Lecture Notes in Computer Science

Lecture Notes in Artificial Intelligence 15497
Founding Editor

Jörg Siekmann

Series Editors

Randy Goebel, *University of Alberta, Edmonton, Canada*
Wolfgang Wahlster, *DFKI, Berlin, Germany*
Zhi-Hua Zhou, *Nanjing University, Nanjing, China*

The series Lecture Notes in Artificial Intelligence (LNAI) was established in 1988 as a topical subseries of LNCS devoted to artificial intelligence.

The series publishes state-of-the-art research results at a high level. As with the LNCS mother series, the mission of the series is to serve the international R & D community by providing an invaluable service, mainly focused on the publication of conference and workshop proceedings and postproceedings.

Amir Hussain · Bo Jiang · Jinchang Ren ·
Mufti Mahmud · Erfu Yang · Aihua Zheng ·
Chenglong Li · Shuqiang Wang · Zhi Gao ·
Zhicheng Zhao
Editors

Advances in Brain Inspired Cognitive Systems

14th International Conference, BICS 2024
Hefei, China, December 6–8, 2024
Proceedings, Part I

Editors
Amir Hussain
Edinburgh Napier University
Edinburgh, UK

Jinchang Ren
Robert Gordon University
Aberdeen, UK

Erfu Yang
The University of Strathclyde
Glasgow, UK

Chenglong Li
Anhui University
Hefei, China

Zhi Gao
Wuhan University
Hubei, China

Bo Jiang
Anhui University
Hefei, China

Mufti Mahmud
King Fahd University of Petroleum
and Minerals
Dhahran, Saudi Arabia

Aihua Zheng
Anhui University
Hefei, China

Shuqiang Wang
Shenzhen Institutes of Advanced Technology,
Chinese Academy of Science
Shenzhen, China

Zhicheng Zhao
Anhui University
Hefei, China

ISSN 0302-9743   ISSN 1611-3349 (electronic)
Lecture Notes in Artificial Intelligence
ISBN 978-981-96-2881-0   ISBN 978-981-96-2882-7 (eBook)
https://doi.org/10.1007/978-981-96-2882-7

LNCS Sublibrary: SL7 – Artificial Intelligence

© The Editor(s) (if applicable) and The Author(s), under exclusive license
to Springer Nature Singapore Pte Ltd. 2025, corrected publication 2025

This work is subject to copyright. All rights are solely and exclusively licensed by the Publisher, whether the whole or part of the material is concerned, specifically the rights of translation, reprinting, reuse of illustrations, recitation, broadcasting, reproduction on microfilms or in any other physical way, and transmission or information storage and retrieval, electronic adaptation, computer software, or by similar or dissimilar methodology now known or hereafter developed.
The use of general descriptive names, registered names, trademarks, service marks, etc. in this publication does not imply, even in the absence of a specific statement, that such names are exempt from the relevant protective laws and regulations and therefore free for general use.
The publisher, the authors and the editors are safe to assume that the advice and information in this book are believed to be true and accurate at the date of publication. Neither the publisher nor the authors or the editors give a warranty, expressed or implied, with respect to the material contained herein or for any errors or omissions that may have been made. The publisher remains neutral with regard to jurisdictional claims in published maps and institutional affiliations.

This Springer imprint is published by the registered company Springer Nature Singapore Pte Ltd.
The registered company address is: 152 Beach Road, #21-01/04 Gateway East, Singapore 189721, Singapore

If disposing of this product, please recycle the paper.

Preface

We are thrilled to welcome you to the 14th edition of the International Conference on Brain-Inspired Cognitive Systems (BICS 2024). Building on the success of previous years, this conference continues to serve as a key platform for exploring the latest advancements in brain-inspired computing, artificial intelligence, and cognitive systems. Since its inception in 2004, BICS has gathered researchers, practitioners, and thought leaders from around the world to share ideas and shape the future of intelligent systems.

Following the success of BICS 2023 in Kuala Lumpur, Malaysia, this year's event promised to continue that tradition of excellence. BICS 2024 covered a wide range of topics, from computational neuroscience and deep learning to brain-machine interfaces and cognitive computing. Our goal was to foster collaboration across disciplines and provide a space for meaningful discussions that will drive the next wave of innovation in these rapidly evolving fields.

After a single-blind review with submissions receiving on average 2.5 reviews each, 57 papers were accepted from 124 submissions. Of these, 56 papers can be found in these proceedings.

We would like to express our sincere gratitude to all contributors, speakers, and attendees who made this conference so impactful. With your continued support, we are confident that BICS 2024 was another milestone in advancing brain-inspired technologies and their applications.

Organization

General Chairs

Amir Hussain   Edinburgh Napier University, UK
Bo Jiang   Anhui University, China
Jinchang Ren   Robert Gordon University, UK

Advisory Board and Publicity Chairs

Cheng-Lin Liu   CAS, China
Bin Luo   Anhui University, China

Program Chairs

Mufti Mahmud   Nottingham Trent University, UK
Erfu Yang   University of Strathclyde, UK
Aihua Zheng   Anhui University, China
Chenglong Li   Anhui University, China
Shuqiang Wang   Chinese Academy of Sciences, China
Zhi Gao   Wuhan University, China

Publication Chairs

Junchi Yan   Shanghai Jiao Tong University, China
Qi Liu   University of Science and Technology of China, China
Zhicheng Zhao   Anhui University, China

Finance Chairs

Zhengzheng Tu   Anhui University, China
Ping Sun   Anhui University, China

Registrations Chairs

Wei Jia   Hefei University of Technology, China
Yun Xiao   Anhui University, China

Local Arrangement Chairs

Haifeng Zhao   Anhui University, China
Futian Wang   Anhui University, China
Yang Zhao   Hefei University of Technology, China
Cunhang Fan   Anhui University, China
Lingma Sun   Hefei University, China

Contents – Part I

A Lightweight Neural Network for SAR Ship Detection Based
on YOLOv8 and Swin-Transformer 1
    Fei Gao, Chen Fan, Tianjin Liu, Jun Wang, and Amir Hussain

RA-BLS: A Sequential BLSs Integrated with Residual Attention
Mechanism 10
    Yanqiang Wu, Jing Wang, and Wei Hu

EEG-Based Emotion Recognition Using Similarity Measures of Brain
Rhythm Entropy Matrix 20
    *Guanyuan Feng, Peixian Wang, Xinyu Wu, Ximing Ren, Chen Ling,
Yuesheng Huang, Leijun Wang, Jujian Lv, Jiawen Li, and Rongjun Chen*

Intensity Controllable Emotional Speech Synthesis Based
on Valence-Arousal-Dominance 30
    Guoping Li and Yanxiang Chen

Unsupervised Person Re-identification with Random Occlusion
and ContrastiveCrop 41
    Yang Jing, Gu Lingkang, Xia Zhouxiang, and Wu Mengqi

Dynamic Prompt Adjustment for Multi-label Class-Incremental Learning 52
    Haifeng Zhao, Yuguang Jin, and Leilei Ma

Using Decision Tree Classification to Identify Cost Drivers
of Hospitalization Expenses for Elderly Patients 62
    Xiaojing Hu, Yudian Liu, and Shixi Liu

Adversarial Attacks on Facial Images Based on Attribute-Conditioned
High-Camouflage Editing 72
    *Jingjing Zhang, Huabin Wang, Dongxu Shang, Hongrui Yuan,
and Liang Tao*

A High Accuracy Text CAPTCHA Recognition Approach Through
Opertimized Vision Transformer 82
    Wei Hao, Shoulai Shang, and Yepeng Zhang

LightMamba-UNet: Lightweight Mamba with U-Net for Efficient Skin
Lesion Segmentation 93
    Wanzhen Hou, Shiwei Zhou, and Haifeng Zhao

Exploiting Memory-Aware Q-Distribution Prediction for Nuclear Fusion
via Modern Hopfield Network 104
  *Qingchuan Ma, Shiao Wang, Tong Zheng, Xiaodong Dai, Yifeng Wang,
  Qingquan Yang, and Xiao Wang*

Multi-modal Fusion Based Q-Distribution Prediction for Controlled
Nuclear Fusion 115
  *Shiao Wang, Yifeng Wang, Qingchuan Ma, Xiao Wang, Ning Yan,
  Qingquan Yang, Guosheng Xu, and Jin Tang*

Deformable Transformer for 3D Medical Image Segmentation 126
  Haifeng Zhao, Tianxia Yang, Minghui Xu, and Yanping Fu

On the Gap Between AI-Generated and Human-Written Patent Texts 136
  *Zhanhao Xiao, Wei Hu, Yanqiang Wu, Weiqi Chen, Huihui Li,
  and Xiaoyong Liu*

MRI-CT Brain Image Registration Based on SuperPCA
and Block-Matching Algorithm 147
  Wannan Zhang

Multi-teacher Knowledge Distillation with Triplet Loss for Cross-Modal
Object Tracking 155
  Yi Li, Lei Liu, Mengya Zhang, and Chenglong Li

Enhanced Comprehensive Competition Network for Domain Adaptive
Palmprint Recognition 166
  Congcong Jia, Xingbo Dong, Zhe Jin, and Lianqiang Yang

MBDR-V2: A Network for MRI Brain Tumor Image Segmentation
with Incomplete Modalities 177
  *Yanqi Hou, Longfeng Shen, Jiacong Chen, Liangjin Diao, Youle Xu,
  and Wei Zhao*

An Innovative Eco-Friendly Weighing System for Reusable Bags
Incorporating K210 and QR Code Technology 187
  Yubin Wei, Yufei Li, and Yiwen Zhang

Focal Consistency Network for Developmental Stage Classification
of Embryos with Time-Lapse Embryo Video Datasets 197
  Yiming Li, Hua Wang, Jingfei Hu, and Jicong Zhang

Chest X-ray Image Rib Segmentation via Disentanglement Enhancement
Network 208
  Lili Huang, Shiqi Li, Lingma Sun, and Chuanfu Li

Instance-Level 3D Model Reassembling from CLuttered Fragments 218
    Longteng Jiang, Yijian Liu, Feixiang Lu, Chenming Wu, and Xin Jin

Brain-Inspired Action Generation with Spiking Transformer Diffusion
Policy Model 229
    Qianhao Wang, Yinqian Sun, Enmeng Lu, Qian Zhang, and Yi Zeng

Single-Stage Dual-Task Joint Learning Framework for Hand Hygiene
Assessment 239
    Sizhe Qin, Zijian Tu, Deyu Su, and Zi Wang

Enhancing Few-Shot Learning in Spiking Neural Networks Through
Hebbian-Augmented Associative Memory 249
    Weiyi Li, Dongcheng Zhao, Yiting Dong, Guobin Shen, and Yi Zeng

Palmprint Texture Fusion Based on TinyViT for Recognition 259
    Fuchuan Huang, Cunyu Sheng, Jian He, and Wei Jia

Novel Device Placement Approach with Neighbor Effect Aware Graph
Mamba Networks 269
    Hao Shu, Wangli Hao, Meng Han, and Fuzhong Li

Research on Improved PointPillars Algorithm Based on Attention
Mechanism and Feature Fusion 280
    RunMei Zhang, AnLong Zhang, and Lei Yin

Correction to: Intensity Controllable Emotional Speech Synthesis Based
on Valence-Arousal-Dominance C1
    Guoping Li and Yanxiang Chen

Author Index 291

Contents – Part II

Multi-modal Dynamic Information Selection Pyramid Network
for Alzheimer's Disease Classification 1
   Yuanmin Ma, Yuan Chen, Yuqing Liu, Jie Chen, and Bo Jiang

Text-Guided Vision Mamba for Alzheimer's Disease Prediction Using
^{18}F-FDG PET 11
   Die Zhou, Yuan Chen, Yuqing Liu, and Bo Jiang

EEG-Based Recognition of Knowledge Acquisition States in Second
Language Learning 21
   Shanlin Xi, Ziyu Li, and Xia Wu

A Study on the Neural Mechanism of the Spatial Position of Speech
in Different Masking Types Affecting Auditory Attention Processing 31
   Dawei Xiang, Yong Ma, and Yiming Yang

DSCF-DE: A Query-Based Object Detection Model via Dynamic
Sampling and Cascade Fusion 41
   Dengdi Sun, Wenhao Liu, and Zhuanlian Ding

MDFNet: Multi-dimensional Fusion Attention for Enhanced Image
Captioning 52
   Dengdi Sun, Xuetao Li, and Chaofan Mu

Dynamic Points Location of Professional Model Pose Based on Improved
Network Stacking Model 62
   Kaizhan Mai, Dazhi Li, Yuefang Gao, Pingping Mi, and Li Hao

A Redundancy Free Facial Acne Detection Framework Based
on Multi-view Face Images Stitching 72
   *Ye Luo, Jianfei Wang, Linglin Zhang, Xinyu Liu, Ji Rao, Wantong Xu,
Jianwei Lu, and Xiuli Wang*

A New Device Placement Approach with Dual Graph Mamba Networks
and Proximal Policy Optimization 82
   *Meng Han, Yan Zeng, Hao Shu, Xiaofei Lu, Jilin Zhang, Yongjian Ren,
and Wangli Hao*

Cross-Generational Contrastive Continual Learning for 3D Point Cloud
Semantic Segmentation 93
   Yuan He, Guyue Hu, and Shan Yu

TGAM-SR: A Sequential Recommendation Model for Long
and Short-Term Interests Based on TCN-GRU and Attention Mechanism 104
   Jiajing Zhang, Zhiya Shen, Jinlan Chen, and Qilang Li

Investigating ChatGPT Translation Hallucination
from an Embodied-Cognitive Translatology Perspective 117
   Hui Jiao, Xinwei Li, Jonathan Ding, and Xiaojun Zhang

A Study on Chinese Acronym Prediction Based on Contextual Thematic
Consistency 127
   Wan Tao, Xiaoran Wang, and Qiang Zhang

Learning Supportive Two-Stream Network for Audio-Visual Segmentation 138
   Hongfan Jiang, Tianyang Xu, Xuefeng Zhu, and Xiaojun Wu

Multi-exposure Driven Stable Diffusion for Shadow Removal 148
   Zheng Yan, Wenhao Tan, and Linbo Wang

Human Disease Prediction Based on Symptoms Using Novel Machine
Learning 159
   Ibukunoluwa Oluwabusayo Efunwoye, Mandar Gogate,
   Adeel Hussain, Bin Luo, Jinchang Ren, Fengling Jiang, Amir Hussain,
   and Kia Dashtipour

CAT-LCAN: A Multimodal Physiological Signal Fusion Framework
for Emotion Recognition 168
   Ao Li, Zhao Lv, and Xinhui Li

A Novel Thermal Imaging and Machine Learning Based Privacy
Preserving Framework for Efficient Space Allocation, Utilisation
and Management 178
   Maria Bruevich, Nilupulee A. Gunathilake, Mandar Gogate,
   Adeel Hussain, Bin Luo, Jinchang Ren, Amir Hussain, Fengling Jiang,
   and Kia Dashtipour

Training Feature-Awared GPU-Memory Allocation and Management
for Deep Neural Networks 188
   Qintao Zhang, Xin Li, Chengchuang Huang, Ying Zhu, Jilin Zhang,
   and Meng Han

TR-LDA: An Improved Potential Topic Recognition Model 201
  Anzhen Li, Shufan Qing, Weijie Qin, Liwen Qin, Jiawei Zhang,
  Meilin Shi, Jinchang Ren, and Mingchen Feng

Brain-Inspired Object Domain Adaptive Segmentation 211
  Mengyin Pang, Song Xu, Lina Wang, Zhenfei Liu, Meijun Sun,
  and Zheng Wang

Task Adaptive Feature Distribution Based Network for Few-Shot
Fine-Grained Target Classification 222
  Ping Li, Hongbo Wang, Jie Ren, Xin Mi, and Chao Shi

ST_TransNeXt: A Novel Pig Behavior Recognition Model 233
  Wangli Hao, Hao Shu, Xinyuan Hu, Meng Han, and Fuzhong Li

A Method for Predicting the RUL of HDDs Based on Bidirectional LSTM
and Transformer 243
  ZeHong Wu, Jinghui Qin, Zhijing Yang, and Yongyi Lu

Spatio-temporal Graph Learning on Adaptive Mined Key Frames
for High-Performance Multi-Object Tracking 252
  Futian Wang, Fengxiang Liu, and Xiao Wang

From Image to the Ground: Recover the Ground Location of Vehicles
from Traffic Cameras Using Neural Networks 262
  Xuzhen Wang, Wenzhong Wang, and Jin Tang

In-Depth Evaluation and Analysis of Hyperspectral Unmixing Algorithms
with Cognitive Models 273
  Shunan Deng, Jinchang Ren, Rongjun Chen, Huimin Zhao,
  and Amir Hussain

Effective Gas Classification Using Singular Spectrum Analysis
and Random Forest in Electronic Nose Applications 283
  Yuntao Wu, Jinchang Ren, Rongjun Chen, Huimin Zhao,
  and Amir Hussain

Author Index 295

A Lightweight Neural Network for SAR Ship Detection Based on YOLOv8 and Swin-Transformer

Fei Gao[1], Chen Fan[1], Tianjin Liu[1(✉)], Jun Wang[1], and Amir Hussain[2]

[1] School of Electronic and Information Engineering, Beihang University, Beijing 100191, China
tianjinliu@buaa.edu.cn
[2] Edinburgh Napier University, Edinburgh EH11 4BN, UK

Abstract. In response to the current challenges of large parameter size and low detection accuracy in SAR ship detection models, this paper proposes an improved YOLOv8 model. The model integrates the Swin-Transformer architecture into YOLOv8 through an adaptive feature fusion method, enhancing the model's global information perception and detection accuracy. Additionally, it employs a top-down unidirectional semantic pyramid and a lightweight detection head structure to achieve model lightweighting. Experimental comparisons on the HRSID dataset show that, compared to the benchmark algorithm, the proposed algorithm increases mAP50 by 1.2% and reduces model parameter size by 75%. Compared to current mainstream algorithms, this algorithm offers better detection performance and a lower model parameter size, effectively meeting the requirements for model lightweighting and high detection accuracy in SAR image detection terminal devices.

Keywords: SAR ship detection · Swin-Transformer · YOLOv8

1 Introduction

Synthetic Aperture Radar (SAR) is an all-weather, all-time active Earth observation system, extensively utilized for natural disaster assessment, urban planning, traffic management, environmental monitoring, and the detection and identiSfication of maritime targets [1]. Among its applications, SAR ship detection holds significant value for navigation safety, fisheries management, ship salvage, and national defense construction [2].

Traditional SAR ship detection methods face challenges in complex scenarios, with low detection performance and poor generalization capabilities, which fail to meet the requirements for high precision, real-time processing, and strong adaptability in detection tasks. In contrast, one-stage SAR ship detection approaches leveraging deep learning have garnered widespread attention in research due to their superior detection capabilities, efficiency, and generalization. CAM-SSD [3] enhances ship detection accuracy by integrating a rotation box detection branch and a Convolutional Block Attention Module feature fusion module with the SSD network. RO-YOLO [4] enhances the detection precision of the model by integrating SPD-Conv into YOLOv8 and employing Dyhead as the head network. Despite these advancements in one-stage detection algorithms, the underlying network structures, predominantly convolutional networks adept at capturing local features, impose inherent limitations on the improvement of detection performance. Currently, the Transformer architecture [5], predicated on attention mechanisms, has been increasingly adopted in the field of computer vision, with numerous researchers exploring its application in SAR ship detection. For instance, CRTransSar [6] constructs a backbone network that amalgamates Swin-Transformer and CNN, employing a cross-attention mechanism to augment feature Fusion within the neck structure, thereby elevating the model's detection performance. ST-YOLOA [7] embeds Swin-Transformer into the STCNet backbone to enhance the extraction of global information and utilizes residual structures and top-down sampling in the neck network to refine the semantic and global information of the output features, successfully enhancing both detection performance and speed. While the incorporation of the Transformer architecture has significantly improved detection accuracy, it has also led to a sharp increase in the number of model parameters and computational load.

To enhance the detection accuracy and reduce the model parameter size, we have implemented the following three innovations based on YOLOv8:

(1) The Swin-Transformer is integrated into the YOLOv8 backbone network through an adaptive feature fusion approach, enhancing the model's feature extraction capability.
(2) A top-down unidirectional semantic pyramid is proposed, which improves the model's feature fusion and enhancement capabilities while reducing the model's parameter count and computational load.
(3) The decoupled head network is designed with differentiation, achieving a lightweight model.

2 Proposed Method

We present a lightweight model design, as depicted in Fig. 1, based on the Swin-Transformer and YOLOv8 architectures. The proposed structure integrates a Swin-Transformer branch into the YOLOv8 backbone, preserving the backbone's ability to extract local information while enhancing its capacity for global feature extraction. By employing a top-down unidirectional semantic feature pyramid, the structure effectively augments the multi-scale information input from the backbone, thereby improving detection performance and achieving model lightweighting. Additionally, a differentiated design of the decoupled head is implemented, which significantly reduces the model's computational load while maintaining its detection capabilities.

2.1 Swin-Transformer and Adaptive Feature Fusion

The Transformer architecture leverages the attention mechanism to model long-range dependencies within the data, which is conducive to enhancing the ability

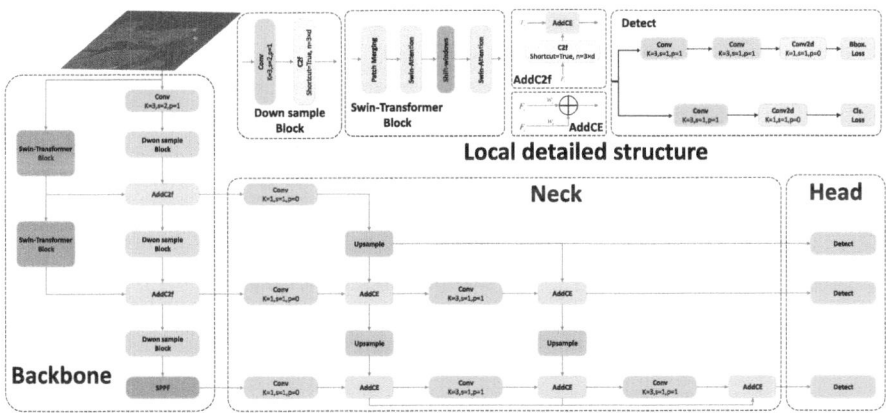

Fig. 1. Overall architecture of the proposed method. The network architecture is composed of three distinct sections: the Backbone, the Neck, and the Head. The Conv block is a feature extraction unit that consists of a basic convolution, batch normalization, and a Sigmoid Linear Unit activation function. The AddCE module serves as an adaptive feature fusion component. The AddC2f module is responsible for channel mapping and adaptive feature fusion within the backbone.

of convolutional networks to extract global information. The Swin-Transformer addresses the issue of excessive computational redundancy in Transformer architectures for image processing by employing a window-based attention mechanism and cyclic shift operations of the windows, thereby enhancing the visual model's capacity to capture long-range information. The window-based attention mechanism segments the feature map into multiple windows and performs attention computations solely among the feature pixels within these windows, alleviating the computational redundancy inherent in the Transformer. We adopt a uniform window partitioning strategy, wherein each window is of identical size, contains no holes, and is arranged in a tight, non-overlapping pattern. The cyclic shift operation entails moving the defined windows as a cohesive unit, with portions exceeding the boundary reemerging on the opposite side. Figure 2 illustrates the uniform window partitioning and cyclic shift operations.

We integrates the Swin-Transformer into the backbone of YOLOv8, enhancing the feature extraction capabilities of the model's backbone by adaptively weighting and summing the local information obtained from CNNs with the global information acquired by the Swin-Transformer.

Figure 3 illustrates the structure of the Swin-Transformer module utilized in this paper. The Patch Merging component employs a scaling and flattening approach to alter the dimensions of the input feature map, followed by a linear layer that maps each feature pixel into a space of the same dimensionality. The Swin-Attention section performs windowed attention calculations on the input feature map, completing the weight allocation of attention among feature pixels within the windows. The Shift Window operation enables inter-window pixel

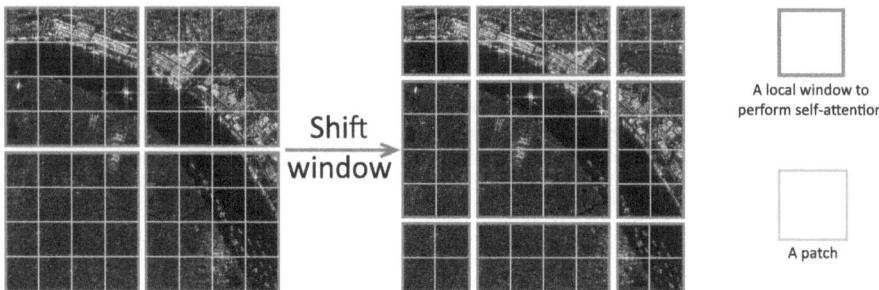

Fig. 2. Uniform window partitioning and cyclic shift operations.

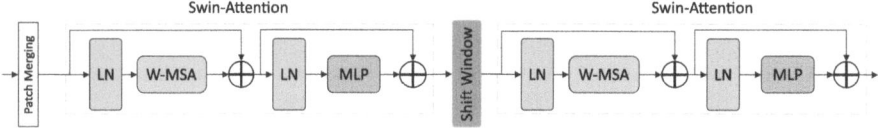

Fig. 3. Swin-Transformer module. LN refers to Layer Normalization, W-MSA denotes the Window Multi-head Attention mechanism, MLP stands for Multi-Layer Perceptron, and Shift Window indicates the cyclic shift operation.

interaction and enhancing the structure's ability to perceive global information. The attention calculation formula is as follows:

$$Attention(K, Q, V) = Softmax(QK^T/\sqrt{d} + B)V \tag{1}$$

K, Q, and V are vectors derived from feature pixels that are mapped into a d-dimensional space through distinct linear layers. B represents the relative positions of different feature pixels within the same window, composed of learnable parameters that are randomly initialized at the time of model creation and are updated during the training process. The Swin-Transformer module utilizes these learnable relative position parameters, enabling the model to autonomously learn the importance of different features based on the training dataset. This capability allows the model to better adapt to the processing requirements of various domain-specific data, thereby enhancing the model's performance.

In Fig. 1, AddCE denotes an adaptive feature fusion method, which employs learnable weight parameters to perform a weighted summation of two features with identical shapes. The formula is expressed as follows:

$$F_{mix} = w_t * F_t + w_c * F_c \tag{2}$$

F_t, F_c, and F_{mix} denote the features extracted by the Swin-Transformer, the convolutional neural network (CNN), and the fused features from both, respectively; w_c and w_t represent the adaptive weighted parameters for the features extracted by the CNN and the Swin-Transformer, respectively. To achieve uniformity in the shape of features extracted by the Swin-Transformer and the

CNN, this paper employs a stacked Swin-Transformer architecture to obtain multi-scale features, thereby standardizing the dimensions of feature width and height. A c2f structure, rich in gradient flow, is selected for channel transformation of the Swin-Transformer's multi-scale features, ensuring the unification of feature channel dimensions while maintaining the gradient backpropagation.

2.2 Unidirectional Semantic Pyramid

In contrast to convolutional structures, the Transformer architecture exhibits only partial translational invariance, which leads to the quality of semantic information in the features surpassing that of spatial information. Leveraging this characteristic, this paper introduces a unidirectional semantic pyramid that amalgamates the high-fidelity semantic information from low-resolution features with the high-fidelity spatial information from high-resolution features, thereby enhancing the quality of features extracted by the model. The unidirectional semantic pyramid adopts the design philosophy of feature pyramids, progressively cascading downscaled feature maps, and generating a series of feature maps across various scales through a process of upsampling and adaptive weighted summation.

In Fig. 1, the "neck" portion delineates the configuration of a unidirectional semantic pyramid. This module comprises three top-down pathways, a series of lateral connections, and three residual connections. The first pathway initially employs a 1×1 convolution to match the channel dimensions of the multi-scale features outputted by the backbone network. Subsequently, it performs resolution matching and feature fusion by sequentially applying nearest neighbor interpolation and weighted summation between adjacent scales of features in a top-down manner. The second pathway shares the same top-down feature processing approach as the first pathway, but for the lateral transfer of multi-scale features, it utilizes a 3×3 convolution to suppress aliasing effects introduced by upsampling. The third pathway receives multi-scale features from both the first and second pathways and obtains semantically enhanced feature maps through weighted summation.

2.3 Lightweight Detection Head

The detection head structure of YOLOv8 is intricate, particularly the feature extraction component, which is composed of two layers of convolutional structures, significantly increases the computational load of the model. Considering that classification tasks are less complex than regression problems, this paper adopts a single-layer Conv structure for the transformation of features output by the neck network in the classification part, and employs the ReLU activation function to simplify computations, aligning with the characteristics of binary classification for SAR ship detection. Additionally, taking into account the single-channel nature of SAR images, this paper adjusts the output channel count of the Conv structures in the regression part, Achieved lightweight optimization of the model's Head section.

Figure 1 illustrates the detailed architecture of a lightweight detection head, wherein the classification branch utilizes cross-entropy loss, while the regression branch employs Distribution Focal Loss (DFL) and Complete Intersection Over Union (CIOU) loss. The computation formulas for DFL and CIOU loss are as follows:

$$LossCIoU = 1 - (IOU - d^2/c^2 - v^2/(1 - IOU + v)) \quad (3)$$

$$DFL = -(y_{i+1} - y)log(S_i) - (y - y_i)log(S_{i+1}) \quad (4)$$

In the CIOU loss function, d represents the Euclidean distance between the centers of the predicted bounding box and the ground-truth bounding box. The variable c denotes the diagonal distance of the smallest enclosing rectangle. The variable v serves as a correction factor, which is employed to further refine the loss function. This factor can be determined based on the width and height of the ground-truth bounding box w_G, h_G and the predicted bounding box w_p, h_p. The specific computation of v is as follows:

$$v = 4 * (arctan(w_G/h_G) - arctan(w_p/h_p))^2/\pi^2 \quad (5)$$

In the DFL, y, y_i, and y_{i+1} denote the true position and the adjacent integer values on the left and right sides, respectively. S_i and S_{i+1} denote the model's predicted probabilities for the left and right positions, respectively.

3 Experiment

3.1 Experiment Data and Experiment Settings

The experimental data is selected from the High Resolution SAR Images Dataset. The dataset comprises 5,604 high-resolution SAR images of size 800 × 800, along with 16,951 ship instances, covering a variety of scenarios. The experimental training set and test set image ratios are 65% and 35%, respectively. The size of the input data is scaled down to 640 × 640 through nearest-neighbor interpolation.

The experiments were conducted on two NVIDIA GeForce RTX 4090 GPUs. The deep learning framework used is PyTorch. The number of iterations for model training is set to 200, and the batch size is set to 32. The model's parameter updates employed the stochastic gradient descent optimizer with a learning rate of 0.01 and a momentum parameter of 0.937.

3.2 Experiment Result

In order to verify the detection performance of the improved algorithm, this paper compares the proposed improved algorithm with mainstream algorithms RT-DETR, YOLOv5, YOLOv10, and YOLOv8 on the HRSID dataset in terms of detection accuracy, model parameter size, and computational load. The results are shown in Table 1, where the bold font indicates the best results in each column.

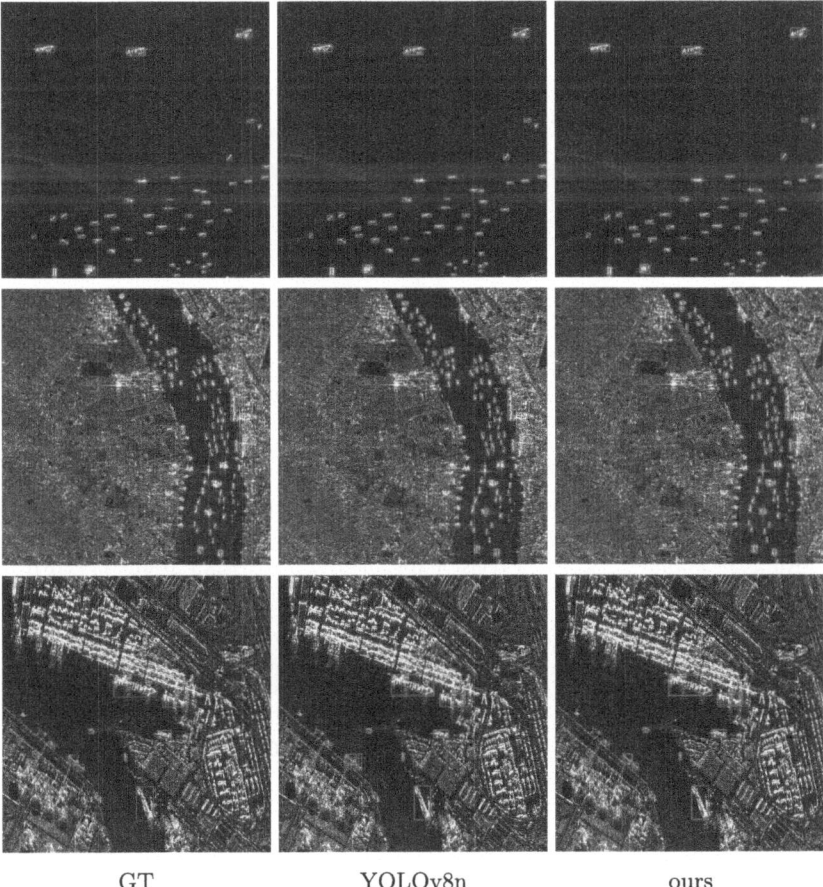

GT YOLOv8n ours

Fig. 4. Visualization Comparison of Detection Results. The red boxes represent the model's detection boxes, and the green boxes represent the ground truth boxes of the targets. (Color figure online)

From the table, it is evident that the algorithm proposed in this paper outperforms other algorithms in terms of recall and mAP. Compared to the second-best results, mAP_{50}, mAP_{50}^{95}, and recall have improved by 1.2%, 0.5%, and 0.5%, respectively. Furthermore, the algorithm presented in this paper can significantly reduce the number of parameters, thereby decreasing the model's memory resource consumption and better adapting to scenarios where terminal devices have limited resources. Due to the use of the Swin-Transformer structure in the algorithm, the computational load of the model has increased compared to YOLOv5.

Figure 4 illustrates the detection performance of the improved YOLOv8 model and YOLOv8n in various scenarios. It can be observed from the figure that the algorithm proposed in this paper effectively reduces the false alarm prob-

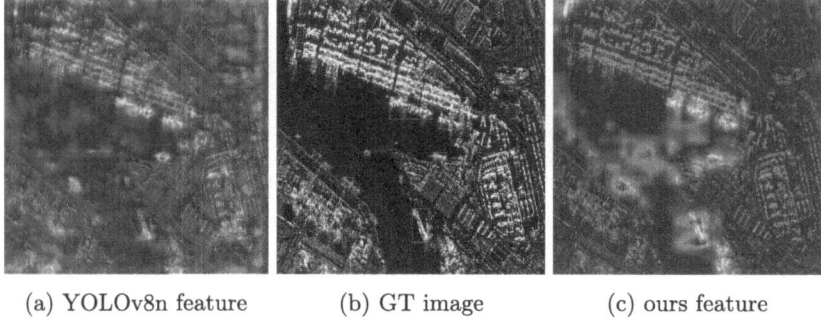

(a) YOLOv8n feature (b) GT image (c) ours feature

Fig. 5. Comparison of Feature Visualization Results. The bluer parts indicate that the features in that area have less influence on the model's output, while the redder parts indicate that the features have a greater influence on the model's output. (Color figure online)

Table 1. Comparison experiments between detection algorithms.

Algorithm	Backbone	Precision	Recall	mAP_{50}	mAP_{50}^{95}	Parameters	GFLOPS
RT-DETR	ResNet50	0.854	0.753	0.857	0.566	4.19 M	125.6
YOLOv5	New CSP-Darknet53	**0.916**	0.809	0.899	0.658	2.17M	**5.5**
YOLOv10	Enhance CSPNet	0.910	0.795	0.892	0.65	2.69 M	8.2
YOLOv8	Darknet	0.908	0.820	0.902	0.666	3.01 M	8.1
ours	Darknet+Swin	0.914	**0.825**	**0.914**	**0.671**	**0.76M**	9.4

ability and enhances the localization accuracy of the model in scenarios with small, dense ship targets and complex backgrounds. Figure 5 shows the visualization results of the feature maps output by the neck network of YOLOv8n and the model proposed in this paper using Grad-CAM. It can be seen from the figure that the model proposed in this paper can better lock onto the target's location when making detection decisions, giving higher attention weight to the locations where targets exist, thereby improving the model's detection accuracy. This also indicates that applying the unidirectional semantic pyramid and Swin-Transformer to the YOLOv8n network can effectively enhance the model's feature extraction and feature enhancement capabilities.

4 Conclusion

In response to the current issues of large parameter size and low detection accuracy in SAR ship detection models, this paper proposes an improved YOLOv8 algorithm for SAR ship detection. This algorithm integrates the Swin-Transformer architecture into the YOLOv8 backbone, enhancing the model's detection performance. By employing a unidirectional semantic pyramid and a lightweight detection head structure, the model has been effectively streamlined.

Experimental results on the HRSID dataset demonstrate that the algorithm significantly reduces the model's parameter volume, while improving detection performance. The parameter volume is reduced to approximately 25% of YOLOv8, and the detection performance metrics mAP50 and mAP50-95 are increased by 1.2% and 0.5%, respectively. Moreover, the algorithm also outperforms current mainstream algorithms in terms of accuracy and model parameter size. In future work, the loss function and the positive and negative sample allocation strategy of the model can be improved to further enhance the stability and generalization capability of detection.

Acknowledgments. This research was supported by the National Natural Science Foundation of China under Grant 62371022.

References

1. Gao, F., et al.: SAR target incremental recognition based on features with strong separability. IEEE Trans. Geosci. Remote Sens. (2024)
2. Correa, V., et al.: Applications of GANs to aid target detection in SAR operations: a systematic literature review. Drones **8**(9), 448 (2024)
3. Wang, J., Lu, C., Jiang, W.: Simultaneous ship detection and orientation estimation in SAR images based on attention module and angle regression. Sensors **18**(9), 2851 (2018)
4. Ying, L., Miao, D., Zhang, Z.: A robust one-stage detector for SAR ship detection with sequential three-way decisions and multi-granularity. Inf. Sci. **667**, 120436 (2024)
5. Vaswani, A.: Attention is all you need. In: Advances in Neural Information Processing Systems (2017)
6. Xia, R., Chen, J., Huang, Z., et al.: CRTransSar: a visual transformer based on contextual joint representation learning for SAR ship detection. Remote Sens. **14**(6), 1488 (2022)
7. Zhao, K., Lu, R., Wang, S., et al.: ST-YOLOA: a swin-transformer-based YOLO model with an attention mechanism for SAR ship detection under complex background. Front. Neurorobot. **17**, 1170163 (2023)

RA-BLS: A Sequential BLSs Integrated with Residual Attention Mechanism

Yanqiang Wu, Jing Wang(✉), and Wei Hu

School of Computer Science, Guangdong Polytechnic Normal University,
Guangzhou 510665, China
wj_adr@163.com

Abstract. Broad Learning System (BLS) has attracted the attention of many researchers because of its excellent performance. However, The number of random nodes becomes very large when the BLS copes with large complex datasets. Based on this problem, a novel residual attention mechanism is designed and introduced into the BLS to form a sequential BLSs integrated with residual attention mechanism (RA-BLS). The goal is to shrink the network structure by correcting random enhancement nodes. The RA-BLS feeds the residuals back to the enhancement nodes to get efficient and compact feature representations, further strengthening the approximation capability. The RA-BLS drastically reduces the node requirements and further improves the performance of BLS. Finally, experimental results show that RA-BLS achieves excellent performance on UCI, MNIST, FASHION-MNIST, and NORB datasets. Notably, an accuracy of 92.33% is achieved on the NORB dataset, marking an improvement of 2.99% with only 17% of the original BLS size. The source code is available all at https://github.com/arolme/RABLS-Residual-Attention-Broad-Learning-System.

Keywords: Broad Learning System · Residual · Classification · Attention Mechanism · Machine Learning

1 Introduction

Broad Learning System [1], introduced by Chen, uses Ridge Regression to build a concise structure with an explicit solution. Because BLS can achieve [2] satisfactory performance quickly, it is widely applied [3] to address time-consuming tasks. The BLS employs a strategy of randomly generating hidden nodes, which necessitates a large number of these nodes [4,5] to capture the information of the input units comprehensively. This approach can cause the BLS to become very fat [6,7]. The fat BLS ensures thorough learning but impacts computational efficiency. Additionally, the fat BLS is also not conducive to practical application.

The core reason [8] for the fat BLS is that the random enhancement nodes are not designed based on the specific task. This means that different random nodes may capture similar or even duplicate information. Some scholars have proposed solutions to the problem of conflict between model size and performance. The practice is to keep the compact network. Ding [9] proposed a greedy BLS that reduces the redundant nodes in the hidden layer, constructing a structure that compromises width and depth. In D&BLS [10], Xie passes the residuals to the next subsystem. Furthermore, multiple lightweight BLSs are stacked to replace the fat BLS.

Our motivation is to address the fat BLS via a compact structure. Inspired by the D&BLS, we sequence multiple lightweight BLSs to replace the fat BLS. We propose a sequential Broad Learning System with residual attention mechanism (RA-BLS) to address fat BLS caused by excessive random nodes. Unlike D&BLS, the lightweight subsystem in RA-BLS does not require the regeneration of hidden nodes. Instead, the hidden nodes of RA-BLS are transferred from lower to upper layers, facilitating their reuse and potentially improving computational efficiency. In addition, inspired by attention mechanism [11], we feed the residuals back into the enhancement nodes of the subsystem, thereby weakening the randomness of the enhancement nodes. The aim is to adapt the enhancement nodes to a specific task. The hidden nodes of the uppermost layer are directly connected to the output, with the output weights calculated by the pseudoinverse. The key contributions are outlined:

1. The new system introduces a residual attention mechanism to correct the enhancement nodes according to the current task. This method leads to the creation of slim and compact subsystems, which improves the system's efficiency.
2. A small-scale network is maintained by forming a sequential model through multiple compact subsystems. The optimal output of the last subsystem is generated after passing through multiple subsystems. This innovative approach corrects the enhancement nodes to adapt to the current task rather than capture random features. The fatness problem of BLS is solved, further improving the superior accuracy on large and complex datasets.

The structure is as follows as follows: Sect. 2 briefly describes the background of BLS and the attention mechanism. The residual attention mechanism and the structure of RA-BLS are detailed in Sects. 3. Experimental results are shown in Sect. 4. The performance and the generalization capability of RA-BLS are compared with other popular methods. The effects of different parameters are also discussed. Finally, Sect. 5 offers conclusions and suggests directions for future research.

2 Related Works

2.1 Brief Introduction of Broad Learning System

The BLS [1] contains (see Fig. 1) three main structures: input unit, hidden layer, and output layer. Two distinct node types are present within the hidden layer: feature nodes and enhancement nodes. The core idea involves a mapping function that maps raw input units into feature nodes. These feature nodes are then augmented to generate enhancement nodes randomly. Finally, each hidden node is directly connected to the output, with the pseudoinverse calculated output weights.

For the original BLS, a given input $\{(X,Y)|X \in \mathbb{R}^{N \times M}, Y \in \mathbb{R}^{N \times C}\}$. Let $\phi(\cdot)$ denote the mapping function for n feature nodes, and $\xi(\cdot)$ denote the activation function

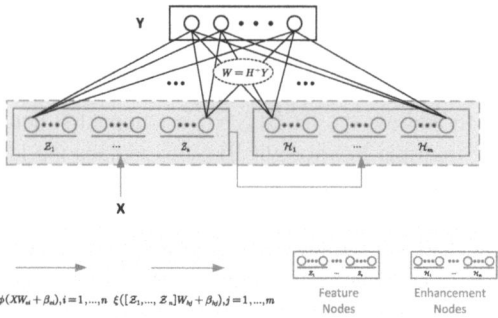

Fig. 1. Illustration of the BLS.

that produces m enhancement nodes. The mapping function $\phi(\cdot)$ and activation function $\xi(\cdot)$ have no explicit restrictions, which means that the common choices such as kernel mappings or nonlinear transformations are acceptable. Specifically, if the function in the feature mapping uses kernel mappings, the BLS has additional connecting nodes called enhancement nodes between the feature mapping layers and the output layer. The feature information hidden in the enhancement nodes can be further explored. The n feature nodes and m enhancement nodes can be collectively represented as hidden nodes:

$$H = [\mathcal{Z}_1, \ldots, \mathcal{Z}_n, \mathcal{H}_1, \ldots, \mathcal{H}_m]. \tag{1}$$

where, using $\mathcal{Z}_i = \phi(XW_{ei} + \beta_{ei}), i = 1, \ldots, n$ to denote the ith $\mathcal{Z}_i$ mapped features, and denoting all feature nodes as $[\mathcal{Z}_1, \ldots, \mathcal{Z}_n]$. And denoting $\mathcal{H}_j = \xi([\mathcal{Z}_1, \ldots, \mathcal{Z}_n]W_{hj} + \beta_{hj}), j = 1, \ldots, m$ as the jth enhancement nodes, where $W_{ei}, \beta_{ei}, W_{hj}$ and β_{hj} are generated randomly with corresponding dimensions.

The output of BLS can be represented as the equation of the form

$$Y = [\mathcal{Z}_1, \ldots, \mathcal{Z}_n, \mathcal{H}_1, \ldots, \mathcal{H}_m]W. \tag{2}$$

W represents the connecting weights between the hidden nodes and the output. The pseudoinverse is expressed as $H^+ = (\lambda I + H^T H)^{-1} H^T$. The regression approximation can be easily computed as W using the equation

$$W = (\lambda I + H^T H)^{-1} H^T Y. \tag{3}$$

2.2 Attention Mechanism

Attention [11] is inspired by the biological system. Humans actively focus their attention on a certain part when processing a large amount of information. Attention mechanisms are widely applied in deep learning.

In deep learning, the attention mechanism is designed according to the specific tasks. The **query** and **key** are designed according to the particular task or network structure. The correlation between the **key** and the **query** reflects the importance of the attention. A matrix Q is packed with all **queries**. Similarly, the **keys** and **values** are also packed together into matrices K and V. The attention mechanism is computed in the matrix of outputs as:

$$\text{Attention}(Q, K, V) = \text{softmax}(\frac{QK^T}{\sqrt{D}})V. \quad (4)$$

where Q and K are computed dot product, divide each by $\sqrt{D}$, and the softmax function is applied to the score of V. D denotes the dimension of the feature. The $\frac{1}{\sqrt{D}}$ is used as a scaling factor to counteract the effect of large scores.

3 The Proposed Methods

3.1 Residual Attention Mechanism

The BLS is a discriminative model. The decision boundary of BLS is searched to separate different categories. The residual [12] represents the distance between the predicted value and the true value. The smaller the residual is, the clearer the decision boundary is. Therefore, we propose residual attention mechanisms to correct the random enhancement nodes to fit a given task.

$$\text{Attention}(Q, K, R) = \text{softmax}(\frac{QK^T}{\sqrt{D}})R. \quad (5)$$

where R represents the residual of the current state, and D represents the dimension of the feature.

3.2 BLS with Residual Attention Mechanism

In RA-BLS, the hidden nodes of the lowest layer are randomly generated. The hidden nodes are passed layer by layer for reuse. Two distinct node types are present within the hidden nodes: feature nodes and enhancement nodes. The feature nodes are retained in the next subsystem. The enhancement nodes have their randomness weakened by the residual attention mechanism. Immediately after that, the feature nodes and the corrected enhancement nodes are combined to form the new hidden nodes. These hidden nodes are connected to the output.

For the RA-BLS, mathematically, we denote the input $\{(X, Y)|X \in \mathbb{R}^{N \times M}, Y \in \mathbb{R}^{N \times C}\}$. First, recall that the original BLS can be represented equivalently by Eq. (1)–(3). Note that the nodes for the first iteration are generated randomly. The feature nodes $F = [\mathcal{Z}_1, \ldots, \mathcal{Z}_n]$, inactive enhancement nodes $E_{inactive}^{(1)} = [(FW_{h1} + \beta_{h1}), \ldots, (FW_{hm} + \beta_{hm})]$, enhancement nodes $E_{active}^{(1)} = \xi(E_{inactive}^{(1)})$, weights $W^{(1)}$ and residual $R^{(1)}$ for the first iteration are easily obtained.

Then, the process of optimizing the enhancement nodes begins. The weights are split into two parts

$$\begin{bmatrix} W_f^{(1)} \\ W_e^{(1)} \end{bmatrix} = W^{(1)}. \quad (6)$$

where $W_e^{(1)}$ is a weight that connects the enhancement nodes to the output. Similarly, $W_f^{(1)}$ is a weight that connects the feature nodes to the output. So that the output is equivalent to $Output^1 = [F, E_{active}^{(1)}] \begin{bmatrix} W_f^{(1)} \\ W_e^{(1)} \end{bmatrix} = FW_f^{(1)} + E_{active}^{(1)} W_e^{(1)}$.

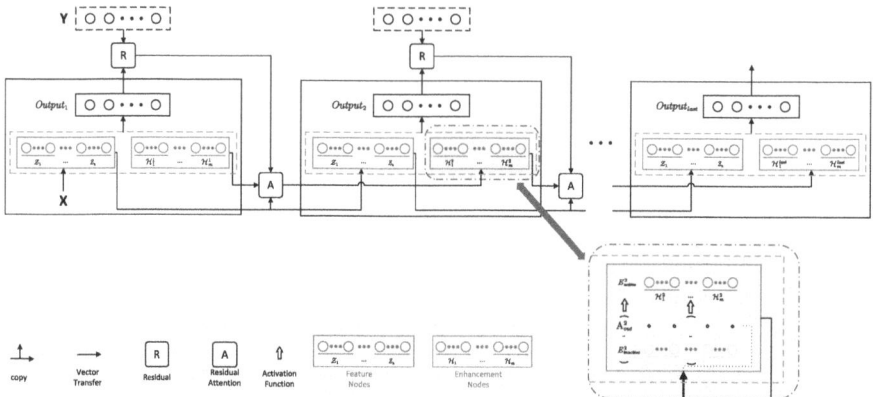

Fig. 2. Lightweight subsystems are sequenced. All nodes are randomly generated in the first subsystem. Each subsystem reuses feature nodes. Only the enhancement nodes of the lower layer have their randomness weakened by the residual attention mechanism. Immediately after, the nodes are fed to the upper layer subsystem.

In RA-BLS (See Fig. 2), the feature nodes F represent the matrix K and matrix Q. $A(Q, K, R, W) = (\nabla \xi) \circ \text{softmax}(\frac{FF^T}{\sqrt{D}}) RW^T$ is the residual attention block. $E_{inactive}$ denotes an inactive enhancement node. The core of correction is to fuse the residual information before the enhancement node is activated. The equation for (t+1)th correction-based residual attention is expressed as:

$$E_{inactive}^{(t+1)} = E_{inactive}^{(t)} - (\nabla \xi) \circ \text{softmax}(\frac{FF^T}{\sqrt{D}}) R^{(t)} W_e^{(t)T}. \tag{7}$$

where $R^{(t)}$ denotes the residuals of (t)th iteration. $\circ$ denotes the Hadamard product. $(\nabla \xi)$ denotes the gradient of the activation function $\xi(\cdot)$. D denotes the dimension of the feature nodes. The purpose of introducing the $(\nabla \xi)$ is to counteract the effect of the activation function in the residual pass. The W_e^t is a weight that connects the enhancement nodes to the output. The role of W_e^T is to communicate the residual back to the corresponding enhancement nodes.

Next, the corrected nodes are activated to form enhancement nodes

$$E_{active}^{(t+1)} = [\mathcal{H}_1^{(t+1)}, \ldots, \mathcal{H}_m^{(t+1)}] = \xi(E_{inactive}^{(t+1)}). \tag{8}$$

The $H^{(t+1)} = [\mathcal{Z}_1, \ldots, \mathcal{Z}_n, \mathcal{H}_1^{(t+1)}, \ldots, \mathcal{H}_m^{(t+1)}]$ denotes as the hidden nodes of the (t+1)th iteration. And the weights $W^{(t+1)} = (\lambda I + H^{(t+1)T} H^{(t+1)})^{-1} H^{(t+1)T} Y$ are calculated using pseudoinverse. Then, the output is obtained

$$Output^{last} = H^{(last)}W^{(last)}. \qquad (9)$$

The RA-BLS is an optimization algorithm based on the residual attention mechanism. The number of subsystems is also a very important hyperparameter. In general, too much attention to residuals can cause overfitting. Therefore, we set the iteration to terminate when the performance drops off.

Algorithm 1. RA-BLS

Require: The samples $\{(X,Y)|X \in \mathbb{R}^{N \times M}, Y \in \mathbb{R}^{N \times C}\}$, mapping function $\phi(\cdot)$, activation function $\xi(\cdot)$, number of Feature Nodes n, number of Enhancement Nodes m and regularization coefficient λ.
Ensure: $Output_{last}$
1: Initialize parameters;
2: Randomly generate n Feature Nodes $\mathcal{Z}_i \Leftarrow \phi(XW_{ei} + \beta_{ei}), i = 1, \ldots, n$;
   % Matrix of all Feature Nodes $F = [\mathcal{Z}_1, \ldots, \mathcal{Z}_n]$.
3: Randomly generate m Enhancement Nodes $\mathcal{H}_j \Leftarrow \xi(FW_{hj} + \beta_{hj}), j = 1, \ldots, m$; % Feature Nodes are mapped to get inactive Enhancement Nodes $E^{(1)}_{inactive} = [(FW_{h1} + \beta_{h1}), \ldots, (FW_{hm} + \beta_{hm})]$. Immediately after that, a nonlinear activation function is used to activate Enhancement Nodes $E^{(1)}_{active} = \xi(E^{(1)}_{inactive})$.
4: Combine the Hidden Nodes $H^{(1)} \Leftarrow [F, E^{(1)}_{active}]$;
5: Calculate the Weight $W^{(1)} \Leftarrow (\lambda I + H^{(1)T}H^{1})^{-1}H^{(1)T}Y$;
6: Calculate the output $Output^{(1)} \Leftarrow H^{(1)}W^{(1)}$;
7: Calculate the Residual $R^{(1)} \Leftarrow (Y - Output^{(1)})$;
8: $t \Leftarrow 1$;
9: **while** $True$ **do**
10:   Split the weight corresponding to Enhancement Nodes $W_e^{(t)} \Leftarrow W^{(t)}$ by Eq. 6;
11:   Calculate **new** Enhancement Nodes $E^{(t+1)}_{active} \Leftarrow \xi(E^{(t)}_{inactive} - (\triangledown \xi) \circ$ softmax$(\frac{FF^T}{\sqrt{D}})R^{(t)}W_e^{(t)T})$;%D denotes the dimension of the feature nodes.
12:   Combine **new** Hidden Nodes $H^{(t+1)} = [F, E^{(t+1)}_{active}]$
13:   Calculate **new** Weight $W^{(t+1)} \Leftarrow (\lambda I + H^{(t+1)T}H^{t+1})^{-1}H^{(t+1)T}Y$;
14:   Calculate **new** output $Output^{(t+1)} \Leftarrow H^{(t+1)}W^{(t+1)}$;
15:   Calculate **new** Residual $R^{(t+1)} \Leftarrow (Y - Output^{(t+1)})$;
16:   $t \Leftarrow t+1$;
17:   **if** performance drops **then**
18:     **Return** $Output_{last} \Leftarrow Output^{(t)}$
19:   **end if**
20: **end while**

---

4 Experiments

This section details the experiments of the RA-BLS. Various methods are also compared on several popular datasets.

In our paper, function $\phi(\cdot)$ uses a linear feature mapping in the proposed RA-BLS. For the enhancement nodes, the Tanh function $\xi(\cdot)$ is chosen to build the RA-BLS. The regularization coefficient λ for RA-BLS, BLS, Stacked BLS, and ER-BLS is set as 2^{-30} by grid search. The W_{hj} and β_{hj}, for $j = 1, \ldots, m$ obey the standard uniform distribution on the interval $[-1,1]$. In addition, we set the iteration to terminate when the performance drops off.

4.1 Experiment Environment and Assessment Metric

The experiments for RA-BLS are given. To make it more convincing, the results are compared on several authoritative public datasets using different methods. The accuracy is averaged over 20 experiments. Our source code will provide frozen random seeds to fix the experimental results. The fixed results are slightly higher or equal to the paper results.

The RA-BLS is evaluated against several methods, including SVM with RBF Kernel [13], BLS [1], Stacked BLS [14], and ER-BLS [15]. The datasets include MNIST [16], FASHION-MNIST [17], NORB [18], and UCI datasets. The RA-BLS is compared with the current popular methods on these datasets. The SVM does not use special feature selection. Regarding experimental settings, all experiments in this paper are implemented using MATLAB 2021a.

Accuracy is the most straightforward and widely adopted metric for evaluating performance in classification tasks. The Accuracy can be represented as the equation of the form:

$$Accuracy = \frac{\text{total correct predictions}}{\text{total samples}} \times 100\%. \tag{10}$$

For the regression task, the evaluating metric is Root Mean Squared Error. The equation is:

$$RMSE = \sqrt{\frac{1}{k}\sum_{t=1}^{k}(\mathcal{Y}_i - \hat{\mathcal{Y}}_i)^2}. \tag{11}$$

4.2 Performance Evaluation on Classification

BLS is experimentally demonstrated on MNIST. In addition to MNIST, our experiments utilize the FASHION-MNIST dataset, which presents a more challenging classification task. The FASHION-MNIST dataset maintains the same structure as MNIST but with increased complexity. The NORB dataset is primarily utilized to assess the efficacy of computer vision algorithms in 3D object recognition tasks. Table 1 presents details regarding the datasets used.

Table 1. Classification dataset details

Datasets	No. of Samples		Features	No. of Categories
	Training	Testing		
MNIST	60000	10000	768(28 × 28)	10
FASHION-MNIST	60000	10000	768(28 × 28)	10
NORB	24300	24300	2048(32 × 32 × 2)	5

The results are presented in Table 2. The number of feature nodes N_f and enhancement nodes N_e are determined through an exhaustive search from the ranges of [1, 2000] and [100, 15000]. To ensure fairness in the results. The feature engineering of the RA-BLS remains the same as that of the BLS and only updates the enhancement node proposed in our method. At the same time, the N_f and the hyperparameter settings are kept the same.

Table 2. Comparative Analysis of SVM, BLS, Stacked BLS, ER-BLS and RA-BLS for Classification

Method	MNIST			FASHION-MNIST			NORB		
	N_f	N_e	Acc(%)	N_f	N_e	Acc(%)	N_f	N_e	Acc(%)
SVM	–	–	97.92	–	–	88.28	–	–	84.49
BLS(2017)	1000	15000	98.72	1500	9000	89.73	1500	9000	89.34
Stacked BLS(2021)	1000	9000	98.91	1000	15000	89.59	1000	3000	89.67
ER-BLS(2023)	100	9000	98.94	200	8000	90.29	1600	2000	89.18
RA-BLS(ours)	1000	2000	**99.02**	1500	2000	**91.07**	1500	300	**92.33**

Table 2 shows that RA-BLS consistently achieves the best accuracy. The Accuracy on MNIST, FASHION-MNIST and NORB datasets are **99.02%**, **91.07%** and **92.33%**. This is because the performance of RA-BLS improves as the randomness of the enhancement nodes is iteratively weakened. Remarkably, the RA-BLS even improves accuracy, while the structure size is only 17% of that of BLS on the NORB dataset. This result indicates that RA-BLS can significantly reduce the model size while maintaining high performance. The RA-BLS demonstrates a significant advantage when higher accuracy is sought in the NORB dataset. The results confirm that RA-BLS outperforms BLS and prove that it is promising and effective.

4.3 Performance Evaluation on Regression

To evaluate the performance of the RA-BLS, three regression datasets from the UCI are selected for analysis. Table 3 presents details regarding the datasets used.

BLS and RA-BLS maintain the same number of nodes for low-complexity regression tasks. The parameters N_f and N_e have specified search ranges of [1,150] and

Table 3. Regression dataset details

Datasets	No. of Samples		Features
	Training	Testing	
Basketball	62	32	8
Mortgage	699	350	15
Housing	337	169	13

Table 4. Performance comparison of SVM, BLS and RA-BLS for Regression

Method	Basketball			Mortgage			Housing		
	N_f	N_e	RMSE	N_f	N_e	RMSE	N_f	N_e	RMSE
SVM	–	–	0.0894	–	–	0.0485	–	–	0.0989
BLS(2017)	21	1	0.0791	50	130	0.0051	145	50	0.0748
RA-BLS(ours)	21	1	**0.0790**	50	130	**0.0047**	145	50	**0.0742**

[1,200]. The RA-BLS performs excellently (See Table 4) on regression datasets and BLS. The proposed RA-BLS demonstrates superior performance on the Mortgage dataset. The results demonstrate the effectiveness of the RA-BLS.

5 Conclusion

This paper proposes a compact network called RA-BLS and performs an exhaustive experimental evaluation. The new system adopts a residual attention mechanism to correct the enhancement nodes to adapt to the current task rather than capture random features. The enhancement nodes without randomness significantly improve the efficiency and performance of BLS. This design replaces the fat BLS with a more compact and efficient structure. In addition, the RA-BLS provides a highly efficient and robust framework for dealing with various complex data tasks. The method for the residual attention mechanism is interchangeable, such as methods related to other attention mechanisms or reducing redundancy. The RA-BLS demonstrates the potential of BLS in modern data science and opens up new possibilities for future research and applications.

However, RA-BLS uses linear mapping to generate feature nodes. It cannot handle ultra-complex datasets such as CIFAR100 or ImageNet well. For ultra-complex datasets, the feature nodes should be generated by a more advanced deep learning feature extractor. Exploring the combination of advanced deep learning and RA-BLS is also a topic of great significance.

Funding. The work was supported by the Research Platforms and Program of the Education Department of Guangdong Province in China [grant numbers 2023ZDZX1011 and 2022ZDZX1013], and Scientic and Technological Planning Projects of Guangdong Province [grant numbers 2021A-0505030074]. It was also supported partly by Guangdong Basic and

Applied Basic Research Foundation [grant numbers 2021A1515011999], and the Guangdong Provincial Key Laboratory of Traditional Chinese Medicine Informatization [grant numbers 2021B1212040007].

References

1. Chen, C.L.P., Liu, Z.: Broad learning system: an effective and efficient incremental learning system without the need for deep architecture. IEEE Trans. Neural Netw. Learn. Syst. **29**(1), 10–24 (2017)
2. Chen, C.L.P., Liu, Z., Feng, S.: Universal approximation capability of broad learning system and its structural variations. IEEE Trans. Neural Netw. Learn. Syst. **30**(4), 1191–1204 (2018)
3. Gong, X., Zhang, T., Chen, C.L.P., Liu, Z.: Research review for broad learning system: algorithms, theory, and applications. IEEE Trans. Cybern. **52**(9), 8922–8950 (2021)
4. Chen, C.L.P.: A rapid supervised learning neural network for function interpolation and approximation. IEEE Trans. Neural Networks **7**(5), 1220–1230 (1996)
5. Chen, C.L.P.: A rapid supervised learning neural network for function interpolation and approximation. IEEE Trans. Neural Network **7**(5), 1220–1230 (1996)
6. Liu, Z., Chen, B., Xie, B., Qiang, H., Zhu, Z.: Feature selection for orthogonal broad learning system based on mutual information. In: 2019 International Joint Conference on Neural Networks (IJCNN), pp. 1–8. IEEE (2019)
7. Ma, J., et al.: Factorization of broad expansion for broad learning system. Inf. Sci. **630**, 271–285 (2023)
8. Asi, H., Duchi, J.C.: The importance of better models in stochastic optimization. Proc. Natl. Acad. Sci. **116**(46), 22 924–22 930 (2019)
9. Ding, W., Tian, Y., Han, S., Yuan, H.: Greedy broad learning system. IEEE Access **9**, 79 307–79 315 (2021)
10. Xie, R., Wang, S.: Downsizing and enhancing broad learning systems by feature augmentation and residuals boosting. Complex Intell. Syst. **6**(2), 411–429 (2020). https://doi.org/10.1007/s40747-020-00139-2
11. Vaswani, A., et al.: Attention is all you need. In: Advances in Neural Information Pprocessing Systems, vol. 30 (2017)
12. Ling, R.F.: Residuals and influence in regression. Technometrics **26**(4), 413–415 (1984)
13. Han, S., Qubo, C., Meng, H.: Parameter selection in SVM with RBF kernel function. World Autom. Congress **2012**, 1–4 (2012)
14. Liu, Z., Chen, C.L.P., Feng, S., Feng, Q., Zhang, T.: Stacked broad learning system: from incremental flatted structure to deep model. IEEE Trans. Syst. Man Cybern. Syst. **51**(1), 209–222 (2020)
15. Gan, J., Xie, X., Zhai, Y., He, G., Mai, C., Luo, H.: Facial beauty prediction fusing transfer learning and broad learning system. Soft Comput. **27**(18), 13 391–13 404 (2023)
16. Le Cun, Y., et al.: Handwritten digit recognition with a back-propagation network. In: Proceedings of the 2nd International Conference on Neural Information Processing Systems, vol. 9. Cambridge, MA, USA: MIT Press, pp. p. 396–404 (1989)
17. Xiao, H., Rasul, K., Vollgraf, R.: Fashion-mnist: a novel image dataset for benchmarking machine learning algorithms (2017). https://arxiv.org/abs/1708.07747
18. LeCun, Y., Huang, F.J., Bottou, L.: Learning methods for generic object recognition with invariance to pose and lighting. In: Proceedings of the 2004 IEEE Computer Society Conference on Computer Vision and Pattern Recognition, 2004. CVPR 2004, vol. 2, pp. 2–104 (2004)

EEG-Based Emotion Recognition Using Similarity Measures of Brain Rhythm Entropy Matrix

Guanyuan Feng[1,2], Peixian Wang[1], Xinyu Wu[1], Ximing Ren[1], Chen Ling[1], Yuesheng Huang[1], Leijun Wang[1], Jujian Lv[1], Jiawen Li[1,2(✉)], and Rongjun Chen[1,3]

[1] School of Computer Science, Guangdong Polytechnic Normal University,
Guangzhou 510665, China
lijiawen@gpnu.edu.cn
[2] Anhui Provincial Key Laboratory of Multimodal Cognitive Computation, Anhui University,
Hefei 230601, China
[3] Guangdong Provincial Key Laboratory of Intellectual Property and Big Data, Guangdong Polytechnic Normal University, Guangzhou 510665, China

Abstract. Constructing the Brain Rhythm Entropy Matrix (BREM) as a feature for emotion recognition is feasible due to the correlation between brain rhythm activities and arousal/valence levels, as well as the effectiveness of entropy in quantifying such activities. Besides, similarity, a vital metric in bioinformatics, measures the degree of resemblance between internal elements. Building on these foundations, this paper introduces an emotion recognition method that leverages the similarity measures of BREM derived from EEG signals. The results using the DEAP database indicate that the proposed method achieves an accuracy range of 75% to 92% by employing the optimal single-channel data for emotion recognition. Furthermore, significant individual differences were observed in selected channels and time windows for emotion recognition. Such findings not only offer an innovative manner for EEG-based emotion recognition through BREM and similarity measures but also provide advances for designing portable emotion computing devices in the future.

Keywords: Emotion Recognition · EEG · Brain Rhythm Entropy Matrix (BREM) · Similarity Measures

1 Introduction

Emotion is a comprehensive response to external stimuli or internal mental activities, encompassing various psychological, behavioral, and physiological changes. For example, emotions such as happiness, anger, and sadness swiftly influence cognitive, decision-making, behavioral, and physiological states, regulating interactions with the external environment accordingly [1]. Generally, emotion can be expressed through non-verbal behaviors, such as facial expressions, speech, and body movements. Nonetheless, social norms and individual preferences influence emotional expression, posing challenges for accurate recognition. To address this issue, physiological measures, including electroencephalography (EEG) signals, heart rate, and skin conductance, are widely utilized as they reflect authentic responses to emotional stimuli. Specifically, EEG signals record the electrophysiological activities occurred in the brain, capturing neural changes and offering direct emotional insights. Therefore, analyzing EEG signals enhances accuracy and mitigates errors resulting from inconsistencies in external behavioral expressions.

Intracranial electrophysiology studies demonstrate that emotion processing engages multiple brain regions. For instance, negative emotions are primarily processed in the subcortical nucleus accumbens, temporal lobe, temporoparietal junction, and inferior frontal gyrus [2]. The forebrain, among these regions, plays a critical role in emotion regulation. Specifically, hyperactivity in the left prefrontal lobe is associated with positive emotions, while hyperactivity in the right prefrontal lobe correlates with negative emotions [3]. Moreover, emotion is associated not only with different brain regions but also with distinct brain rhythms reflected in EEG signals. For example, happy stimuli are positively correlated with delta rhythm energy near the prefrontal cortex midline [4], while gamma rhythm marks emotional stimuli in arousal ratings [5]. Consequently, the two classic dimensions of emotion assessment, arousal and valence, are strongly related to the changes in brain rhythms.

Although brain rhythms are complex, entropy, as a metric capable of analyzing EEG signals, sensitive to non-linear dynamics, and less sensitive to artifacts and noise, is an appropriate tool for analyzing rhythmic activity as a reflection of emotion. Besides, dynamic entropy analysis can uncover hidden information not acquired by averaging or spectral analyses [6], enriching the feature set for classifying emotional states accordingly. Furthermore, in bioinformatics, similarity is commonly employed to classify biological sequences. Based on that, we propose the similarity measures to the Brain Rhythm Entropy Matrix (BREM) for emotion recognition in this paper.

During the experiment, EEG signals from various channels and time windows are analyzed to identify the optimal single-channel data and most suitable time window for emotion recognition with the highest accuracy. This process reduces data redundancy and the computational cost of EEG-based emotion recognition, which contributes a novel perspective for the development of portable emotion computing devices in the future.

2 Database

The EEG signals utilized in this paper were sourced from a publicly accessible DEAP database [7], widely assessed in emotion recognition. In this database, 32 participants (17 male and 14 female) aged 19 to 37 years were recruited. During the experiment, the participants were presented with 40 music videos from various genres, each lasting 60 s, to elicit different emotional responses. Afterward, the participants rated the arousal, valence, like, dominance, and familiarity of the 40 music videos based on the Self-Assessment Manikin (SAM) scale (1–9). In this paper, only arousal and valence were considered, i.e., High Valence (HV) was defined as valence ≥ 5, while Low Valence (LV) as valence <5. Similarly, the same criteria were adopted to classify High Arousal (HA) and Low Arousal (LA). Meanwhile, the 10–20 EEG acquisition system recorded data during each trial, creating a database of 32 participants $\times 40$ trials $\times 32$ channels. Lastly, the EEG signals were sampled at 128 Hz and filtered within 0.1–100 Hz.

3 Methodology

The proposed method is organized into four main stages: brain rhythm extraction, entropy extraction, matrix construction, and similarity classification. First, considering that brain rhythms beneficially denote emotional arousal and valence levels, we employ the Discrete Wavelet Transform (DWT) to extract five brain rhythms, i.e., delta, theta, alpha, beta, and gamma. In this regard, DWT builds on the localization concept derived from the Short-time Fourier Transform (STFT), offering a time-frequency window that adapts with frequency, which helps to avoid the data redundancy potentially generated by the Continuous Wavelet Transform (CWT). Hence, DWT is appropriate for time-frequency analysis in EEG signals processing.

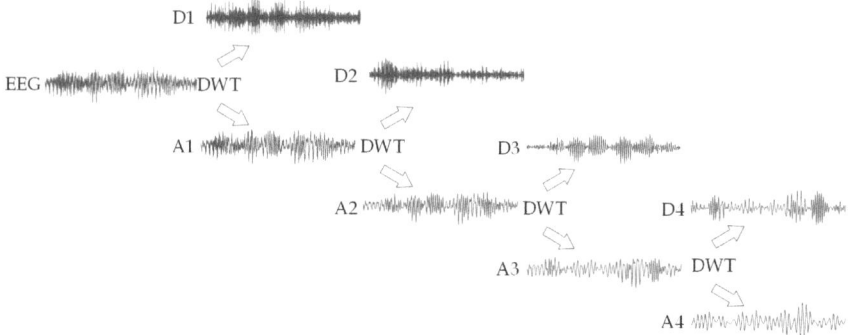

Fig. 1. DWT decomposes EEG signals into high-frequency and low-frequency components, where D1, D2, D3, D4, and A1 correspond to gamma, beta, alpha, theta, and delta, respectively.

Mathematically, DWT is implemented via multi-level high-frequency and low-frequency filter banks, as expressed in (1) and (2), which can be conceptualized as a discrete wavelet tree, depicted in Fig. 1. Particularly, the orthogonality of the DB-4 wavelet basis function used in this paper improves both the efficiency and accuracy of signal processing, as its smoothness makes it suitable for analyzing signals with smooth transitions and intricate details. Such characteristics are vital for EEG signals analysis because they contain physiological information and subtle fluctuations.

$$a_{j+1}[k] = \sum_n h[n - 2k]a_j[n] \quad (1)$$

$$d_{j+1}[k] = \sum_n g[n - 2k]d_j[n] \quad (2)$$

Subsequently, entropy, as a measure of complexity and uncertainty, effectively denotes the dynamic changes in EEG signals. By using the entropy of brain rhythms, dynamic variations in brain activity can be captured, which, in turn, reflect emotional states and avoid extracting meaningless data properties due to insufficient size [8]. To achieve this, we adopt four entropies: approximate entropy (AE) [6], sample entropy

(SE) [9], fuzzy entropy (FE) [10], and permutation entropy (PE) [11], to characterize the complex variations in five brain rhythms. By calculating the entropy for each brain rhythm and integrating these values, the BREM is constructed, reflecting the complexity and uncertainty of brain activity in different emotional states, as presented in (3). Here, BREM integrates five brain rhythms with four entropies, leveraging both the characteristics of brain rhythm changes during emotional processes and entropy's sensitivity to non-stationary EEG signals. This way establishes a more comprehensive feature to indicate the dynamic characteristics in EEG signals, enhancing the accuracy and interpretability of emotion recognition.

$$\text{BREM} = \begin{bmatrix} AE_\gamma & FE_\gamma & SE_\gamma & PE_\gamma \\ AE_\beta & FE_\beta & SE_\beta & PE_\beta \\ AE_\alpha & FE_\alpha & SE_\alpha & PE_\alpha \\ AE_\theta & FE_\theta & SE_\theta & PE_\theta \\ AE_\delta & FE_\delta & SE_\delta & PE_\delta \end{bmatrix} \quad (3)$$

Next, a high similarity between the two matrices indicates that the features they represent are quite similar. Conversely, a low similarity between matrices suggests significant differences in the features they represent. Based on that, measuring the similarity of BREM is a vital step for identifying different emotional states. Given the close linkage between brain rhythm activities and emotional states, and considering that entropy quantifies the complexity and uncertainty of these activities, similar BREMs reflect similar brain rhythm entropy features, which imply the same emotional state. Thus, matrix similarity is employed as a core in the proposed method for accomplishing emotion recognition. Specifically, Dynamic Time Warping (DTW) is utilized to measure the similarity level between two BREMs. Unlike traditional Euclidean distance, DTW allows for stretching to find the optimal alignment path, making it more suitable for addressing the nonlinear alignment needs of brain rhythm entropy features. Using DTW to compute the optimal alignment path between two BREMs and determine the minimal distance, we can measure their similarity level and perform a comparison for classification.

Before classification through similarity measures, it is essential to establish a standard template. Considering individual differences and the emotional variations induced by watching music videos in each trial, we employ a leave-one-out cross-validation approach for determining the standard template. For example, Subject 1 (S1) participated in 40 trials, with 16 in LA and 24 in HA. We select the BREMs from the first LA trial and the first HA trial as the initial standard templates, while the remaining trials served as test samples for similarity measures to evaluate their emotional states (LA or HA). Similarly, the BREMs from the second LA trial and the first HA trial were chosen as standard templates, and the process was repeated. As a result, a total of 16×24 classifications were conducted for S1.

Lastly, to investigate the optimal time window, we segment the EEG data into non-overlapping 10-s intervals. This segmentation not only aids in analyzing the specific characteristics of brain activities across various periods but also reduces data size and improves computational efficiency. For instance, for S1, there were 16 LA trials and 24 HA trials, with each divided into six 10-s time windows. Therefore, the total number of classifications for S1 amounted to 16 (LA) × 24 (HA) × 32 (channels) × 6 (time windows). By fully analyzing each classification result, we prioritize maximizing accuracy and decide the key factors, including the optimal channel and the suitable time window for each subject in the DEAP database.

4 Results and Discussion

After applying the proposed method, we can identify the optimal channels and their corresponding time windows for 32 subjects of the DEAP database, along with the respective classification accuracies, as listed in Table 1. As seen, in both the arousal and valence classifications, all accuracies exceeded 75%. Particularly, the highest reached 92.11% for S13 across 32 channels and 6 time windows (0–10 s, 10–20 s, 20–30 s, 30–40 s, 40–50 s, and 50–60 s) in the arousal classification, as illustrated in Fig. 2. Furthermore, our analysis revealed that the optimal channels for arousal and valence varied among the subjects, indicating that different dimensions of emotion recognition may correspond to different EEG channels, consistent with findings in the previous work [12]. Besides, in arousal classification, 11 subjects demonstrated the best performance within the 0–10 s window, while in valence classification, 10 subjects exhibited a similar trend. It can be said that about 30% of the subjects exhibited heightened emotional awareness during the initial stages (0–10 s) when watching music videos, with the impact of the music video on emotions diminishing as the experiment progressed. Meanwhile, it was found that certain subjects, such as S1 and S24, had the same optimal channel and time segment (CP6 and 0–10 s) for arousal classification but entirely different ones for valence classification. Similar results were observed in S18 and S19, as well as S9 and S29. Next, in valence classification, S2 and S10, as well as S6 and S11, displayed discrepancies similar to those observed in arousal classification. These findings highlighted significant individual differences in emotion recognition, which is expected given the influence of factors such as personal experiences, cultural background, and history on emotional responses [13]. Finally, for about 75% of the subjects, their time windows for arousal and valence classifications were inconsistent. Therefore, it is not advisable to rate arousal and valence within the same time window when employing music videos as stimuli. That implies a personalized classification method for acquiring a suitable time window is preferred.

Table 1. Optimal channels and suitable time windows for EEG-based emotion recognition of 32 subjects in the DEAP database.

Subject	Arousal classification			Valence classification		
	Channel	Time window (s)	Accuracy (%)	Channel	Time window (s)	Accuracy (%)
S1	CP6	0–10	84.21	FC1	20–30	86.84
S2	FZ	50–60	78.95	T7	0–10	84.21

(*continued*)

Table 1. (*continued*)

Subject	Arousal classification			Valence classification		
	Channel	Time window (s)	Accuracy (%)	Channel	Time window (s)	Accuracy (%)
S3	AF4	0–10	89.47	T8	20–30	81.58
S4	CP5	0–10	86.84	F8	50–60	81.58
S5	PO4	50–60	81.58	O1	0–10	78.95
S6	O1	0–10	78.95	FC5	0–10	86.84
S7	C3	50–60	84.21	FP2	50–60	84.21
S8	F4	20–30	81.58	P3	0–10	76.32
S9	PO3	20–30	84.21	F7	30–40	81.58
S10	CZ	0–10	76.32	T7	0–10	78.95
S11	P4	20–30	84.21	FC5	0–10	78.95
S12	CP5	20–30	92.11	P3	30–40	78.95
S13	F8	10–20	92.11	P4	10–20	76.32
S14	F7	0–10	81.58	AF3	10–20	78.95
S15	FC2	0–10	76.32	FC5	20–30	84.21
S16	CP1	30–40	84.21	OZ	10–20	78.95
S17	PO3	10–20	76.32	CZ	50–60	81.58
S18	F3	20–30	81.58	FP1	10–20	81.58
S19	F3	20–30	81.58	T8	20–30	78.95
S20	F7	40–50	86.84	F8	0–10	81.58
S21	FC5	0–10	89.47	F8	20–30	78.95
S22	F3	30–40	81.58	F8	0–10	78.95
S23	C4	40–50	81.58	FC1	10–20	81.58
S24	CP6	0–10	89.47	PZ	30–40	78.95
S25	C3	20–30	89.47	FP2	0–10	81.58
S26	C3	10–20	78.95	CP2	40–50	84.21
S27	P8	0–10	86.84	C4	0–10	86.84
S28	F8	20–30	78.95	F8	10–20	78.95
S29	PO3	20–30	84.21	T8	30–40	84.21
S30	PO3	30–40	78.95	F4	30–40	84.21
S31	FC2	20–30	78.95	C4	20–30	84.21
S32	O2	0–10	86.84	F7	30–40	78.95

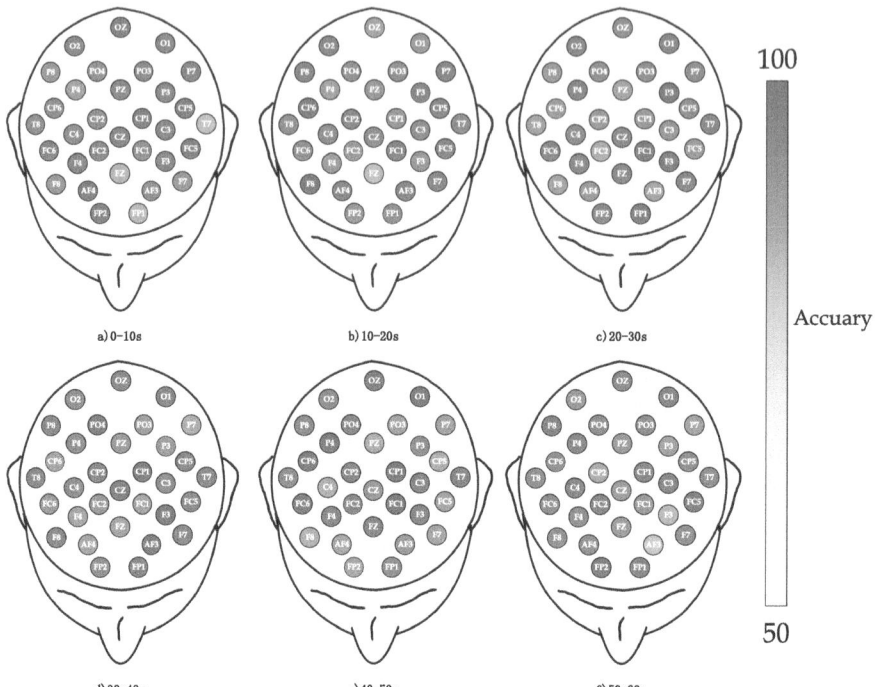

Fig. 2. The arousal classification accuracies of S13 across 32 channels and 6 time windows using the proposed method. The color intensity indicates the level of accuracy, with deeper colors representing higher accuracy: a) 0–10 s; b) 10–20 s; c) 20–30 s; d) 30–40 s; e) 40–50 s; f) 50–60 s.

Table 2 presents a comparative study focusing on parameters such as the number of channels, the duration of time windows, the methodologies employed, and the classification accuracy for arousal and valence. Previous studies have focused on multi-channel analysis, which is a logical approach since an increased number of channels usually provides more comprehensive data, allowing for finer details to be captured. Nevertheless, using fewer channels involves fewer electrodes, and single-channel methods push this simplification to its limits. Specifically, single-channel data significantly reduces the complexity of EEG input, which facilitates the development of portable devices. Regarding the time window, although the work [14] segmented EEG signals to increase the number of samples and enhance classification performance, it did not conduct a comprehensive analysis of various time windows to identify the suitable one. Besides, previous work did not employ similarity measures of BREM. Therefore, the overall performance of the proposed method indicates it can be viewed as a cost-effective EEG-based emotion recognition solution.

Table 2. Comparative study of recent EEG-based emotion recognition works.

Work	Number of channel	Time window (s)	Methodology	Accuracy(%)	
				Arousal	Valence
Zhu et al. [14]	32	60	Dynamic energy features with Bidirectional Long Short-term Memory (Bi-LSTM)	77.34	75.78
Tiwari et al. [15]	32	60	Joint Emotion Class Learning with Self-Similarity Learning (JEC-SSL)	69.98	70.30
Liu et al. [16]	32	2	Emotion Recognition Transformer Network (ERTNet)	80.99	73.31
Rajamanickam et al. [17]	32	60	Fractal Dimension with Classification and Regression Tree (FD-CART)	79.90	78.18
Kannadasan et al. [18]	32	60	Convolutional Neural Network (CNN) features with Support Vector Machine (SVM)	58.44	58.13
This work	1	10	**BREM with similarity measures**	**83.39**	**81.33**

5 Conclusion

This paper validated the proposed similarity-based classification method using the BREMs derived from EEG signals in the DEAP database. The analysis included data from 32 participants, 32 channels, and 40 trials. The findings revealed that, compared to previous works, the classification performance remained high when utilizing single-channel data, with an accuracy of approximately 75% to 92%, despite a reduction in data input. Specifically, the experiments analyzed optimal channels for different subjects, demonstrating individual differences in emotion recognition, and also found that arousal and valence time windows often do not align in most cases. Consequently, a personalized approach for EEG-based emotion recognition is desired, which can be achieved through the proposed similarity measures of BREMs.

Although this paper exhibited the potential to develop portable devices by utilizing optimal single-channel data sources and selecting appropriate time windows, ongoing efforts should focus on enhancing accuracies when classifying more emotional dimensions and improving generalization for other EEG-based applications.

Acknowledgments. The authors extend their appreciation to the financial support in part by the Special Projects in Key Fields of Ordinary Universities of Guangdong Province under Grant 2021ZDZX1087, in part by the Guangzhou Science and Technology Plan Project under Grants 2024B03J1361, 2023B03J1327, and 2023A04J0361, in part by the Graduate Education Innovation Project of Guangdong Province under Grant 2020JGXM079, in part by the Open Research Fund of Anhui Provincial Key Laboratory of Multimodal Cognitive Computation under Grant MMC202305, in part by the Guangdong Province Higher Vocational Education Teaching Quality and Teaching Reform Project under Grant 2023JG296, in part by the Guangdong Province Ordinary Colleges and Universities Young Innovative Talents Project under Grant 2023KQNCX036, in part by the Key Discipline Improvement Project of Guangdong Province under Grants 2022ZDJS015 and 2021ZDJS025, in part by the Scientific Research Capacity Improvement Project of the Doctoral Program Construction Unit of Guangdong Polytechnic Normal University under Grant 22GPNUZDJS17, in part by the Graduate Education Innovation Program of Guangdong Polytechnic Normal University under Grants 2024XJSFKC006, 2023YJSY01008, and 2023YJSY04002, in part by the Graduate Education Teaching Achievement Cultivation Project of Guangdong Polytechnic Normal University under Grant 2024XJJXCG002, in part by the Engineering Construction project of Science, Industry, and Education Integration Practice Teaching Base of Guangdong Province under Grant 2021-31, in part by the Special Fund for Science and Technology Innovation Strategy of Guangdong Province (Climbing Plan) under Grant pdjh2024a226, and in part by the Research Fund of Guangdong Polytechnic Normal University under Grant 2022SDKYA015.

Disclosure of Interests. The authors have no competing interests to declare that are relevant to the content of this article.

References

1. Houssein, E.H., Hammad, A., Ali, A.A.: Human emotion recognition from EEG-based brain-computer interface using machine learning: a comprehensive review. Neural Comput. Appl. **34**(15), 12527–12557 (2022)
2. Li, X., Zhang, Y., Tiwari, P., et al.: EEG based emotion recognition: a tutorial and review. ACM Comput. Surv. **55**(4), 1–57 (2022)
3. Mir, M., Nasirzadeh, F., Bereznicki, H., et al.: Investigating the effects of different levels and types of construction noise on emotions using EEG data. Build. Environ. **225**, 109619 (2022)
4. Iyer, A., Das, S.S., Teotia, R., et al.: CNN and LSTM based ensemble learning for human emotion recognition using EEG recordings. Multimedia Tools Appl. **82**(4), 4883–4896 (2023)
5. Li, D., Xie, L., Chai, B., et al.: Spatial-frequency convolutional self-attention network for EEG emotion recognition. Appl. Soft Comput. **122**, 108740 (2022)

6. Cacciotti, A., Pappalettera, C., Miraglia, F., et al.: Complexity analysis from EEG data in congestive heart failure: a study via approximate entropy. Acta Physiol. **238**(2), e13979 (2023)
7. Koelstra, S., Mühl, C., Soleymani, M., et al.: DEAP: a database for emotion analysis using physiological signals. IEEE Trans. Affect. Comput. **3**(1), 18–31 (2012)
8. Hasan, M.J., Kim, J., Kim, C.H., et al.: Health state classification of a spherical tank using a hybrid bag of features and k-nearest neighbor. Appl. Sci. **10**, 2525 (2020)
9. Zhang, M., Zhang, J., Hou, A., et al.: Aerodynamic system instability identification with sample entropy algorithm based on feature extraction. Propul. Power Res. **12**(1), 138–152 (2023)
10. Xu, X., Tang, J., Xu, T., et al.: Mental fatigue degree recognition based on relative band power and fuzzy entropy of EEG. Int. J. Environ. Res. Public Health **20**, 1447 (2023)
11. Boaretto, B.R.R., Budzinski, R.C., Rossi, K.L., et al.: Spatial permutation entropy distinguishes resting brain states. Chaos Solitons Fractals **171**, 113453 (2023)
12. Akbarnia, Y., Daliri, M.R.: EEG-based identification system using deep neural networks with frequency features. Heliyon **10**, 4 (2024)
13. Lim, N.: Cultural differences in emotion: differences in emotional arousal level between the East and the West. Integrat. Med. Res. **5**(2), 105–109 (2016)
14. Zhu, M., Wang, Q., Luo, J.: Emotion recognition based on dynamic energy features using a Bi-LSTM network. Front. Comput. Neurosci. **15**, 741086 (2022)
15. Tiwari U., Chakraborty R., Kopparapu S.K.: Joint class learning with self similarity projection for EEG emotion recognition. In: 7th Joint International Conference on Data Science & Management of Data, pp. 207–211. ACM, Bangalore, India (2024)
16. Liu, R., Chao, Y., Ma, X., et al.: ERTNet: an interpretable transformer-based framework for EEG emotion recognition. Front. Neurosci. **18**, 1320645 (2024)
17. Rajamanickam, Y., Thagavel, P., Thomas, J., et al.: Comprehensive analysis of feature extraction methods for emotion recognition from multichannel EEG recordings. Sensors **23**, 915 (2023)
18. Kannadasan, K., Miraj, M.T.I., Bheekharry, K.S., et al.: Analysis of feature extraction models for emotion recognition using EEG Signals. In: 2022 IEEE 3rd Global Conference for Advancement in Technology, pp. 1–6. IEEE, Bangalore, India (2022)

Intensity Controllable Emotional Speech Synthesis Based on Valence-Arousal-Dominance

Guoping Li(✉) and Yanxiang Chen

Hefei University of Technology, Hefei, China
`2022111035@mail.hfut.edu.cn, chenyx@hfut.edu.cn`

Abstract. Speech spoofing technologies have advanced significantly, enabling the creation of fake audio that closely mimics authentic human voices. Nevertheless, these synthetic speeches often lack precise control over emotional intensity. This paper introduces a technique to modulate the intensity of emotions in synthesized speech using a three-dimensional (3D) emotion representation: Valence-Arousal-Dominance (VAD). The process entails mapping emotion embeddings onto a continuous 3D emotion continuum and fine-tuning the dimensionality values within specific ranges to regulate emotional intensity. Leveraging a feature fusion network grounded on an emotion2vec pre-trained model, we devise a transformation model from labeled data to convert VAD vectors into emotion embeddings. Experimental results confirm that our method enhances the quality of synthetic speech production and affords superior command over emotional intensity.

Keywords: Feature fusion · Emotion intensity control · Valence-Arousal-Dominance · Emotion speech synthesis

1 Introduction

Currently, speech synthesis technology can generate smooth and natural speech, but the synthesized speech lacks emotional expression, and it is difficult to simulate the rich emotional expression ability that real human speech has. There have been many emotional speech synthesis related studies in recent years. One direct method is to use speech data with emotion labels to train a TTS model, as done by Lee et al. [1], who embedded emotion labels into the decoder to obtain speech with corresponding emotions. However, speech data with emotion labels are more difficult to obtain, to overcome this difficulty, emotional speech synthesis techniques based on style migration have been proposed to migrate the emotional style of the reference speech to the target speech, Global style token and its derived models [2, 3] use a set of tokens that can learn different style representations to model the emotions in the speech, and then through these tokens to migrate emotions to the target speech. In order to model more rhyming embeddings and obtain higher dimensional emotion embedding vectors, works such as [4, 5] use a pre-trained speech emotion recognition model to extract speech features as deep emotion features.

The original version of the chapter has been revised. The Chapter 4 corresponding author name has been corrected and both the author's affiliation has been corrected. A correction to this chapter can be found at https://doi.org/10.1007/978-981-96-2882-7_29

© The Author(s), under exclusive license to Springer Nature Singapore Pte Ltd. 2025, corrected publication 2025
A. Hussain et al. (Eds.): BICS 2024, LNAI 15397, pp. 30–40, 2025.
https://doi.org/10.1007/978-981-96-2882-7_4

While the aforementioned methodologies can indeed facilitate speech synthesis across diverse emotional categories, they nonetheless fall short in modulating the intensity of any singular emotion. The varying intensity of an emotion profoundly impacts the perception of that emotion by an individual. Current mainstream methods for controlling emotion intensity include scaling factors [6], relative attributes [7], and interpolation [8]. In [6], relative attributes are used to rank different emotional audio samples, and the authors also introduce a scaling factor for regulating emotional intensity. In [7], a relative attribute-based ranking function is proposed, where the authors consider the sentiment strength as the relative gap between neutral speech and other emotional speech, and the strength and relative gap of emotional speech are numerically equal if the sentiment strength of neutral speech is zero. The ranking function is obtained by learning the features between each pair of neutral and emotional speech to predict the intensity value of the corresponding emotion. [8] proposed a nonlinear interpolation technique by gradually changing the weight values of the emotional embedding vectors, which in turn leads to the gradual increase in intensity from neutral speech with intensity 0 to emotional speech with gradually increasing intensity.

In this study, we aim to develop a high-performance Text-to-Speech (TTS) system that can produce varied emotional expressions with adjustable intensity. We present a novel method consisting of two-stage training and single-stage inference. Initially, a feature processing network, guided by the pre-trained emotion2vec model [9], is used to extract and refine the emotional features from reference audio, resulting in deep emotion embeddings for the acoustic model. The emotion2vec model employs a combination of sentence-level and frame-level loss in an online distillation strategy on unlabeled datasets for self-supervised training to effectively capture emotional nuances. In the second training stage, we introduce a three-dimensional emotion representation, Valence-Arousal-Dominance (VAD) [10], enabling precise manipulation of emotional intensity through adjustments in the VAD components. These components are then transformed into high-dimensional emotion embeddings using a specialized transformation model, serving as critical inputs to the acoustic model. The VAD vectors act as crucial latent variables, linking the encoder and decoder during training. To achieve this transformation, we design an auto-encoder-like model using wav2vec [11] as the encoder to convert input speech into 3D VAD vectors. These vectors are then converted into deeper emotion embeddings by a decoder comprising linear layers. During inference, we can synthesize speech matching the reference audio's emotion category but with varied intensity or directly generate sentiment speech. Distinct from prior work, our VAD emotion intensity control modifies the emotion embedding directly without utilizing an intensity-dependent weight vector. The paper's key contributions are as follows:

1. An emotion2vec-based feature extraction network is designed to distill both global and local emotional features from speech. These features are then seamlessly integrated using a weighted feature fusion approach.
2. We introduce a three-dimensional emotion representation, VAD, and propose a transformation model with VAD vectors as latent variables. The emotion intensity can be controlled by fine-tuning the VAD vectors. In practice, the vectors representing different emotion strengths can be directly input to generate the target speech, and the synthesis step is simpler.

3. We conducted extensive experiments to validate the effectiveness and superiority of the proposed method in generating emotional speech and emotional intensity control.

2 The Proposed Method

Our model builds upon the core architecture of VITS [12], known for its superior performance in speech synthesis. As a complete end-to-end system, VITS produces natural and smooth speech without requiring a vocoder. In this research, we enhance the VITS framework by incorporating an emotion feature extractor, emotion intensity control mechanisms, and the original VITS structure, as shown in Fig. 1. The emotion feature extractor, based on emotion2vec, captures emotional nuances from the audio. This information is then processed by various feature modules, fused, and fed into VITS alongside phoneme data. The emotion intensity control mechanism modifies the VAD vector of the speech signal and includes three main parts: a wav2vec pre-trained model, the VAD representation, and a transfer model. In this configuration, wav2vec acts as an encoder to convert audio inputs into latent VAD variables, which are then transformed into deep emotion embeddings by the transfer model acting as a decoder. The following sections describe each of these modules.

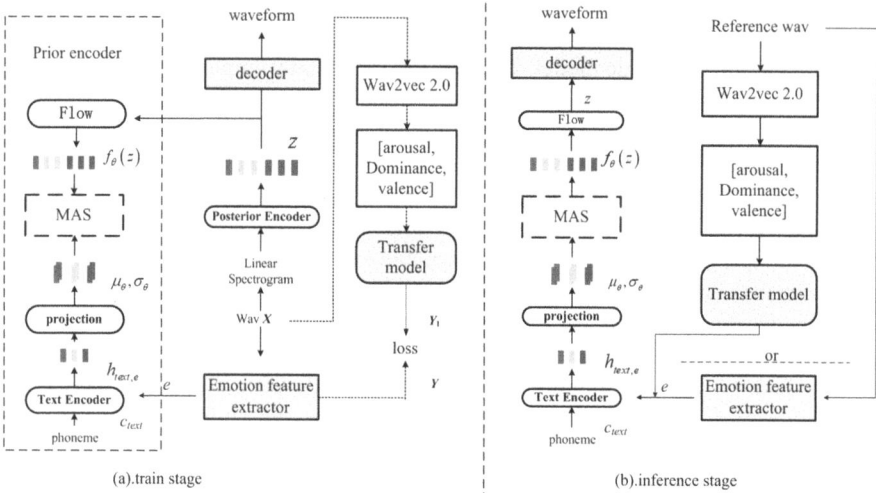

Fig. 1. The figure shows the overall architecture of the model, which is divided into training stage (a) and inference stage (b). The model consists of a prior encoder, a posterior encoder, an emotion feature extractor and a decoder, where the posterior encoder is not involved in the inference.

2.1 Emotion Feature Extractor

Inspired by [13], the emotion feature extractor consists of a pre-trained emotion2vec model, a global feature processing module, a local feature processing module, and the structure is shown in Fig. 2. The specific process is as follows: the audio input is first extracted by emotion2vec to extract the emotion features, which are fed into the subsequent linear layer to change the dimensionality. Then the transformed feature sequence is input into the global feature processing module and local feature processing module respectively, and the emotional features are divided into coarse-grained and fine-grained outputs, and finally the two obtained features are fused.

The global feature extraction module is used to extract the sentiment information contained in the entire utterance. First the features obtained from emotion2vec pass through a linear layer with a Relu activation function. Then it is passed through a LSTM network [14] which remains the contextual information to a large extent because of its good memory. Finally, the feature sequences are aggregated to form a feature vector of length 768 through a MaskAvg module.

In contrast, local features focus on changes in expression in a sentence. The output obtained from emotion2vec is passed through a linear layer with a Relu activation function, fed into a maximum pooling layer, and finally passed through the linear layer. The maximum pooling layer expands the perceptual field while still preserving locally salient features. The final linear layer is to ensure that the local feature vectors have the same dimensions as the global features to facilitate feature fusion.

The global features and local features of the audio information are obtained through the global feature processing module and the local feature processing module, respectively. Inspired by [15], we use weighted summation for the fusion of global and local features. The fusion formula is as follows:

$$h_{fusion} = \alpha h_{global} + (1 - \alpha) h_{local} \tag{1}$$

where h_{global}, h_{local} denote global and local features, respectively. α is a weight parameter, which can be determined by model training. h_{fusion} is the fused feature vector which is used as the input to the prior encoder together with the phoneme information.

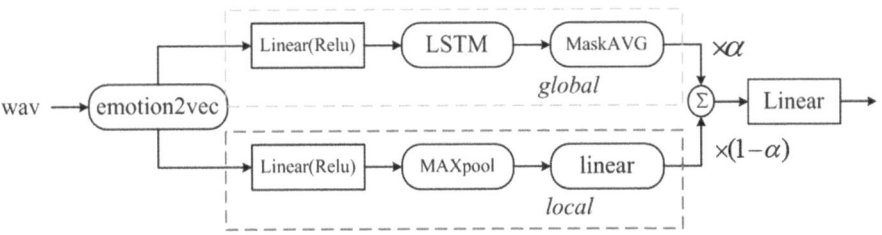

Fig. 2. Emotion feature extraction and fusion

2.2 Emotion Intensity Control

According to the VAD theory of affective space, human emotions can be described by three dimensions: valence, arousal, and dominance. We represent emotions in speech as a three-dimensional vector corresponding to these dimensions. Our findings indicate that this vector can encapsulate the same emotion within a certain range, with variations in dimension components reflecting different intensities. Consequently, we developed an emotion control module based on this concept.

In Fig. 1, wav2vec is used to extract the VAD feature vectors of wav X when the model is trained. The obtained VAD vectors are used as inputs to the transfer model, which consists of a stack of linear layers, and its goal is to transform the dimensionality of the VAD vectors from 3 to 768 as the deep emotion embedding.

In the training stage, the VAD vectors of the wav X are input to the transfer model to get the L1 paradigm between the emotion embedding Y_1 and the target emotion embedding Y as a loss function for training. In the inference stage, the VAD feature vector of the reference audio is first obtained by wav2vec, and the VAD component is adjusted to control the intensity of the emotion, and the emotion embedding e carrying the intensity information is obtained by the transfer model. The whole process can be represented as Fig. 3.

Training stage	Inference stage
Input: training dataset ESD Repeat: 1: for (X, Y) in ESD: 2: e_VAD = wav2vec(X) 3: Y_1 = transfer_model(e_VAD) 4: loss = $\Vert Y_1 - Y \Vert_1$ 5: backward and update parameters 6: end for Until: loss stabilizing **Output**: save current model	**Input**: Reference audio X 1: VAD = wav2vec(X) 2: Adjust the value of VAD 3: e = transfer_model(VAD) 4: end for **Output**: e

Fig. 3. Transformation model training and inference process

3 Experiment

3.1 Experiment Setup

In our experiments, the dataset we used is the ESD dataset [16], a multi-task emotion dataset, which consists of two parts in English and Chinese, 10 speakers in each part, and five emotions for each speaker: anger, happiness, sadness, surprise, and neutral. Each emotion contains 350 utterances, 300 for training, 30 for testing, and 20 for validation. We use the English part, totaling about 750 min of training data. The audio is sampled at 16 kHz and feature extraction is performed on each audio to obtain the corresponding VAD feature vector and 768-dimensional emotion embedding, respectively. An 80-dimensional linear spectrum is extracted by short-time Fourier transform every 12.5 ms with a frame size of 50 ms, which is used as input to the posterior encoder.

In training, the model was trained using the Adam optimizer [17], with the initial learning rate set to 0.0002 and the exponentially weighted mean decay coefficient of 0.8 and 0.99 respectively. We used NVIDIA RTX3090 GPU for training with batch-size set to 64.

3.2 Result and Analysis

Speech Quality. In order to validate the effectiveness of the proposed feature extraction method, three different methods (GST [2], wav2vec2 [13] and emotion2vec) were used as feature extractors and the acoustic model was unified using VITS. We also performed an ablation study to validate the effectiveness of the feature fusion strategy. The global or local feature processing networks in the feature extractors were removed for comparison, respectively. We will compare the quality of the generated speech of these three methods in terms of objective and subjective evaluation respectively. Objective evaluation metrics include Mel-cepstral distance (MCD) [18], emotion accuracy (Emotion accuracy is obtained by using a BiLSTM-based emotion recognition model [19] to recognize the emotions in the synthesized speech and by calculating the classification accuracy) and subjective evaluation metrics are dominated by mean opinion score (MOS) [20]. The result of evaluation is shown in Tables 1 and 2.

Table 1. Compare the MCD and emotion recognition accuracy (Emo.Acc.) of different feature extraction modules.

Method	MCD(dB)	Emo.Acc.(%)
Ground-truth	–	98.98
GST	4.97	85.35
Wav2vec	4.33	92,84
Emo2vec(g)	4.62	86.66
Emo2vec(l)	5.06	83.96
Emo2vec	**4.02**	**96.48**

Table 1 shows the objective evaluation results of proposed (emo2vec) with emo2vec(g), emo2vec(l), emo2vec and two baselines. Emo2vec(g) and emo2vec(l) denote the emotion2vec pre-trained model plus the global and local processing modules, respectively. The MCD value of emo2vec is the smallest, which indicates that the speech generated by the feature extractor based on emotion2vec is the closest to the ground-truth, while a single local or global feature extractor is not as effective as emo2vec, which suggests that the fusion feature of global and local information that we have proposed is better than a single feature.

For Emotion accuracy, except for ground-truth, emo2vec has the highest accuracy, which indicates that the synthesized speech emotion based on the model of emo2vec has higher recognition compared to GST and wav2vec. Emo2vec(g) and emo2vec(l) have lower accuracy than emo2vec, which It shows that fused features perform better than single features. From the experimental results, our proposed feature fusion strategy based on emo2vec outperforms GST and wav2vec in terms of quality and emotion of the generated speech.

Table 2. The MOS scores are presented with 95% confidence intervals

Method	MOS
GT	4.75 ± 0.010
GST	3.73 ± 0.073
Wav2vec	4.36 ± 0.109
Emo2vec(g)	4.02 ± 0.060
Emo2vec(l)	3.95 ± 0.133
Emo2vec	**4.71 ± 0.065**

We selected 20 texts corresponding to 100 speeches from the test set and synthesized 600 emotional speeches using six models for evaluation. Ten testers scored the speeches based on quality and emotional similarity, with each voice assessed by a different tester. The average MOS and 95% confidence interval were calculated. Table 2 displays the MOS scores, showing that Emo2vec outperforms GST and wav2vec by margins of 0.98 and 0.35 points, respectively. This indicates the superiority of the emotion2vec-based approach over GST and wav2vec. Models emo2vec(g) and emo2vec(l), which use only global or local vectors, scored lower than the fused feature vectors of emo2vec.

Emotion Intensity Control. To control the intensity of emotional expression in synthesized speech, we adjust the values within the VAD vector. We examine the impact of various components on emotion intensity using synthesized speech samples from Speaker 13 of the ESD dataset. In this study, we choose four distinct emotions and select 20 speech samples with identical text for each intensity level, totaling 240 instances. The VAD ranges for different emotions and intensities are outlined in Tables 3, 4, 5 and 6, with the interval endpoints representing the maximum and minimum VAD values across all audio at the same intensity.

Table 3. VAD range of anger

Intensity	Valence	Arousal	Dominance
weak	−0.65–0.60	**0.31~0.47**	0.54~0.60
medium	−0.78–0.64	**0.45~0.62**	0.57~0.66
strong	−0.80–0.60	**0.82~ .88**	0.60~0.71

The findings suggest that valence is negative for anger and sadness but positive for happiness and surprise, underscoring its key role in determining emotional positivity or negativity. Arousal levels vary significantly across intensities for anger and sadness, while valence varies with intensity for happiness and surprise. Dominance tends to rise with increasing emotional intensity. The data indicate that arousal primarily regulates the intensity of anger and sadness, while valence largely governs happiness and surprise.

Table 4. VAD range of happiness

Intensity	Valence	Arousal	Dominance
weak	**0.41~0.50**	0.61~0.70	0.50~0.55
medium	**0.49~0.66**	0.65~0.76	0.64~0.70
strong	**0.60~.74**	0.64~0.75	0.65~0.78

Table 5. VAD range of sadness

Intensity	Valence	Arousal	Dominance
weak	−0.30~−0.21	**0.34~0.50**	0.53~0.65
medium	−0.27~−0.19	**0.21~0.44**	0.54~0.68
strong	−0.25~−0.11	**0~0.32**	0.57~0.69

Table 6. VAD range of surprise

Intensity	Valence	Arousal	Dominance
weak	**0.61~0.67**	0.64~0.70	0.70~0.85
medium	**0.68~0.73**	0.67~0.80	0.74~0.88
strong	**0.79~0.88**	0.70~0.81	0.82~0.97

In order to evaluate the emotional control of VAD, we visualize the pitch and energy attributes of different strengths of speech with the same content and the same emotion, and show the trend of each attribute under different strengths more intuitively. Figure 4 show the energy and f0 contours for four emotions, anger, happiness, sadness and surprise, respectively, under three intensities. The energy is related to the volume of the audio, and f0 can represent the change in pitch of the sound. As can be seen in Fig. 4 (left), the energy of the anger emotion increases with increasing intensity, the energy of the sadness emotion decreases, and the energies of the happy and surprised emotions do not have a similar trend. Figure 4 (right) shows that when the intensity of the emotion increases, the pitch also increases, especially for happy and surprised emotions. It can be concluded that the intensity of anger and sadness emotions is more affected by changes in energy, and the intensity of happiness and surprise is more closely related to changes in pitch. Combined with the previous conclusions it can be concluded that arousal regulates energy to control emotional intensity and valence controls intensity by influencing pitch. This suggests that our method can effectively control the intensity of emotions by changing pitch and energy.

The Best-Worst Scaling (BWS) [21] test was employed to assess our emotion intensity control methods against alternatives such as scaling factors [6], relative attributes [7], and interpolation [8]. Each method was classified by emotion type: anger, happiness, and sadness, each with three intensities—weak, medium, and strong—and five audios per intensity, totaling 180 test audios. Testers evaluated these in a controlled environment, selecting the best and worst based on emotional expressiveness and speech quality across different intensities. Table 7 presents the results, indicating that the VAD method outperforms the others in emotion intensity control, producing higher-quality speech. This highlights the superiority of our method in controlling emotion intensity.

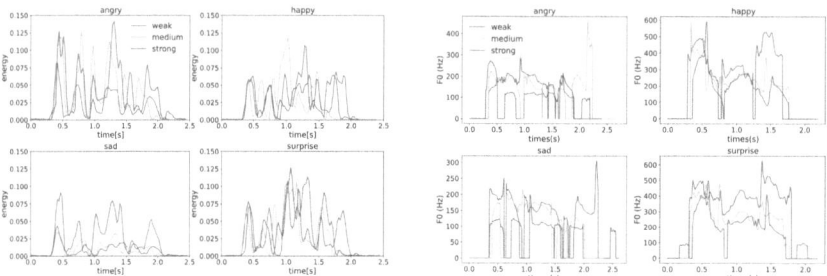

Fig. 4. Comparison of energy (left) and pitch (right) at different emotional intensities (weak, medium, strong).

Table 7. Comparison of BWS results under different emotion control methods

model	weak		medium		strong	
	B	W	B	W	B	W
scaling	14%	39%	0%	63%	0%	77%
relative	20%	18%	25%	10%	18%	9%
interpolation	21%	30%	17%	21%	11%	14%
VAD	**45%**	**13%**	**58%**	**6%**	**71%**	**0%**

4 Conclusion

This paper presents a speech synthesis model equipped with controllable emotion intensity, grounded on the VAD emotion representation. We harness an emotion2vec-driven feature extractor, paired with a nuanced feature fusion approach, to distill emotion features. By integrating the VAD model, we adeptly map emotions into a three-dimensional space. During speech synthesis, these emotional dimensions are fine-tuned, transforming them into emotion embeddings enriched with intensity cues. Comprehensive experiments ascertain that our approach excels in modulating emotion intensity and concurrently produces superior quality speech.

References

1. Lee, Y., Rabiee, A., Lee, S.-Y.: Emotional end-to-end neural speech synthesizer. arxiv preprint arxiv:1711.05447 (2017)
2. Wang, Y., Stanton, D., Zhang, Y., et al.: Style tokens: unsupervised style modeling, control and transfer in end-to-end speech synthesis. In: International Conference on Machine Learning, PMLR, pp. 5180–5189 (2018)
3. Wu, P., Ling, Z., Liu, L., Jiang, Y., Wu, H., Dai, L.: End-to-end emotional speech synthesis using style tokens and semi-supervised training. In: 2019 Asia-Pacific Signal and Information Processing Association Annual Summit and Conference (APSIPA ASC), pp. 623–627. IEEE, November 2019
4. Tang, H., Zhang, X., Wang, J., Cheng, N.: Emomix: emotion mixing via diffusion models for emotional speech synthesis. arxiv preprint arxiv:2306.00648 (2023)
5. Li, T., et al.: Cross-speaker emotion transfer based on prosody compensation for end-to-end speech synthesis. arxiv preprint arxiv:2207.01198 (2022)
6. Zhu, X., et al.: Controlling emotion strength with relative attribute for end-to-end speech synthesis. In: 2019 IEEE Automatic Speech Recognition and Understanding Workshop (ASRU), IEEE (2019)
7. Zhou, K., et al.: Emotion intensity and its control for emotional voice conversion. IEEE Trans. Affect. Comput. **14**(1), 31–48 (2022)
8. Um, S.Y., Oh, S., Byun, K., et al.: Emotional speech synthesis with rich and granularized control. In: ICASSP 2020–2020 IEEE International Conference on Acoustics, Speech and Signal Processing (ICASSP), IEEE (2020)
9. Ma, Z., et al.: emotion2vec: Self-supervised pre-training for speech emotion representation. arxiv preprint arxiv:2312.15185 (2023)
10. Nandini, D., et al.: Design of subject independent 3D VAD emotion detection system using EEG signals and machine learning algorithms. Biomed. Sig. Process. Control **85**, 104894 (2023)
11. Baevski, A., et al.: wav2vec 2.0: a framework for self-supervised learning of speech representations. Adv. Neural Inf. Process. Syst. 33, 12449–12460 (2020)
12. Kim, J., Kong, J., Son, J.: Conditional variational autoencoder with adversarial learning for end-to-end text-to-speech. In: International Conference on Machine Learning, PMLR (2021)
13. Zhao, W., Yang, Z.: An emotion speech synthesis method based on vits. Appl. Sci. **13**(4), 2225 (2023)
14. Sherstinsky, A.: Fundamentals of recurrent neural network (RNN) and long short-term memory (LSTM) network. Physica D **404**, 132306 (2020)
15. Vielzeuf, V., et al.: Centralnet: a multilayer approach for multimodal fusion. In: Proceedings of the European Conference on Computer Vision (ECCV) Workshops (2018)
16. Zhou, K., et al.: Seen and unseen emotional style transfer for voice conversion with a new emotional speech dataset. In: ICASSP 2021–2021 IEEE International Conference on Acoustics, Speech and Signal Processing (ICASSP), IEEE (2021)
17. Ilya, L., Frank, H.: Decoupled weight decay regularization. arxiv preprint arxiv:1711.05101 (2017)
18. Robert, K.: Mel-cepstral distance measure for objective speech quality assessment. In: Proceedings of IEEE pacific Rim Conference on Communications Computers and Signal Processing, vol. 1, IEEE (1993)
19. Sajjad, M., Kwon, S.: Clustering-based speech emotion recognition by incorporating learned features and deep BiLSTM. IEEE Access **8**, 79861–79875 (2020)

20. Streijl, R.C., Winkler, S., Hands, D.S.: Mean opinion score (MOS) revisited: methods and applications, limitations and alternatives. Multimedia Syst. **22**(2), 213–227 (2016)
21. Svetlana, K., Mohammad, S.M.: Best-worst scaling more reliable than rating scales: a case study on sentiment intensity annotation. arxiv preprint arxiv:1712.01765 (2017)

Unsupervised Person Re-identification with Random Occlusion and ContrastiveCrop

Yang Jing$^{(\boxtimes)}$, Gu Lingkang$^{(\boxtimes)}$, Xia Zhouxiang, and Wu Mengqi

Anhui Polytechnic University, Wuhu 241000, Anhui, China
`amx56@foxmail.com, glk@ahpu.edu.cn`

Abstract. In order to alleviate the problem of person re-identification task's dependence on labeled data and low accuracy in occlusion scenarios, an unsupervised person re-identification method combining random occlusion and ContrastiveCrop is proposed. Firstly, the input image is randomly occluded with various patterns according to the real scene, and the contrast cropping method is used to generate pedestrian image samples with greater differences while ensuring the semantic consistency of the positive sample pair, so as to alleviate the impact of occlusion and complex background on the network. Then, the SC-CAResNet network was designed in the contrastive learning model for multi-granularity feature extraction, so that it can pay more attention to the important areas of the image; finally, the network was trained by combining multiple loss functions. Experimental results show that this method outperforms traditional classical methods on two public datasets of person re-identification, Market-1501 and DukeMTMC-reID, has stronger robustness, and significantly improves the ability of person re-identification in model occlusion scenes.

Keywords: Random occlusion · ContrastiveCrop · SC-CAResNet · Unsupervised · Person re-identification

1 Introduction

Person re-identification (person re-id) aims to solve the association and matching of target pedestrians across cameras and scenes. This is like a smaller issue within tasks involving finding images. It can help make up for the fact that fixed cameras can't see everything and can work together with technologies that detect and follow people. [1]. Due to the urgent need for social stability and the increasing number of surveillance cameras in residential areas, stations, intersections, etc., the application of person re-identification in intelligent security is crucial. It has important research implications and practical significance, and is widely used in intelligent security, human-computer interaction, and other fields. Person re-identification faces many uncontrollable factors in complex environments such as significant changes in posture, variations in viewpoint, changes in lighting conditions, occlusions, and low image resolution.

Today, deep learning algorithms have replaced traditional metric learning methods and taken the lead in person re-identification.Yan et al. [2] introduced a lightweight locally enhanced representation network that aggregates unique information of local features using local correlations. Wang et al. [3] presented a pose-guided feature disentanglement method based on transformers, which uses pose information to clearly separate semantic components and selectively match non-occluded parts. She et al. [4] introduced an attention mechanism module based on the residual network to handle person re-identification under occlusion.

This paper proposes an unsupervised person re-identification method that combines random occlusion and ContrastiveCrop to optimize the re-id network from multiple dimensions. It aims to extract features at a deeper level for different granularity information, allowing samples to achieve better clustering effects in the feature space. This approach alleviates the problem of gradient vanishing and enhances the model's generalization ability, effectively improving the accuracy of person re-identification in occluded scenarios.

2 Basic Principles

The network architecture involved in this method mainly consists of several parts: random masking [5], ContrastiveCrop [6], Inter-instance Contrastive Encoding (ICE) [7], and the SC-CAResNet extraction network. The overall framework of the method is shown in Fig. 1.

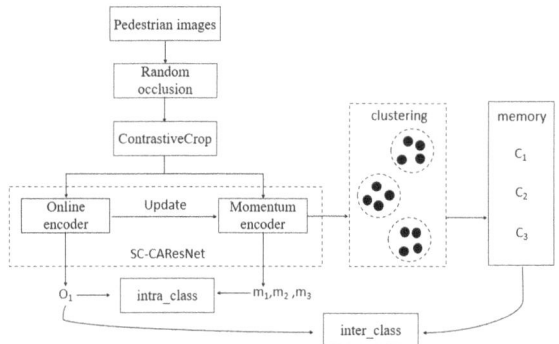

Fig. 1. Method framework. The center points of different clusters C1, C2, C3; after the encoder operation, the feature vectors O1 and m1, m2, m3 are obtained.

Firstly, the random erasing module is introduced to generate occluded pedestrian images by randomly masking the original input images. The occluded images are then subjected to ContrastiveCrop, which differs from conventional random cropping (RandomCrop) in that it utilizes a contrastive learning approach and samples from a non-uniform distribution with central suppression, increasing the variance of multiple samplings and enhancing the differences between different samples. These two methods are combined with the baseline model ICE. This method is an optimized version of the MOCO [8] model, which adds clustering algorithms based on the MOCO model to generate category labels for samples, thereby calculating inter-class loss and intra-class loss to obtain more compact clustering results. The momentum encoder and online encoder are designed based on ResNet50 network [9], with both having identical structures; where the momentum encoder (with weights set as θ_m) is updated by the online encoder (with weights set as θ_m).

$$\theta_m^t = \alpha \theta_m^{t-1} + (1-\alpha)\theta_o^t \tag{1}$$

α is like a speed control knob for momentum encoding updates, while t just shows which iteration we're on.

Before each training round starts, a momentum encoder picks out feature vectors from the training data. After that, a clustering method called DBSCAN [10] is used to create pseudo-labels. The centroids C1, C2, C3… Are calculated using formula (2).

$$p_a = \frac{1}{N_a} \sum_{c_i \in y_a} c_i \tag{2}$$

By utilizing online encoders and momentum encoders to calculate feature vectors, and combining the class information generated by clustering, the inter-class loss and intra-class loss can be computed [7].

Considering the difficulty of extracting multi-scale pedestrian features in the ResNet50 network of the encoder, we replaced the regular convolution with Spatial and Channel Reconstruction Convolution (SCConv) [11] to enhance feature adaptability. Additionally, we added Coordinate Attention (CA) [12] in the residual structure to optimize feature extraction. The SC-CAResNet network is designed to more effectively extract key features and focus on the multi-scale information required for pedestrian re-identification tasks. The combination of these two approaches guides the network to simultaneously consider multiple dimensions, allowing it to pay more comprehensive attention to significant areas of pedestrians in occluded scenes. This approach alleviates gradient vanishing issues and enhances generalization capabilities. Finally, by jointly using inter-class loss, cross-camera difference loss, hard sample contrastive loss, and consistency loss for training, we aim to improve sample collection risks such as lens differences and data distribution changes while mitigating the problem of low accuracy in pedestrian re-identification under occluded scenes.

2.1 Random Occlusion

Random erasing is an easily implemented data augmentation method, which belongs to the category of information deletion methods. It not only reduces the risk of overfitting but also enhances the generalization ability of the network model to deal with occlusion issues. Most occlusion operations involve randomly boxing out a regular area on the image for random erasure. While this approach may yield good results, it does not reflect the diverse and irregular shapes of occlusions in actual scenarios. In this study, we set a 50% probability for randomly obscuring pedestrian images by randomly boxing out rectangular and circular areas across the entire input image for random obscuration; when two overlapping obstructions form multiple random patterns. This better reflects real-world occlusion scenarios as compared to traditional approaches. The proportion range of randomly obscured areas relative to input images is set at 0.02~0.4, which better prepares models to handle real-world occlusion scenarios (Fig. 2 illustrates the effect of obscuring pedestrian images).

Original Image   Regular shape occludes the image   Irregular shapes occlude image

Fig. 2. Pedestrian images

2.2 ContrastiveCrop

Contrastive learning is a form of self-supervised learning method, commonly used to learn the general features of a dataset by letting the model learn the similarity or dissimilarity of data points in the absence of labels. It is considered effective in many downstream tasks [13]. Random cropping is one of the common image augmentation operations in contrastive learning, as shown in Eq. (3).

$$(x, y, h, w) = \mathbb{R}_{crop}(s, r, I) \tag{3}$$

In the formula, s represents the size of cropping, r represents the aspect ratio of cropping, I represents the input pedestrian image, and R_{crop} represents the random cropping function, which obtains a quadruple (x, y, h, w) representing the center coordinates, height and width of the crop. However, this direct cropping may generate incorrect positive sample pairs, as shown in Fig. 3(a). Alternatively, it may generate overly similar positive sample pairs that are difficult to improve model recognition accuracy as shown in Fig. 3(b). Especially in scenarios where pedestrians are occluded, occluding objects are often cropped out and treated as positive samples. Therefore, it is necessary and practical to design a cropping method that can enhance images by discerning semantic information of pedestrian images in occluded scenes.

 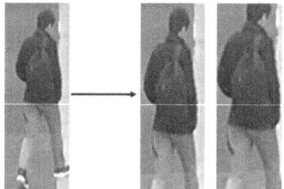

False positive examples (a)   Too similar positive samples (b)

Fig. 3. Randomly cropped pedestrian images

Considering the importance of data augmentation in contrastive learning, and in order to avoid the cropping errors mentioned above, Peng et al. [6] conducted a more in-depth study. They used a semantic-aware approach to reduce the impact of erroneous samples generated during random cropping by incorporating localization information.

In addition, in order to trim out positive sample pairs with greater differences, obtain a more comprehensive global view and finer local information, the method of "central suppression sampling" is introduced. This method uses two Beta distributions β(α,α) with the same parameter α as the probability distribution for sampling. By setting α < 1, the distribution presents a U-shaped pattern with low in the middle and high around it. This method not only reduces the probability of concentrating the trimming area in the center of the image but also decreases the overlap between trimming areas. The final formula for comparativeCrop is:

$$(\dot{x}, \dot{y}, \dot{h}, \dot{w}) = \mathbb{C}_{crop}(s, r, B) \tag{4}$$

The function C_{crop} represents the central suppression sampling. Figure 4 illustrates the contrast between random cropping and center cropping.

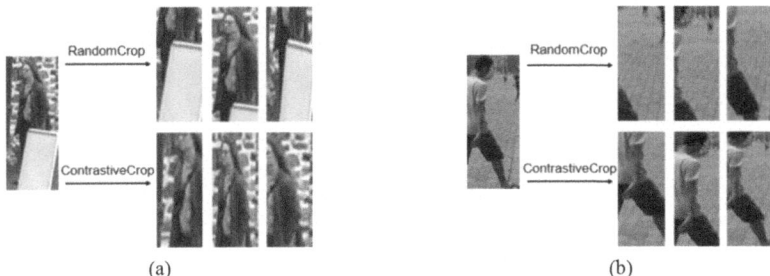

Fig. 4. Comparison of different cutting methods

2.3 SC-CAResnet Network Model

Inspired by the literature [11, 12], this paper designs a new extraction network, named SC-CAResnet, which uses spatial and channel reconstruction convolutions to replace the standard convolutions in ResNet50 and incorporates a coordinate attention mechanism. This design aims to ensure broader hardware support and higher inference speed while maintaining training speed. The SC-CAResnet pays full attention to both local and global information of pedestrian images, resulting in higher recognition accuracy, making it more suitable for person re-identification tasks.

Figure 5 displays where the convolution module is placed for spatial and channel reconstruction. This setup includes two parts: the spatial reconstruction unit (SRU) and the channel reconstruction unit (CRU), which are carried out one after the other in a sequence.

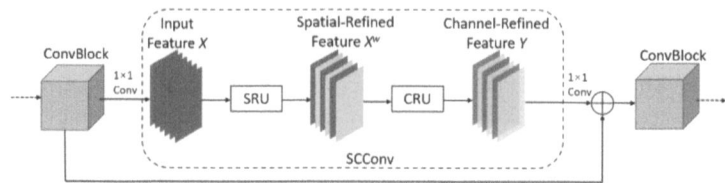

Fig. 5. SCConv insertion position

First, the input feature X gets fine-tuned for space to become X^W, and then it gets fine-tuned for channels to become Y. The combination of these two modules can effectively enhance the feature representation of CNN networks and reduce the redundancy between spatial and channel features. The spatial reconstruction unit is used to handle the spatial redundancy of features, while the channel reconstruction unit is used to handle the channel redundancy. The specific processing process is shown in Fig. 6.

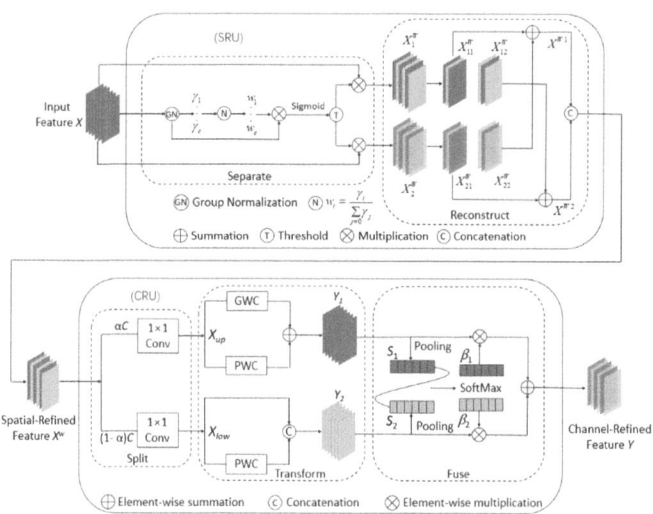

Fig. 6. SCConv Processing Flow

The attention mechanism was initially applied to machine translation tasks and has now been widely used in various fields of deep learning, such as image classification, object detection, and categorization. The coordinate attention mechanism used in this paper is able to simultaneously consider the relationships between channels and long-distance positional information. It not only captures inter-channel information and efficiently processes the relationships between channels, but also includes direction-related positional information, by making full use of the positional details we've gathered, the model can improve its ability to find and recognize targets. In Fig. 7, you can see how the coordinate attention mechanism works. The key step involves turning global pooling into two 1D feature representations.

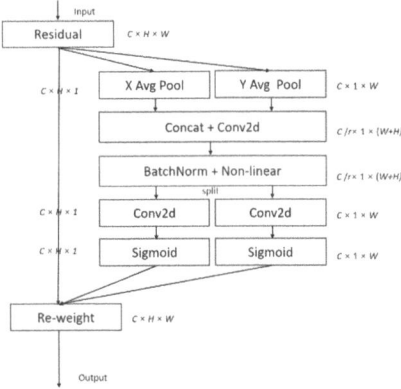

Fig. 7. Coordinate Attention Flowchart

This paper embeds attention into the later parts of conv2, conv3, and conv4 of ResNet50 to improve the network's accuracy. This attention mechanism significantly enhances the performance of the network in handling tasks.

2.4 Loss Function

Interclass Contrast Loss. The definition of inter-class contrastive loss is as follows:

$$L_{\text{inter\_class}} = \mathbb{E}\left[-\log \frac{\exp(o_a \cdot c_a / \tau_{\text{inter\_class}})}{\sum_{i=1}^{|p|} \exp(o_a \cdot c_i / \tau_{\text{inter\_class}})}\right] \quad (5)$$

Cross-Camera Discrepancy Loss. The definition of cross-camera discrepancy loss is as follows, with the aim of reducing the distance between different samples from the same class and minimizing intra-class camera style differences.

$$L_{\text{camera\_loss}} = \mathbb{E}\left[-\frac{1}{|p|}\sum_{i \neq b \cap i \in C} \log \frac{\exp(s(o_{ab}, c_{ai})/\tau_{\text{camera}})}{\sum_{j=1}^{Nneg+1} \exp(s(o_{ab}, c_j)/\tau_{\text{camera}})}\right] \quad (6)$$

Contrastive Loss for Challenging Samples. The definition of contrastive loss for difficult samples is as follows, aiming to reduce the generation of outliers during the clustering process and strengthen the compactness of clusters.

$$L_{hard} = \mathbb{E}\left[-\log \frac{\exp(s(o^i, m_k^i)/\tau_{hard})}{\sum_{j=1}^{J+1} \exp(s(o^i, m_j)/\tau_{hard})}\right] \quad (7)$$

Consistency Loss. To prevent the similarity between instances from changing after data augmentation, it is necessary to ensure consistency in the data distribution before and after data augmentation. An consistency loss is introduced.

Distribution after data enhancement:

$$P_1 = \frac{\exp(s(O_A, m)/\tau_{aug})}{\sum_{j=1}^{N} \exp(s(O_A, m_j)/\tau_{aug})} \quad (8)$$

Use KL divergence to narrow the distance between the distribution before and after data enhancement:

$$L_{aug} = \mathcal{D}_{KL}(P_1 \| P_2) \tag{9}$$

When the above loss functions are combined, the overall loss function is as follows:

$$L = \lambda_1 L_{\text{inter\_class}} + 0.5 L_{\text{camera\_loss}} + \lambda_2 L_{hard} + \lambda_3 L_{aug} \tag{10}$$

3 Experimental Results and Analysis

3.1 Data Set

This paper verifies the proposed method on the datasets Market-1501 [14] and DukeMTMC-reID [15]. The specific information of the datasets is shown in Table 1.

Table 1. Dataset details.

Characteristic	Market-1501	DukeMTMC-reID
Collection location	Tsinghua University	Duke University
Number of pedestrians	1501	1404
Number of cameras	6	8
Number of training set IDs	751	702
Number of training set images	12936	16522
Number of test set IDs	750	702
Number of test set images	19732	17661
Total number of images	32668	36411

The experiment uses Rank-k and mean average precision (mAP) to verify the method proposed in this paper (Table 2).

Table 2. Ablation experiments on the dataset.

Characteristic	Market-1501	
	Rank-1(%)	mAp(%)
Baseline	90.4	78.6
+ Random Occlusion	91.8	81.7
+ ContrastiveCrop	91.3	80.6
+ SC-CAResNet	94.6	84.4
ours	95.6	85.7

3.2 Ablation Experiment

3.3 Visual Comparison

To validate the effectiveness of the model, pedestrian heat maps were used for visualization. Figure 8 illustrates the visual comparison of heat maps for pedestrian image recognition before and after modifying ResNet50 to SC-CAResNet.

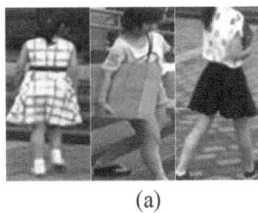 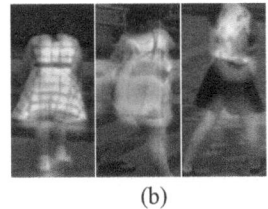 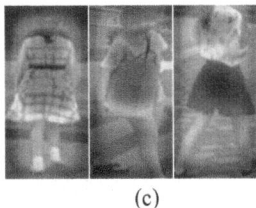

(a)　　　　　　　　　　(b)　　　　　　　　　　(c)

Fig. 8. Heatmap Visualization

3.4 Comparative Experiment

The performance of the method in this paper is compared with the current mainstream methods. The Rank-1 and mAP of each method are shown in Table 3.

Table 3. Comparison of results of different advanced methods on the dataset

Characteristic	Market-1501		DukeMTMC-reID	
	Rank-1	mAp	Rank-1	mAp
HG [16]	95.6	86.1	87.1	77.5
ICE [7]	93.8	82.3	83.3	69.9
MCC-DCL [17]	94.8	85.8	84.8	74.4
PAN [18]	94.9	87.6	89.5	78.2
HQP [19]	92.3	80.3	82.6	68.0
OCNet [20]	95.0	89.3	90.5	80.2
MHSA-Net [21]	94.3	82.5	87.0	72.6
Ours	95.6	85.7	86.7	76.0

References

1. Wang, S., Xiao, S.: Review of person re-identification. J. Beijing Univ. Technol. **48**(10), 1100–1112 (2016)

2. Yan, G., Wang, Z., Geng, S., Yu, Y., Guo, Y.: Part-based representation enhancement for occluded person re-identification. IEEE Trans. Circ. Syst. Video Technol. (2023)
3. Wang, T., Liu, H., Song, P., Guo, T., Shi, W.: Pose-guided feature disentangling for occluded person re-identification based on transformer. In: Proceedings of the AAAI Conference on Artificial Intelligence, vol. 36, pp. 2540–2549 (2022)
4. She, X., Li, R., Ye, O.: Pedestrian re-identification combining random erasing and residual attention network. Comput. Eng. Appl. **58**(03), 215–221 (2022)
5. DeVries, T., Taylor, G.W.: Improved regularization of convolutional neural networks with cutout, arXiv preprint arXiv:1708.04552 (2017)
6. Peng, X., Wang, K., Zhu, Z., Wang, M., You, Y.: Crafting better contrastive views for Siamese representation learning. In: Proceedings of the IEEE/CVF Conference on Computer Vision and Pattern Recognition, pp. 16031–16040 (2022)
7. Chen, H., Lagadec, B., Bremond, F.: Ice: inter-instance contrastive encoding for unsupervised person re-identification. In: Proceedings of the IEEE/CVF International Conference on Computer Vision, pp. 14960–14969 (2021)
8. He, K., Fan, H., Wu, Y., Xie, S., Girshick, R.: Momentum contrast for unsupervised visual representation learning. In: Proceedings of the IEEE/CVF Conference on Computer Vision and Pattern Recognition, pp. 9729–9738, (2020)
9. He, K., Zhang, X., Ren, S., Sun, J.: Deep residual learning for image recognition. In: Proceedings of the IEEE Conference on Computer Vision and Pattern Recognition, pp. 770–778 (2016)
10. Ester, M., Kriegel, H.-P., Sander, J., Xu, X., et al.: A density-based algorithm for discovering clusters in large spatial databases with noise. In: kdd, vol. 96, pp. 226–231 (1996)
11. Li, J., Wen, Y., He, L.: Scconv: spatial and channel reconstruction convolution for feature redundancy. In: Proceedings of the IEEE/CVF Conference on Computer Vision and Pattern Recognition, pp. 6153–6162 (2023)
12. Hou, Q., Zhou, D., Feng, J.: Coordinate attention for efficient mobile network design. In: Proceedings of the IEEE/CVF Conference on Computer Vision and Pattern Recognition, pp. 13713–13722 (2021)
13. Zhang, J., Wang, Y., Zhou, Z., Luan, T., Wang, Z., Qiao, Y.: Learning dynamical human-joint affinity for 3d pose estimation in videos. IEEE Trans. Image Process. **30**, 7914–7925 (2021)
14. Zheng, L., Shen, L., Tian, L., Wang, S., Wang, J., Tian, Q.: Scalable person re-identification: a benchmark. In: Proceedings of the IEEE International Conference on Computer Vision, pp. 1116–1124 (2015)
15. Zheng, Z., Zheng, L., Yang, Y.: Unlabeled samples generated by gan improve the person re-identification baseline in vitro. In: Proceedings of the IEEE International Conference on Computer vision, pp. 3754–3762 (2017)
16. Kiran, M., Praveen, R.G., Nguyen-Meidine, L.T., Belharbi, S., Blais-Morin, L.-A., Granger, E.: Holistic guidance for occluded person re-identification, arXiv preprint arXiv:2104.06524, Contribution Title 5 (2021)
17. Tian, Q., Du, X.: Multi-class center dynamic contrastive learning for unsupervised domain adaptation person re-identification. Comput. Electr. Eng. **116**, 109155 (2024)
18. Ye, P., Zeng, H., Zhang, W., Chen, D.: Part-aware network: a simple but efficient method for occluded person re-identification. In: International Conference on Computer Application and Information Security (ICCAIS 2021), vol. 12260, pp. 158–166. SPIE (2022)
19. Li, Y., Zhu, X., Sun, J., Chen, H., Li, Z.: Unsupervised person re-identification based on high-quality pseudo labels. Appl. Intell. **53**(12), 15112–15126 (2023)

20. Kim, M., Cho, M., Lee, H., Cho, S., Lee, S.: Occluded person re-identification via relational adaptive feature correction learning. In: ICASSP 2022–2022 IEEE International Conference on Acoustics, Speech and Signal Processing (ICASSP), pp. 2719–2723. IEEE (2022)
21. Tan, H., Liu, X., Yin, B., Li, X.: Mhsa-net: Multihead self-attention network for occluded person re-identification. IEEE Trans. Neural Networks Learn. Syst. (2022)

Dynamic Prompt Adjustment for Multi-label Class-Incremental Learning

Haifeng Zhao[1,2], Yuguang Jin[1,2], and Leilei Ma[1,2(✉)]

[1] Anhui Provincial Key Laboratory of Multimodal Cognitive Computation,
Anhui University, Hefei, Anhui, China
xiaomylei@163.com
[2] School of Computer Science and Technology, Anhui University,
Hefei, Anhui, China

Abstract. Significant advancements have been made in *single*-label incremental learning (SLCIL), yet the more practical and challenging *multi*-label class-incremental learning (MLCIL) remains understudied. Recently, visual language models such as CLIP have achieved good results in classification tasks. However, directly using CLIP to solve MLCIL issue can lead to catastrophic forgetting. To tackle this issue, we integrate an improved data replay mechanism and prompt loss to curb knowledge forgetting. Specifically, our model enhances the prompt information to better adapt to multi-label classification tasks and employs confidence-based replay strategy to select representative samples. Moreover, the prompt loss significantly reduces the model's forgetting of previous knowledge. Experimental results demonstrate that our method has substantially improved the performance of MLCIL tasks across multiple benchmark datasets, validating its effectiveness.

Keywords: Multi-Label Class Incremental Learning · Prompt Tuning · Data Replay · Prompt Regularization

1 Introduction

Class-Incremental Learning (CIL) [16,23] focuses on gradually introducing new categories during the training process while maintaining the ability to recognize old categories. This learning approach simulates the ever-changing environment of the real world, where new object categories might appear at any time.

Current research in CIL predominantly addresses the single-label classification challenge [1,4,26], assuming that each image contains single object, as depicted in Fig. 1(a). However, real-world scenarios often involve multiple objects per image, highlighting the necessity for Multi-Label Class-Incremental Learning (MLCIL). In MLCIL, the model learns with limited class information during each session while managing the absence of labels for previously learned classes. As shown in Fig. 1(b), during the training phase, image containing {person;cat;dog} is labeled only for person in the first session. In session 2, this image is re-labeled to include dog, with person now classified as negative class. However, the model is still expected to recognize the person class in test images. This situation aggravates the forgetting of knowledge in MLCIL. Some SLCIL methods [1,29] dynamically expand the model's capacity to handle an increasing number of categories, which may introduce biases, particularly with uneven sample distribution between new and old categories. iCaRL [16], for example, selects representative classes for subsequent training but relies on global average pooling, which favors larger objects and may cause sampling imbalance.

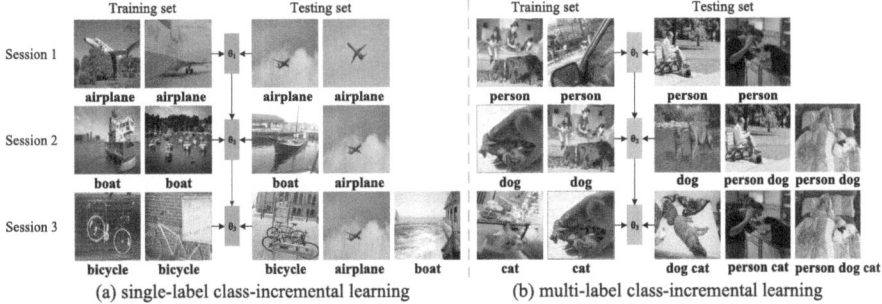

Fig. 1. Illustration of incremental learning for both single-label and multi-label classification tasks. It is assumed that three classes are to be learned, with models θ_1, θ_2, and θ_3 being trained in consecutive sessions.

To address aforementioned challenge, we intruduce a novel method, which consists of two key modules: the Incremental-Context Prompt (ICP) and the Selective Confidence Cluster Replay (SCCR). The ICP employs dual-context prompt mechanism to reduce bias, ensuring that the model remains balanced across different classes. The SCCR utilizes clustering and model confidence to identify and replay significant samples, effectively combating knowledge forgetting. Additionally, we introduce the Textual Prompt Consistency Loss to maintain consistency in textual prompts. Our approach has been validated on the MS COCO and PASCAL VOC datasets, demonstrating its effectiveness in MLCIL. Overall, our contributions are summarized as follows: (**1**) We have utilized a new MLCIL framework based on text prompts. As far as we know, this is the first to employ image-text matching to solve the MLCIL problem; (**2**) We introduce ICP for incremental learning and SCCR for selecting representative samples to alleviate knowledge forgetting; (**3**) Extensive experiments demonstrate that our method achieves competitive results in addressing the MLCIL problem.

2 Related Work

2.1 Class-Incremental Learning

CIL has made significant progress. Approaches in CIL mainly include regularization-based, rehearsal-based, and architecture-based methods. **Regularization-based** methods [11,19] introduce regularization term into the loss function to constrain the variation of model parameters. For example, EWC [10] uses a Fisher information matrix to mitigate changes in important parameters associated with old tasks. **Rehearsal-based** methods [1,4,18,20,22] prevent catastrophic forgetting by replaying a small number of samples or features from old tasks. Xiang et al. [28] generates pseudo-features of old categories using generative adversarial networks which reduces memory usage. **Architectural-based** method [2,5] provides independent parameters for each task. PackNet [14] proposes to isolate the old and new task parameters for knowledge retention. However, this approach can lead to substantial increase in the total number of parameters.

2.2 Prompt Tuning for Incremental Learning

Prompt tuning [8] enables models to achieve high efficiency on downstream tasks with minimal parameter adjustments. Key innovations in this area include L2P's [26] introduction of prompt pools; DualPrompt's [25] distinction between general and specific knowledge through G-Prompt and E-Prompt; and AttriCLIP's [24] use of textual prompts to refine model understanding. However, these methods are primarily designed for SLCIL. We tackle MLCIL by integrating prompt learning techniques inspired by CoOp [30] and leveraging the strengths of CLIP [15].

3 Methodology

3.1 Framework

As shown in Fig. 2, our model mainly comprises two key components: *Incremental Context Prompting* (ICP) and *Selective Confidence Cluster Replay* (SCCR) module. During session S^n, the SCCR module selects samples from the previous $n-1$ sessions, which are combined with the current training set $\mathcal{D}_{tr}^n$ for joint training. Simultaneously, ICP learns category-specific prompts ($\mathcal{P}_c$) and supplementary context prompts ($\mathcal{P}_s$). These prompt features are aligned with image features to produce classification results.

3.2 Incremental Context Prompting Learner

Inspired by CoOp [30], we introduce ICP for MLCIL. Within session n, we devise learnable prompts $\boldsymbol{p}_c$ encompassing all existing categories:

$$\boldsymbol{t}_c = [\omega_1, \omega_2, \ldots, \omega_L, \mathrm{CLS}_j] \ , \tag{1}$$

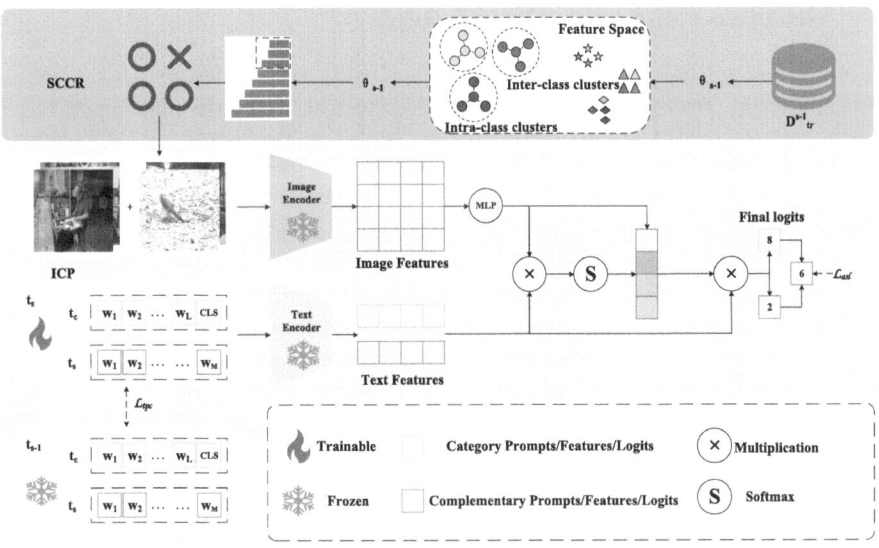

Fig. 2. Overview of our model.

where ω_i denoting the learnable tokens and $[\text{CLS}_j]$ representing the word embedding specific to the j-th category. The length of learnable tokens is L. Motivated by [21], we integrate contextual prompt $\boldsymbol{t}_s$ to complement $\boldsymbol{t}_c$. The form is as follows:

$$\boldsymbol{t}_s^j = [\omega_1, \omega_2, \cdots, \omega_M] \ , \tag{2}$$

where M denotes the count of learnable tokens that is set equal to $L+1$. $\boldsymbol{t}_s^j$ excludes class word embedding to assist $\boldsymbol{t}_c$ capture context and reduce class-specific bias. To achieve classification outcomes, we utilize image $\boldsymbol{x}$ and prompt pair $\boldsymbol{t} = \{\boldsymbol{t}_c, \boldsymbol{t}_s\}$, which are input into the respective encoders $\mathcal{F}(\cdot)$ and $\mathcal{G}(\cdot)$ to extract distinctive features: $\boldsymbol{f}_x = \mathcal{F}(\boldsymbol{x})$, $\boldsymbol{g}_c = \mathcal{G}(\boldsymbol{t}_c)$ and $\boldsymbol{g}_s = \mathcal{G}(\boldsymbol{t}_s)$. The final predicted score is formulated as follows:

$$\boldsymbol{p} = \text{CFA}(\boldsymbol{f}_x, \boldsymbol{g}_c) \odot \boldsymbol{g}_c - \text{CFA}(\boldsymbol{f}_x, \boldsymbol{g}_c) \odot \boldsymbol{g}_s \ , \tag{3}$$

where $\odot$ symbolizes Hadamard product, $\text{CFA}(\cdot, \cdot)$ denotes the class-specific region feature aggregation function [21] and its calculation process is as follows:

$$\begin{aligned} \boldsymbol{f}_{i \to t} &= \text{Proj}_{i \to t}(\boldsymbol{f}_x) \\ \boldsymbol{f}_c &= \text{softmax}(\boldsymbol{f}_{i \to t} \cdot \boldsymbol{g}_c^\top) \cdot \boldsymbol{f}_{i \to t} \ , \end{aligned} \tag{4}$$

where $\text{Proj}_{i \to t}(,)$ projects the image features into the textual feature space. Within $\boldsymbol{f}_c \in \mathbb{R}^{n \times d}$, each vector $\boldsymbol{f}_c^i \in \mathbb{R}^{1 \times d}$ specifically encapsulates the features related to category i.

3.3 Selective Confidence Cluster Replay

Recent rehearsal-based methods [3,16] rely on using average features for category representation, resulting in blurred distinct category features and obscure less salient targets in MLCIL. To overcome this limitation, we introduce the Selective Confidence Cluster Replay (SCCR) strategy.

Initially, we employ Eq. (4) to distill category-related features f from each individual sample. Here, we set $f = f_c$, because we primarily sample based on the ROI of positive samples. We then employ the K-means clustering algorithm to ensure sample diversity by partitioning the feature set of each category into m distinct clusters. To fully leverage the limited old samples and enhance the model's robustness, we further consider sample selection based on the model's performance. To this end, we introduce a confidence-based cluster sampling approach. In detail, within each cluster, we select k hard samples for retraining. The sample set of samples for each category can be obtained by the following process(omitting category identification for brevity):

$$\mathcal{R}_i = \text{Top}_k^{low}\left(\{x_j \mid x_j \in G_i, p_j(x_j \mid y_j, \theta_{s-1})\}\right), \tag{5}$$

where $\mathcal{R}_i$ denotes the set of samples selected from cluster G_i, with x_j representing each sample and p_j representing the predicted probability of x_j by model θ_{s-1}. The Top_k^{low} identifies the k samples with the lowest p_j values from G_i. The final set of samples is $\mathcal{R} = \mathcal{R}_1 \cup \mathcal{R}_2 \cdots \cup \mathcal{R}_m$. Then $\mathcal{R}$ is combined with the training set $\mathcal{D}_{tr}^s$, facilitating the training process in session s: $\widetilde{\mathcal{D}}_{tr}^s = \mathcal{D}_{tr}^s \cup \mathcal{R}$.

3.4 Optimization Objective

In this study, we utilize Asymmetric Loss [17] (ASL) that is widely used in multi-label classification tasks, to optimize the parameters of the prompts.

$$\mathcal{L}_{asl} = \frac{1}{M}\sum_{m=1}^{M}\begin{cases}(1-p_m)^{\gamma+}\log(p_m), & \text{if } y_m = 1\\(p_m)^{\gamma-}\log(1-p_m), & \text{if } y_m = 0\end{cases}, \tag{6}$$

where y_m is a binary label that signifies the presence of category m within the sample. Additionally, $\gamma+$ and $\gamma-$ serve as the attention weights for positive and negative samples. Furthermore, we introduce $\mathcal{L}_{tpc}$ to ensure the consistency of the prompt. The computation process is as follows:

$$\mathcal{L}_{tpc} = 2 - \cos(g_c^s, g_c^{s-1}) - \cos(g_s^s, g_s^{s-1}), \tag{7}$$

where $\cos(\cdot,\cdot)$ denotes cosine similarity. $g_{type\in\{c,s\}}^s$ and $g_{type\in\{c,s\}}^{s-1}$ represent the textual features of the old categories extracted by models θ_s and θ_{s-1} respectively. Finally, the overall optimization objective of our model is: $\mathcal{L} = \mathcal{L}_{asl} + \alpha\mathcal{L}_{tpc}$, where α is balancing factor.

4 Evaluation

4.1 Experiment Setup

Datasets and Metrics. We conduct experiments on the **MS-COCO** [13] and **PASCAL VOC** 2007 [7] datasets to evaluate the effectiveness of our approach. MS-COCO is annotated with 80 distinct categories. PASCAL VOC comprises 20 categories. We adopt the BiCj [3,27] setup for each dataset, where i denotes the number of classes in the initial training session, and j represents the classes incorporated in each incremental session. All categories are arranged alphabetically. We use two key metrics in MLCIL: average accuracy and last accuracy. Average accuracy is the mean mAP score across all sessions, while last accuracy is the mAP score from the final session. We also report per-class F1 (CF1) and overall F1 (OF1) measures.

Implementation Details. We leverage the pre-trained CLIP model, specifically its ViT-B/16 variant as backbone. The number of learnable prompt embeddings L is set to 16. Images are resized to 224 × 224. The model is trained for 20 epochs with batch size of 64 in each session. Optimization is performed using the Adam [9] and the OneCycleLR scheduler with a weight decay of 1e-4. During the incremental session, learning rates are set to 2e-4 for MS-COCO and 1.6e-3 for PASCAL VOC. Moreover, base sessions for all experiments initiate with learning rate of 1.6e-3. We report the upper bounds of model performance: Joint. Joint denotes the results obtained by training the model using all categories from the entire dataset within single session.

Table 1. Experimental results on MS COCO dataset. The best results are shown in **bold**. "*" represents the results reported in [3], and the same applies to the following.

Methods	Buffer Size	MS-COCO B0-C10				MS-COCO B40-C10			
		Avg. Acc	Last Acc			Avg. Acc	Last Acc		
		mAP(%)	CF1	OF1	mAP(%)	mAP(%)	CF1	OF1	mAP(%)
Joint	–	–	–	77.8	81.2	83.9	-	77.8	81.2	83.9
Lwf* [11]	0	47.9	9.0	15.1	28.9	48.6	9.5	15.8	29.9
KRT [3]		74.6	55.6	56.5	65.9	77.8	64.4	63.4	74.0
MULTI-LANE [2]		79.1	65.1	62.8	**74.5**	78.8	66.0	66.6	76.6
Ours		**80.7**	65.1	**65.1**	73.3	**81.1**	**68.2**	**68.5**	**77.2**
iCaRL [16]	20/class	59.7	19.3	22.8	43.8	65.6	22.1	25.5	55.7
ER [18]		60.3	40.6	43.6	47.2	68.9	58.6	61.1	61.6
Der++ [1]		72.7	45.2	48.7	58.8	71.0	46.6	42.1	64.2
AGCN-R [6]		73.2	59.5	60.3	66.0	75.2	64.1	65.2	71.7
KRT-R* [3]		76.5	63.9	64.7	70.2	78.3	67.9	68.9	75.2
Ours		**81.7**	**69.6**	**72.0**	**76.1**	**81.5**	**71.8**	**74.7**	**78.4**
OCDM [12]	1000	49.5	9.3	14.9	28.7	51.5	10.0	16.2	34.9
KRT-R [3]		75.7	61.6	63.6	69.3	78.3	67.5	68.5	75.1
Ours		**81.9**	**69.6**	**71.8**	**78.5**	**81.6**	**71.7**	**74.5**	**78.4**

4.2 Comparsion Results

Table 1 and Table 3 summarize the results of our experiments conducted on the MS-COCO and PASCAL VOC datasets. Notably, the "Buffer Size" column represents the quantity of samples retained for replay.

Results on MS-COCO. Table 1 presents the experimental results of our method compared with other approaches under B40-C10 and B0-C10. Firstly, when training is conducted without rehearsal buffer, our average accuracy surpasses MULTI-LANE [2] by **1.6%**, and achieves last accuracy of **73.3%**. When the buffer size is set to 20 per class, our model achieve last accuracy of **78.4%** under B40-C10, which is **3.2%** higher than the second-best method. Notably, when the buffer size reaches 1000, our method outperforms other methods, with final accuracy that is **3.3%** higher than KRT [3]. This indicates that our *SCCR* module demonstrates superior performance in effective sample acquisition. Table 2 reveal that our method's performance discrepancy relative to Joint is **5.5%**, positioning it closer to Joint's outcome than the KRT. This underscores our method's enhanced capability to mitigate forgetting.

Table 2. Parameter quantity and performance gap with respect to the Joint on MS-COCO dataset under B40-C10 setting. "()" represents the performance gap.

Methods	Backbone	Param.	Avg.mAp(%)	Last.mAP(%)
Joint	TResNetM	29.4M	–	81.8
KRT-R [3]			78.3	75.2($\downarrow$6.6)
Joint	ViT-B/16	28.5M	-	83.9
Ours			81.5	78.4($\downarrow$5.5)

Table 3. Experimental results on PASCAL VOC dataset.

Methods	Buffer Size	VOC B0-C4		VOC B10-C2	
		Avg.Acc	Last Acc	Avg.Acc	Last Acc
Joint	–	–		94.41	–	94.41
AGCN [6]	0	84.3	73.4	79.4	65.1
KRT [3]		89.5	74.8	82.9	67.9
MULTI-LANE [2]		**93.5**	**88.8**	**93.1**	**88.3**
Ours		**93.5**	88.1	88.5	79.0
TPCIL [22]	2/class	87.6	77.3	80.7	70.8
PODNet [4]		88.1	76.6	81.2	71.4
Der++ [1]		87.9	76.1	82.3	70.6
AGCN [6]		86.5	76.0	82.8	69.3
KRT [3]		90.7	83.4	87.7	80.5
Ours		**93.8**	**88.1**	**90.9**	**84.1**

Results on PASCAL VOC. Table 3 presents the results of our method compared to other strategies on the PASCAL VOC dataset under B10-C2 and B0-C4 settings. Specifically, with a buffer size of 0, we achieve best average precision under B0-C4. With a buffer size of 2 samples per class, the mean accuracy increases from **88.5%** to **90.9%** under the B10-C2 scenario. In the final session of the B10-C2 setup, our method leads the second-best by **3.6%**. These results indicate strong performance and potential to approach upper bound performance.

4.3 Qualitative Results

In incremental learning, attention region maps indicate what the model retains or discards. Figure 3 illustrates the evolution of these maps across training sessions, showing consistent focus on base categories {bicycle;bottle;elephant}. As training progresses, the attention regions for each category remain consistent, suggesting the model effectively retains knowledge of existing categories while integrating new information.

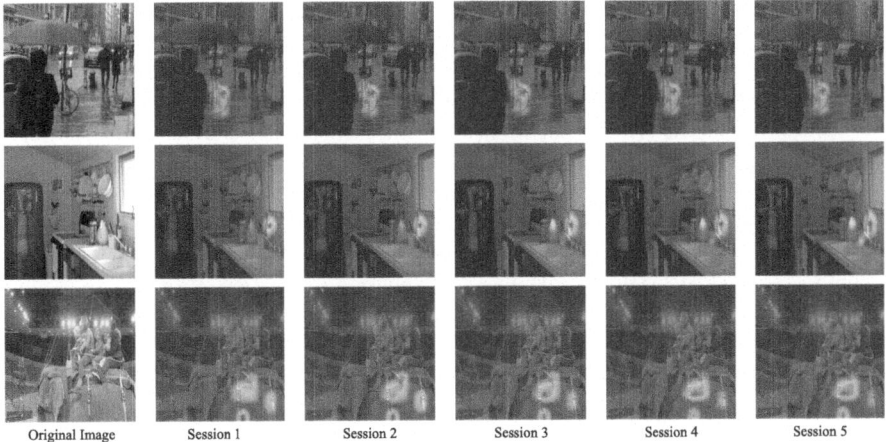

Fig. 3. Visualization of the attention maps on MS-COCO under B40-C10 setting.

Table 4. The results of the ablation study on the effectiveness of different components.

Model	ICP	SCCR	$\mathcal{L}_{tpc}$	Avg.Acc	Last Acc
Baseline	✗	✗	✗	79.7	75.9(+**0.0**)
(1) w/ ICP	✓	✗	✗	80.8	76.7(+**0.8**)
(2) w/ SCCR	✗	✓	✗	80.6	77.0(+**1.1**)
(3) w/o $\mathcal{L}_{tpc}$	✓	✓	✗	81.1	77.9(+**2.0**)
(4) w/o SCCR	✓	✗	✓	81.1	77.2(+**1.3**)
Ours	✓	✓	✓	81.5	78.4(+**2.5**)

4.4 Ablation Studies

Table 4 presents the results of the ablation study on MS-COCO under B40-C10 setting, with a buffer size of 20 per class. We integrate the CoOp with the CFA$(\cdot,\cdot)$ function and concurrently employ DPL [3] for anti-forgetfulness as the baseline approach. The results indicate that both ICP and SCCR significantly improve performance compared to the baseline. SCCR alone improves last accuracy by **1.1%**, while adding ICP increases this to **2.0%**, demonstrating the strong anti-forgetting capabilities of both components. Additionally, we observe that incorporating $\mathcal{L}_{tpc}$ further boosts performance, with **1.3%** increase in last accuracy when combined with ICP. Ultimately, combining all three components yields **2.5%** improvement over the baseline, demonstrating the effectiveness of our approach in MLCIL.

5 Conclusion

In summary, we introduce a novel method for addressing Multi-Label Class-Incremental Learning (MLCIL) by integrating Incremental Context Prompting (ICP) and Selective Confidence Cluster Replay (SCCR). ICP learns a pair of prompts for each category to enhance the model's focus on all categories within an image, while SCCR uses confidence-based sampling to ensure sample diversity and efficient use of limited samples. Our approach advances the use of image-text matching in MLCIL. Experimental results on the MS COCO and PASCAL VOC datasets demonstrate its effectiveness and competitive performance.

Acknowledgements. This work was supported in part by the National Natural Science Foundation of China (No.62472004), Natural Science Foundation of Anhui Province (No. 2308085MF214), University Synergy Innovation Program of Anhui Province (No. GXXT-2022–029), Key Natural Science Project of Anhui Provincial Education Department (No.2023AH050065). We also thank the High-performance Computing Platform of Anhui University for providing computational resources for this project.

References

1. Buzzega, P., Boschini, M., Porrello, A., Abati, D., Calderara, S.: Dark experience for general continual learning: a strong, simple baseline. NeurIPS **33**, 15920–15930 (2020)
2. De Min, T., Mancini, M., Lathuilière, S., Roy, S., Ricci, E.: Less is more: Summarizing patch tokens for efficient multi-label class-incremental learning. In: CoLLAs, PMLR (2024)
3. Dong, S., Luo, H., He, Y., Wei, X., Cheng, J., Gong, Y.: Knowledge restore and transfer for multi-label class-incremental learning. In: ICCV, pp. 18711–18720 (2023)
4. Douillard, A., Cord, M., Ollion, C., Robert, T., Valle, E.: Podnet: pooled outputs distillation for small-tasks incremental learning. In: ECCV, pp. 86–102. Springer (2020)

5. Du, K., et al.: Agcn: augmented graph convolutional network for lifelong multi-label image recognition. In: ICME, pp. 01–06. IEEE (2022)
6. Du, K., et al.: Multi-label continual learning using augmented graph convolutional network. IEEE Trans. Multimedia (2023)
7. Everingham, M., Van Gool, L., Williams, C.K., Winn, J., Zisserman, A.: The pascal visual object classes (VOC) challenge. IJCV **88**, 303–338 (2010)
8. Jia, M., et al.: Visual prompt tuning. In: ECCV, pp. 709–727. Springer (2022)
9. Kingma, D.P., Ba, J.: Adam: a method for stochastic optimization. arXiv preprint arXiv:1412.6980 (2014)
10. Kirkpatrick, J., et al.: Overcoming catastrophic forgetting in neural networks. PNAS **114**(13), 3521–3526 (2017)
11. Li, Z., Hoiem, D.: Learning without forgetting. IEEE TPAMI **40**(12), 2935–2947 (2017)
12. Liang, Y.S., Li, W.J.: Optimizing class distribution in memory for multi-label online continual learning. arXiv preprint arXiv:2209.11469 (2022)
13. Lin, T.Y., et al.: Microsoft coco: common objects in context. In: ECCV, pp. 740–755. Springer (2014)
14. Mallya, A., Lazebnik, S.: Packnet: adding multiple tasks to a single network by iterative pruning. In: CVPR, pp. 7765–7773 (2018)
15. Radford, A., et al.: Learning transferable visual models from natural language supervision. In: ICML, pp. 8748–8763. PMLR (2021)
16. Rebuffi, S.A., Kolesnikov, A., Sperl, G., Lampert, C.H.: ICARL: incremental classifier and representation learning. In: CVPR, pp. 2001–2010 (2017)
17. Ridnik, T., et al.: Asymmetric loss for multi-label classification. In: ICCV, pp. 82–91 (2021)
18. Riemer, M., Cases, I., Ajemian, R., Liu, M., Rish, I., Tu, Y., Tesauro, G.: Learning to learn without forgetting by maximizing transfer and minimizing interference. arXiv preprint arXiv:1810.11910 (2018)
19. Schwarz, J., et al.: Progress & compress: a scalable framework for continual learning. In: ICML, pp. 4528–4537. PMLR (2018)
20. Shin, H., Lee, J.K., Kim, J., Kim, J.: Continual learning with deep generative replay. In: NeurIPS, vol. 30 (2017)
21. Sun, X., Hu, P., Saenko, K.: Dualcoop: fast adaptation to multi-label recognition with limited annotations. NeurIPS **35**, 30569–30582 (2022)
22. Tao, X., Chang, X., Hong, X., Wei, X., Gong, Y.: Topology-preserving class-incremental learning. In: ECCV, pp. 254–270. Springer (2020)
23. Van de Ven, G.M., Tolias, A.S.: Generative replay with feedback connections as a general strategy for continual learning. arXiv preprint arXiv:1809.10635 (2018)
24. Wang, R., et al.: Attriclip: a non-incremental learner for incremental knowledge learning. In: CVPR, pp. 3654–3663 (2023)
25. Wang, Z., et al.: Dualprompt: complementary prompting for rehearsal-free continual learning. In: ECCV, pp. 631–648. Springer (2022)
26. Wang, Z., et al.: Learning to prompt for continual learning. In: CVPR, pp. 139–149 (2022)
27. Wu, Y., et al.: Large scale incremental learning. In: CVPR, pp. 374–382 (2019)
28. Xiang, Y., Fu, Y., Ji, P., Huang, H.: Incremental learning using conditional adversarial networks. In: ICCV, pp. 6619–6628 (2019)
29. Yan, S., Xie, J., He, X.: Der: Dynamically expandable representation for class incremental learning. In: CVPR, pp. 3014–3023 (2021)
30. Zhou, K., Yang, J., Loy, C.C., Liu, Z.: Learning to prompt for vision-language models. IJCV **130**(9), 2337–2348 (2022)

Using Decision Tree Classification to Identify Cost Drivers of Hospitalization Expenses for Elderly Patients

Xiaojing Hu[1], Yudian Liu[2(✉)], and Shixi Liu[1]

[1] School of Computer and Information Engineering, Chuzhou University, Chuzhou 239000, China
[2] Guangji Hospital, Wuhu 241000, China
13093638033@163.com

Abstract. This study aims to identify key factors influencing hospitalization costs for elderly patients undergoing laparoscopic surgery and to provide theoretical support for the refined management of hospital expenses through a grouping model. A retrospective analysis was conducted on the medical records of 1,010 elderly patients who underwent laparoscopic surgery between 2018 and 2023 at a specific hospital. Descriptive statistical analysis, univariate analysis, linear regression, and a regression decision tree model were employed to thoroughly examine the factors affecting hospitalization costs. The decision tree model was further used to categorize patients into distinct groups. The primary factors identified as influencing hospitalization costs were age, length of hospital stay, number of comorbidities, disease outcomes, preoperative hospital stay, and surgical site. Using the decision tree model, with age, number of comorbidities, and preoperative hospital stay as key indicators, patients were effectively categorized. Healthcare institutions can focus on these modifiable factors for targeted treatment and care. Applying the classification standards derived from the decision tree model to standardize diagnostic and treatment processes can assist in managing hospitalization expenses, thereby reducing the economic burden on patients and serving as a reference for healthcare payment system reforms.

Keywords: Decision Tree · Elderly Patients · Laparoscopic Surgery · Hospitalization Costs

1 Introduction

Over recent decades, the acceptance and utilization of laparoscopic surgery have markedly increased, reflecting its rising prominence in medical practice. Laparoscopic surgery is progressively being recognized as the gold standard in the treatment of numerous prevalent diseases, especially those that have a profound impact on the elderly [1]. The '2022 China Health and Wellness Statistical Yearbook' further indicates that the elderly, those aged 60 and above, accounted for 42% of these discharges, signifying that a substantial proportion of surgical patients are elderly.

In recent years, research has begun to focus on the factors determining the hospitalization costs associated with laparoscopic surgery. Some studies have concentrated on single cost-influencing factors for specific laparoscopic procedures, such as the impact of medical instruments used during laparoscopic cholecystectomy on the patients' hospital expenses [2]. Other studies have taken a more comprehensive approach, evaluating multiple cost components of certain surgeries, like the analysis of hospitalization costs and their determinants for patients undergoing laparoscopic sleeve gastrectomy. Some studies have also focused on the cost analysis of different surgical methods. For example, there are comparative analyses of the costs of robotic surgery, laparoscopic surgery, and open abdominal surgery [3].

The current research on the hospitalization costs of surgical patients reveals certain gaps. Primarily, the majority of studies have focused on a general patient population encompassing all age groups, with relatively scarce research dedicated specifically to elderly patients [4]. This oversight is critical because cost-driving factors in the elderly may differ significantly from those in the general population. Secondly, most existing studies examine single factors in isolation, with few addressing the combined effects of multiple factors [5]. Given the unique characteristics and needs of elderly patients, it is essential to conduct targeted investigations into the factors affecting hospitalization costs for elderly patients undergoing laparoscopic surgery.

This study aims to investigate the factors influencing hospitalization costs following laparoscopic surgery in elderly patients. Utilizing statistical analysis tools such as univariate analysis and decision tree models, this research evaluates the hospitalization cost data of 1,010 elderly patients who underwent laparoscopic surgery at a hospital in Anhui Province between 2018 and 2023. The objective is to systematically identify the key factors that affect costs and, based on these findings, offer precise and broadly applicable insights. The outcomes of this study are expected to provide significant guidance and theoretical support for optimizing the structure of hospitalization costs, improving the allocation of health resources, enhancing geriatric medical services, and refining the basic health insurance system.

2 Materials and Methods

2.1 Data Collection and Sample Determination

In this study, through the establishment of a surgical patient hospitalization model, a meticulous screening of key data generated during the patient's hospital stay was conducted. Variables lacking in representativeness were excluded, with a focus on collecting highly indicative data. The gathered data encompassed critical information from the patient's first medical record, including gender, age (60 years and above), diagnosed diseases, surgical sites, disease outcomes, duration of hospital stay, and total hospitalization costs.

The data set analyzed in this study encompasses a total of 3899 patients who underwent laparoscopic surgery from 2018 to 2023, of which 1010 were elderly patients aged 60 and above, accounting for 25.90% of the total sample. The study reveals that the average hospitalization cost for patients under 60 was 7626.02 yuan, with an average hospital stay of 5.89 days. In contrast, the average hospitalization cost for elderly patients was higher at 9298.72 yuan, with an average stay of 7.04 days, both significantly greater than that of patients under 60. This finding reveals the particularity of elderly patients undergoing laparoscopic surgery and highlights the importance of analyzing the factors affecting hospitalization costs in elderly patients.

Within the elderly patient group, there were 464 male and 546 female patients. In total, the overall hospitalization costs for these elderly patients reached 93.917 million yuan. The average age was 68.66 years with a standard deviation of 5.82 years, indicating variability in age distribution. The average length of hospital stay was 7.04 days with a standard deviation of 4.53 days, suggesting a relatively concentrated distribution of hospital stays. Despite the average hospitalization cost being 9298.72 yuan, the standard deviation of cost distribution was quite high at 6197.24 yuan, reflecting considerable variability in hospitalization expenses.

2.2 Method

Definitions of Factors

After the data collection was completed, we defined the variables and clarified their meanings and measurement methods. Additionally, we described the operations performed on these variables, which included the construction of new indicators and the calculation of correlations between variables [6]. We established evaluation indicators to facilitate quality control and validation of the data.

The variable definitions are as follows: (1) Gender: Categorized as male or female. (2) Age: The actual age of the elderly inpatients at the time of their hospitalization is recorded. (3) Length of Hospital Stay: The actual number of days the elderly inpatients are hospitalized is documented. (4) Preoperative Hospital Stay: This refers to the duration from the patient's admission to the hospital until the day before surgery; the day of surgery is excluded. (5) Surgical Site: This is categorized into gallbladder, inguinal hernia, appendix, large intestine (including rectum and colon), kidney, and other parts. (6) Number of Comorbidities: Refers to the number of additional diseases treated within the scope of the patient's hospitalization, other than the primary condition for which they were admitted. This is based on other diagnoses listed on the first page of the patient's hospital record and excludes factors affecting the patient's health status and contacts with healthcare facilities. (7) Disease Outcome: Classified according to the outcome of the disease into cured, improved, or other [7]. For the assignment of variables related to factors affecting hospitalization costs, refer to Table 1.

Table 1. Variable assignment table.

Variable	Variable Name	Assignment
X1	Gender	1 = Male, 2 = Female
X2	Age	Actual Value
X3	Length of Hospital Stay	Actual Value
X4	Preoperative Hospital Stay	Actual Value
X5	Surgical Site	1 = Gallbladder 2 = Inguina hernia 3 = Appendix 4 = Large intestine 5 = Kidney 6 = Other parts
X6	Number of Comorbidities	Actual Value
X7	Disease Outcome	1 = Cured; 2 = Improved; 3 = Other
Y	Hospitalization Cost	Actual Value

Statistical Method

Upon completion of data collection, a database was established using Excel. Statistical analysis was then conducted using Python For quantitative data, arithmetic means and standard deviations were employed for descriptive purposes, and linear regression was used to analyze relationships. Qualitative data were described using proportions and analyzed using Analysis of Variance (ANOVA). A significance level of $P<0.05$ was set for all tests in this study.

To further analyze the data, this study utilized a regression decision tree model. The regression decision tree is a tree-structured machine learning method that is applied to classification and regression tasks [8]. A typical decision tree model comprises a root node, several internal nodes, and a number of leaf nodes. Herein, the leaf nodes represent the predictive outcomes of the model, while each internal node corresponds to a decision criterion based on an attribute. By following a set of decisions from the root to the leaf nodes, all samples are ultimately allocated to different leaf nodes, thereby facilitating classification or regression analysis.

The prioritization of attributes in a decision tree is typically determined by information gain:

$$Gain(D,a) = Ent(D) - \sum_{v=1}^{V} \frac{|D^v|}{|D|} Ent(D^v) \qquad (1)$$

In this context, D^v represents the dataset, denotes the dataset for the v-th branch node, "a" stands for attribute, and "Ent(D)" refers to the information entropy.

$$Ent(D) = -\sum_{k=1}^{|y|} p_k \log_2 p_k \qquad (2)$$

Within this context, p_k refers to the proportion of the k-th category of samples within the dataset D, and "y" represents the number of samples. Generally speaking, the greater the information gain, the greater the increase in purity obtained by using this attribute. Additionally, the Gini index and the gain ratio are also commonly used to determine the optimal attribute. The gain ratio is defined as follows:

$$Gain\,ratio(D,a) = \frac{Gain(D,a)}{IV(a)} \qquad (3)$$

Herein, represents the Information Gain, and is referred to as the Intrinsic Value:

$$IV(a) = -\sum_{v=1}^{V} \frac{|D^v|}{|D|} \log_2 \frac{|D^v|}{|D|} \qquad (4)$$

Herein, D^v denotes the dataset, and represents the dataset for the v-th branch node. The Gini index is defined as:

$$Gini_{index(D,a)} = Ent(D) - \sum_{v=1}^{V} \frac{|D^v|}{|D|} Gini(D^v) \qquad (5)$$

wherein, D represents the dataset, denotes the dataset for the v-th branch node, and indicates the Gini value, which is defined as:

$$Gain(D) = 1 - \sum_{k=1}^{|y|} p_k^2 \qquad (6)$$

Herein, p_k represents the proportion of the k-th category of samples in the dataset D, and y is the number of samples.

3 Results

3.1 Descriptive Statistics

Through the analysis of scatter plots and fitting curves for age, length of hospital stay, and preoperative hospital stay in relation to hospitalization costs, it was found that for patients across different age groups, hospitalization costs tend to rise gradually with age. High costs of hospitalization were observed across all age groups; however, a larger proportion of older patients incurred higher expenses. Conversely, within the oldest age group (≥ 80 years), hospitalization costs showed a tendency to decrease.

Most patients had a preoperative hospital stay of less than 5 days. The fitting curve indicated that hospitalization costs increase with longer preoperative stays. Hospitalization costs also rose with an increase in the total number of inpatient days, especially for patients with longer stays, who incurred substantially higher costs. This suggests a strong correlation between the length of hospital stay and hospitalization costs. The scatter plot illustrating the relationships between age, preoperative hospital stays, total hospital stays, and hospitalization costs can be seen in Fig. 1. The study findings indicate a significant rise in average hospitalization costs with an increase in the number of comorbidities, with patients with higher expenses typically presenting a greater number of comorbid conditions. Moreover, patients with outcomes classified as 'other' at discharge generally incurred higher hospitalization costs. This suggests that adverse prognoses in elderly patients may lead to increased treatment complexity and, consequently, higher hospitalization expenses.

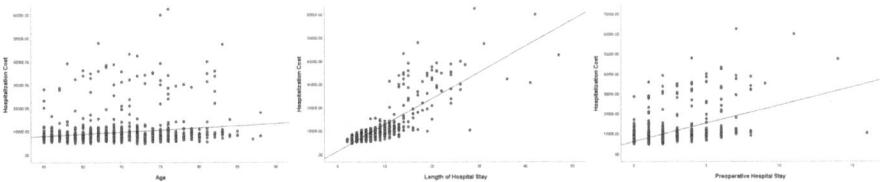

Fig. 1. Scatter plot of the relationship between age, preoperative hospitalization days, hospitalization days, and hospitalization costs.

Analysis of surgical sites revealed that, while hospitalization costs for patients undergoing the same type of laparoscopic surgery tend to be clustered, those who underwent laparoscopic colorectal surgery faced markedly higher costs. Such procedures are usually more complex and involve more extensive resections, often resulting in prolonged hospital stays and increased costs [9]. The relationships between the number of comorbidities, disease outcomes, specific surgical sites, and hospitalization costs are depicted in the box plots shown in Fig. 2.

3.2 Decision Tree Analysis

In this study, we initially examined the linear relationships between the independent variables, and the results indicated no significant linear correlations among them. Consequently, we further explored the relationships between these independent variables and hospitalization costs. Following the preliminary analysis, we identified several key factors affecting hospitalization costs, including both numerical and categorical variables. Utilizing these key variables, we segmented the data into different groups and implemented a stratified management approach to categorize patient groups meticulously. Therefore, we opted to construct a decision tree model for our analysis, as conventional methods such as

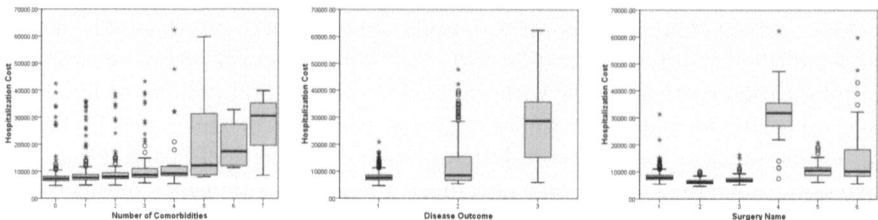

Fig. 2. Box plot of comorbidities, disease outcomes, different surgical sites, and hospitalization costs.

weighted regression analysis are ineffective in addressing the complexity of such data structures and relationships [10].

Experience has demonstrated that the decision tree model is effective in analyzing patients' hospitalization expenses, which is shown in Fig. 3. As a non-parametric model, it has the advantage of not requiring distribution-based assumptions about the data, which enables it to adapt well to the actual data distribution. This is particularly applicable in situations where the data distribution is unknown or does not meet certain statistical prerequisites.

In our study, hospitalization costs were treated as the dependent variable, with age, comorbidity count, and preoperative hospital stay as the categorical node variables. The specific parameters used were as follows: a maximum tree depth of 3, a minimum sample size of 100 for parent nodes, and a minimum sample size of 30 for child nodes. In this study, the decision tree model facilitated the creation of 10 patient groups, with each group demonstrating a P-value less than 0.05, indicating high homogeneity within the groups. The first-level categorical node identified by the decision tree model is patient age, which suggests that age is the most significant factor affecting hospitalization costs. Following age, the preoperative hospital stay and the number of comorbidities are ranked in terms of their impact on hospitalization expenses. Using the decision tree model, nine patient groups were constructed, with each group exhibiting a within-group P-value of less than 0.05, indicating statistically significant homogeneity within the groups. The model identified patient age as the primary classification node in the first tier, suggesting that age is the most significant factor influencing hospitalization costs. This is followed in importance by the number of preoperative hospitalization days and comorbidity count, ranked according to their respective impact on hospitalization expenses. The decision tree model identified several key nodes affecting hospitalization costs: age, preoperative hospitalization duration, and the number of comorbidities. Specifically, nodes 10, 11, and 12 are noteworthy as they are influenced by all three variables. In contrast, nodes 4, 5, 7, 8, and 9 are affected only by two factors, particularly age and preoperative hospitalization duration. Among these critical nodes, nodes 1, 5, 9, and 12 have a proportion exceeding 10%, with each node accounting for more than 100 cases. The specific distribution of these nodes is detailed in Table 2 of the paper.

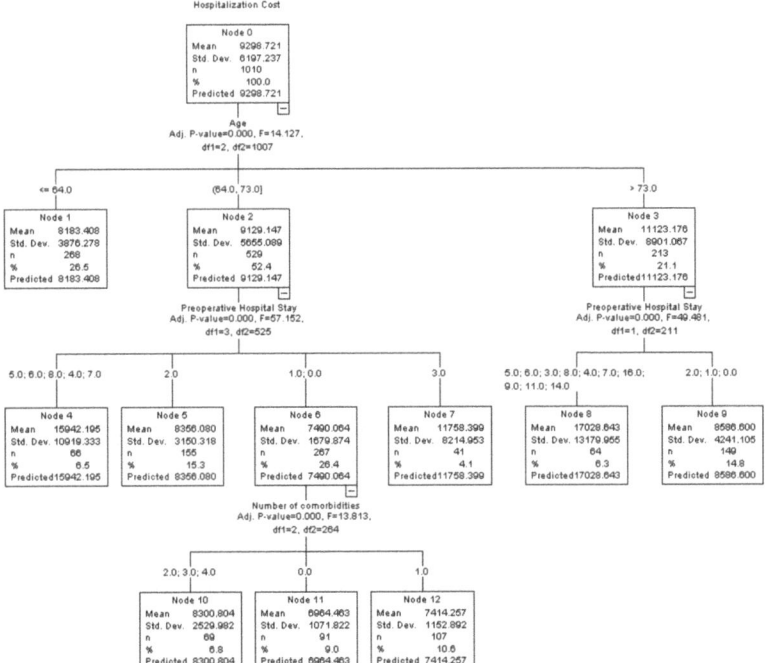

Fig. 3. Decision Tree Model.

Table 2. Decision Tree Grouping.

Node	Mean	Std. Deviation	N	Percent	F	CV
1	8183.41	3876.28	268	26.50%	14.127	0.47
4	15942.19	10919.33	66	6.50%	57.152	0.68
5	8356.08	3150.32	155	15.30%	57.152	0.38
7	11758.40	8214.95	41	4.10%	57.152	0.70
8	17028.64	13179.95	64	6.3%	49.481	0.77
9	8586.60	4241.11	149	14.80%	49.481	0.49
10	8300.80	2529.98	69	6.80%	13.813	0.30
11	6964.46	1071.82	91	9.00%	13.813	0.15
12	7414.26	1152.89	107	10.60%	13.813	0.16

By calculating the coefficient of variation (CV) for each direct point in the hierarchy, our study observed that the CV values for the first and second layers' subnodes were higher compared to those of the third layer, although all CV values were less than 1. This suggests that the data dispersion for subnodes in the first and second layers is relatively high. Consequently, the stratification of patients based solely on age and preoperative hospital stay in these layers does not yield as effective a grouping as the third layer's stratification.

4 Discussion

Compared with previous studies, this research demonstrates several advantages. First, the data was directly extracted from the first-page hospitalization records, which not only made the data more accessible but also ensured its accuracy. Second, the factors under study were comprehensive and representative, effectively reflecting the overall condition of elderly patients and their entire treatment process. Third, considering the varying emphases among different influencing factors, the research refined the primary determinants through variable assignment and the construction of a regression decision tree model, thereby lending greater persuasiveness and scientific validity to the study's conclusions. There are certain limitations to this study. As is well known, medical costs are influenced by many factors such as the patient's basic situation and individual characteristics, and are also affected by objective factors such as regional consumer price index, hospital size and scale, and regional socioeconomic characteristics. Therefore, when setting cost standards, factors such as clinical diagnosis and treatment, patient demand, and regional particularities should be comprehensively considered. In addition, there is no linear relationship between the various influencing factors in this study, but other relationships between the influencing factors have not been explored.

5 Conclusions

This study employed non-parametric tests and regression decision tree methodologies to thoroughly investigate the key factors affecting hospitalization costs for elderly patients undergoing laparoscopic surgery. The variables identified include age, length of hospital stay, number of comorbidities, and disease outcomes. Specifically, age, the number of comorbidities, and preoperative hospitalization days have been proven to be the three primary influencing factors in the decision tree analysis, providing a solid foundation for the creation of a cost prediction model.

It is worth noting that in previous related studies, the application value of the decision tree model in analyzing hospitalization costs has received less attention, with researchers often focusing more on disease grouping studies. However, this study proves that the decision tree model can effectively analyze and predict the factors influencing hospitalization costs. Future research needs to further understand the patient's diagnostic and treatment process, introduce more potential influencing variables for analysis, in order to more accurately reflect various situations and cost changes during the treatment process. Through such research, we can further optimize medical services, improve treatment efficiency, and also help further reduce the hospitalization costs of elderly patients.

Acknowledgement. This study was funded by the Key Research Projects of Higher Education Institutions in Anhui Province 2023AH051594; in part by the Open Research Fund of Anhui Province Engineering Laboratory for Big Data Analysis and Early Warning Technology of Coal Mine Safety CSBD2022-ZD06.

References

1. Chen, J., et al.: How heavy is the medical expense burden among the older adults and what are the contributing factors? A literature review and problem-based analysis. Front. Public Health **11**, 1165381 (2023)
2. Malhotra, L., et al.: Cost analysis of laparoscopic appendectomy in a large integrated healthcare system. Surg. Endosc. **1**, 1–8 (2022)
3. Manole, F., et al.: Review of the effect of aging on health costs. Arch. Pharm. Pract. **14**(3), 58–61 (2023)
4. Zhang, Y., et al.: Diagnosis-intervention packet-based Pareto chart of the proportion of high-cost cases and the analysis of the structure of hospitalization expenses. Technol. Health Care **31**(4), 1355–1364 (2023)
5. Ou, W., et al.: Hospitalization costs of injury in elderly population in China: a quantile regression analysis. BMC Geriatr. **23**(1), 143 (2023)
6. Watrowski, R., et al.: Complications in laparoscopic and robotic-assisted surgery: definitions, classifications, incidence and risk factors-an up-to-date review. Videosurg. Other Miniinvasive Tech. **16**(3), 501–525 (2021)
7. Vieira, B.B., et al.: An integrated cost model based on real patient flow: exploring surgical hospitalization. Healthcare **10**(8), 1458 (2022)
8. Desai, R.J., et al.: Comparison of machine learning methods with traditional models for use of administrative claims with electronic medical records to predict heart failure outcomes. JAMA Netw. Open **3**(1), e1918962 (2020)
9. Feretzakis, D., et al.: Hiding decision tree rules in medical data: a case study. Stud. Health Technol. Inform. **262**, 362–371 (2019)
10. Hao, S.: Modeling hospitalization medical expenditure of the elderly in China. Econ. Anal. Policy **79**, 450–461 (2023)

Adversarial Attacks on Facial Images Based on Attribute-Conditioned High-Camouflage Editing

Jingjing Zhang, Huabin Wang(✉), Dongxu Shang, Hongrui Yuan, and Liang Tao

Anhui Provincial Key Laboratory of Multimodal Cognitive Computation,
School of Computer Science and Technology, Anhui University, Hefei 230601, China
{e22201024,e22201102,Y02214291}@stu.ahu.edu.cn,
{wanghuabin,taoliang}@ahu.edu.cn

Abstract. Research on adversarial attacks for facial recognition identifies and mitigates vulnerabilities in facial recognition systems, enhancing their security. Current adversarial attacks targeting facial attributes lack precision in controlling specific attributes, often altering multiple facial features instead of a single target attribute. Such imprecise modifications reduce the attack's effectiveness and stealthiness. To address these issues, this paper proposes a method for generating adversarial examples based on multi-layer feature map fusion. First, we design a framework called StarAdv, which uses fused multi-layer feature maps with high concealment to achieve realistic facial image transformations under different attributes. This approach generates adversarial images visually similar to the originals while possessing specific misleading properties, enabling transferable attacks on facial recognition systems. Second, we introduce a Multi-layer Fusion module that captures weight information from each layer of the network's residual blocks to adaptively fuse feature maps from different layers, producing high-concealment feature maps. Finally, we propose a Bilinear Feature Map Interpolation (BFMI) algorithm to interpolate the high-concealment features with the original image features, ensuring the final decoded adversarial samples maintain a natural appearance, further reducing the likelihood of detection. We conduct comprehensive experiments on the CelebA dataset, demonstrating that the adversarial facial images generated by our method possess semantic plausibility and authenticity in appearance and achieve a high attack success rate in both white-box and black-box settings.

Keywords: Adversarial attack · high camouflage · black-box transferability · multi-layer feature map fusion · bilinear interpolation

1 Introduction

With the advancement of deep learning, facial recognition systems are increasingly applied in real-world scenarios, bringing greater security risks related to biometric identification. Research shows [1] that deep learning-based facial recognition systems exhibit vulnerabilities to imperceptible or naturally appearing adversarial input images, leading to incorrect predictions [2]. Adversarial attacks are categorized into untargeted attacks (evasion attacks) and targeted attacks (impersonation attacks), with untargeted attacks being easier to implement. This paper generates high-concealment adversarial examples by altering attributes to achieve evasion attacks [3], assessing the robustness and interference resistance of the model. By inputting modified facial images and observing changes in model output, we can understand the model's sensitivity to disturbances and misguidance from different attributes.

Existing methods for evading facial recognition adversarial attacks can be broadly categorized into three types: gradient-based methods, patch-based methods, and invisible methods. The goal of adversarial facial examples is to generate adversarial samples with strong transferability to successfully attack the target model, while ensuring that these adversarial images have high-quality, subtle perturbations that are visually hard to detect.

Recent research [4] introduced the Adv-Attribute method, which generates adversarial noise based on the difference between the target facial recognition features and the actual features, using the StyleGAN model [5]. This noise is embedded into various attributes to create adversarial examples. This method attacks facial recognition models by modifying multiple attributes to predict changes in key attributes. However, altering multiple attributes can compromise the stealthiness of the face, making the generated adversarial examples less camouflaged and more detectable to the human eye. In contrast, our proposed StarAdv method employs a novel strategy by introducing fused high-camouflage feature maps, which not only accurately modify the attributes of facial images but also retain the overall similarity to the original image. This approach ensures that the constructed adversarial samples have high camouflage, effectively disrupting the target recognition model's identification to achieve evasion attacks, thereby enhancing the stealthiness and strength of the adversarial attacks.

To address these issues, we propose a novel facial recognition adversarial attack method, referred to as StarAdv, as shown in Fig. 1. Our contributions are summarized as follows:

- We designed a framework called StarAdv, which generates semantically meaningful adversarial examples by introducing high-camouflage target attribute features. Compared to existing adversarial attacks, our method can produce more imperceptible adversarial faces with higher camouflage, significantly improving the transferability of black-box attacks.
- We proposed a multi-layer feature map fusion strategy to obtain high camouflage features. In the StarGAN generator's residual network, we incorporated a multi-layer feature map fusion mechanism. This fusion method better balances the contributions of feature maps from different layers and helps establish associations between different regions. By flexibly integrating information from different layers, we obtain high-camouflage features that maintain consistency and reasonableness with the original facial features, making the

generated images more similar to the original ones, enhancing the model's overall performance.
- We introduced bilinear interpolation to interpolate between the original image features and high-camouflage features, generating visually smooth adversarial examples. When used in facial images, bilinear interpolation helps ensure smooth transitions of facial features during generation, reducing the perceptual smoothness loss in the objective function. The resulting adversarial examples are more similar to real images, making them easier to deceive the target model without raising suspicion, with higher camouflage and better visual quality.

2 Model and Method

2.1 Problem Definition

A face recognition model (FR model) is primarily defined as a machine learning model trained on a dataset $D(x, y)$ consisting of given image-label pairs to recognize input images. Here, $x \in R^{H*E*D_I}$ is a facial image sampled according to an underlying distribution, and $y \in R^{D_I}$ represents the corresponding ground-truth label. H, W, D_I, and D_L denote the image height, image width, number of image channels, and label dimensions, respectively. Given $f(x) : x \to y$, the model M can predict the label $y = M(x) \in R^{D_L}$ for each input face image x. The primary goal of evasion attacks is to generate an adversarial example x^{adv} that is visually similar to the original image x by adding pixel-level perturbations or applying spatial transformations to x, such that $M(x^{adv}) \neq y$.

2.2 Attribute Image Editing

The generator can reconstruct $y = G(x, c)$ when there are no attribute changes ($c = c_{new}$), where $c_{new} \in R^{D_c}$ is new attribute and y is close to x. Next, we use attribute-conditioned image generator G for semantic image editing. The synthesized image is represented as $x_{new} = G(x, c_{new})$. When our attribute representation is disentangled and the change in attribute values is minimal, our synthesized image x_{new} is expected to be close to the data manifold. Additionally, we can generate many similar images by linearly interpolating between the images x and x_{new} in the feature space of the image-conditioned generator G.

We use the generator to assess the authenticity of the images, the discriminator to classify the labels for each category, and adversarial loss to balance the performance of both. The generator structure is as described in Sect. 3.1, and the formula for designing the reconstruction loss function is given in Eq. (1).

$$L_{rec} = E_{x, \, c_{new}, \, c}[\| x - G(G(x, c_{new}), c) \|_1] \qquad (1)$$

Here, G takes the transformed image $G(x, c_{new})$ and the original domain label c as input and attempts to reconstruct the original image x. We use the L1 norm, which reflects the degree of difference in image reconstruction.

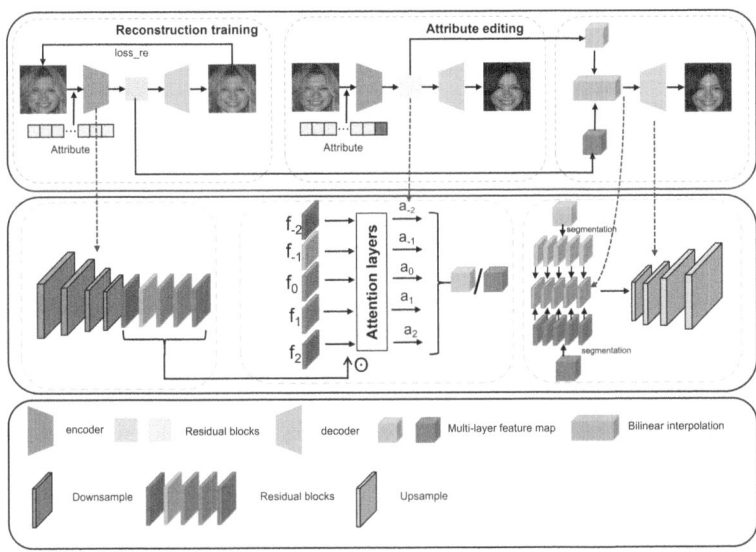

Fig. 1. Main Model. We provides an overview of the framework for attribute-conditioned editing of highly camouflaged adversarial face identity attacks based on multi-layer feature map fusion. The blue dashed lines point to the detailed architecture diagram. (Color figure online)

We use the discriminator to assess the authenticity of the input images, determining whether they are generated or real. Additionally, the discriminator classifies domain labels based on the features of the input image. Its structure is similar to that of the generator, and its loss functions are given by Eq. (2).

$$L^f_{cls} = E_{x,\ c}[-\log D_{cls}(c \mid G(x, c_{new}))] \qquad (2)$$

Among them, $D_{cls}(c|x)$ represents the probability distribution over domain labels computed by D.

2.3 Multi-layer Feature Map Fusion Module

We designed a multi-layer feature map fusion mechanism to extract high concealment feature maps. An adaptive weight is assigned to each layer within residual network, allowing for adaptive fusion of feature maps from different layers of generator's residual network to produce final feature map f_{out}. This adaptive fusion method better balances the contributions of feature maps from different layers, enhances integration of multi-level feature information, and generates more realistic images, improving the model's deception capability and performance.

First, we initialize the weight parameters to 1, indicating each different residual block has an equal impact on the output result, i.e., $a_{-2}, a_{-1}, a_0, a_1, a_2 = 1$.

Then, the attention coefficient for each layer is calculated using Eq. (3).

$$a = \sigma(ln(Conv(x))) \quad (3)$$

Here, $Conv$ represents the convolution operation, used for feature map extraction. In denotes an instance normalization technique, which standardizes the extracted features. The specific explanation of In is given in Eq. (4).

$$InstanceNorm(y) = \frac{y - mean(y)}{\sqrt{var(y) + \varepsilon}} \times \gamma + \beta \quad (4)$$

Here, y represents the feature representation extracted from x after the convolution operation, γ is a learnable scaling parameter, β is a learnable shift parameter, and ε is a small constant.

After the above process, the attention coefficient for each layer is obtained and adaptively fused using Eq. (5). a^k is the adaptive coefficient corresponding to the k-th layer, and f^k is the feature map output from the k-th layer.

$$f_{out} = \sum_{k=0}^{n} f^k \odot a^k \quad (5)$$

Among them, f_{out} represents the result obtained after adaptive fusion through multi-layer fusion, f denotes corresponding feature after convolution, n indicates total number of layers, and a represents the corresponding attention coefficient.

The output result is then upsampled to obtain the synthesized fake image, which is compared with the real image and the target image to calculate the loss. The loss is backpropagated to update the attention parameters.

2.4 Bilinear Interpolation

To better preserve details and textures in generated images and enhance the naturalness and realism of interpolated images. We use bilinear interpolation to reduce blurring or artifacts in the generated images. By incorporating bilinear interpolation, we can blend the original image features with high camouflage features, resulting in visually more continuous and smooth adversarial examples.

We propose using the generator $G = G_D \cdot G_E$ for interpolation. G_E is the encoder module taking the image as input and outputs the feature map, while G_D is the decoder module that takes feature map as input and outputs the synthesized image x* using Eq. (6) and Eq. (7). The feature map $f_{out} = G_E(x, c) \in R^{H_F * W_F * C_F}$ is generated by multi-layer fusion in the generator, where H_F, W_F, and C_F denote the height, width, and number of channels in the feature map.

$$x^* = G_D(f^*) \quad (6)$$

Here, x^* represents the generated countermeasure example picture, G_D represents the decoder, and f^* represents the feature map.

$$f^* = G_E(x, c) + (1 - \gamma) \cdot G_E(x, c^{new}) \quad (7)$$

Among them, γ stands for interpolation coefficient.

2.5 Constructing the Objective Function

Existing methods generate adversarial images x^{adv} by adding perturbations or directly transforming the input image x, as shown in Eq. (8). The first term of our objective function is the adversarial metric, while the second term is a smoothness constraint to ensure perceptual quality. Since concealment and attack strength are somewhat contradictory, it is crucial to design a parameter μ in adversarial attacks to control the balance between these two terms. The decoder G_D takes the feature maps f^* from multi-layer feature fusion interpolation as input and generates the synthesized facial image x* as output.

$$L(x^*) = L_{adv}(x^*;\ M) + \mu \cdot L_{smooth}(x^*) \tag{8}$$

Here, L_{adv} represents confrontation loss, μ represents balance coefficient, and L_{smooth} represents smoothness loss.

As we see in Eq. (9), measures the distance between two identity embeddings from the model M, where in our setup, the normalized L_2 distance (Euclidean distance) is used to measure the similarity between the two identity embeddings, as shown in Eq. (4). Adversarial facial images evade detection by maximizing the distance between $x*$ and x in the feature space. Additionally, we introduce a parameter κ, which represents a constant related to the false positive rate(FPR) threshold computed from the development set.

$$\phi_M^{id}(\ x^*,\ x) = Max_{x^*} \|\ M(x^*) - M(x)\|_2^2 \tag{9}$$

Among them, ϕ_M^{id} stands for distance function and $Max()$ stands for maximum function.

3 Experiments

In the experimental section, we use the CelebA dataset to generate attribute-conditioned adversarial examples to attack face verification models ResNet-101 and ResNet-50. The CelebA dataset is a public dataset and does not involve risks related to personal privacy or ethical controversy; therefore, there are no direct ethical issues. To validate the effectiveness of the StarAdv algorithm, we first quantitatively compared the quality of adversarial examples generated using FaceNet and ArcFace models with those generated by other methods under SSIM and MSE metrics, demonstrating that our method produces more realistic and camouflaged adversarial examples. Second, to evaluate attack transferability, we further demonstrated the effectiveness of our method through the success rate of black-box attacks with no-query on an online facial verification platform.

3.1 Similarity to Original Images

In the experiment, we first attacked the ResNet-101/ResNet-50 models to achieve a white-box attack. The images that were successfully attacked in the white-box

Table 1. MSE and SSIM for Original and Adversarial Samples Generated by Different Methods on FACENET and ARCFACE Models

Attacks Methond	Year	FaceNet		ArcFace	
		MSE(↓)	SSIM(↑)	MSE(↓)	SSIM(↑)
FGSM [6]	2014	0.034	0.388	0.036	0.374
BIM [7]	2018	0.031	0.596	0.032	0.585
PGD [8]	2018	0.026	0.715	0.026	0.700
SemanticAdv [9]	2020	0.023	0.813	0.023	0.813
Sticker [10]	2021	0.958	0.009	0.959	0.008
AMT-GAN [11]	2022	–	0.820	–	–
SAA-StarGan-CS [12]	2023	0.022	0.822	0.020	0.821
SAA-StarGan-CS-M [12]	2023	0.040	0.729	0.036	0.735
MTADV-ST [13]	2024	–	0.855	–	–
Ours	**2024**	**0.017**	**0.882**	**0.019**	**0.828**

test were then used for black-box testing on FaceNet and ArcFace. FaceNet and ArcFace represent two different facial recognition technologies. Our goal is to measure the quality of adversarial facial images relative to the original images using MSE and SSIM. MSE measures absolute error, while SSIM assesses the similarity between two images based on perceived quality. We compared the adversarial facial images generated by our method with baseline methods. As shown in Table 1, our method achieves the highest SSIM values and the lowest MSE values. The main reason is that our model framework introduces fused high-camouflage multi-layer feature maps, enabling more realistic face image transformations by modifying semantic information across different facial attributes. This results in adversarial images that are visually more similar to the original images and possess specific misleading features, providing high camouflage and better visual quality. In contrast, adversarial facial images generated by FGSM, BIM, PGD, and MI-FGSM are more easily perceived by the human eye. Therefore, our method is recommended for generating adversarial facial images.

3.2 Query-Free Black-Box API Attack

As shown in Fig. 2, we conducted a query-free black-box attack. In this experiment, we generated adversarial examples targeting R-101-S with G-FPR = 10^{-3} ($\kappa = 1.24$), G-FPR = 10^{-4} ($\kappa = 0.60$), and G-FPR < 10^{-4} ($\kappa = 0.30$). For the ResNet-101 model trained with softmax loss, the false positive rate (FPR) is an important metric for evaluating the performance of the face recognition model. We used four different FPRs: 10^{-3} ($\kappa = 1.24$), 3×10^{-4} ($\kappa = 1.05$), 10^{-4} ($\kappa = 0.60$), and < 10^{-4} ($\kappa = 0.30$). We evaluated our algorithm on two industry-level face verification APIs, Face++ and AliYun, as shown in Table 2.

Table 2. Quantitative Analysis of Query-Free Black-Box Attacks.

Model	Year	FaceNet		ArcFace	
		T-FPR = 10^{-3}	T-FPR = 10^{-4}	T-FPR = 10^{-3}	T-FPR = 10^{-4}
CW [14]	2017	37.24	20.41	18.00	9.50
StarGAN+CW [15]	2018	47.45	26.20	20.00	8.50
M-DI^2-FGSM [16]	2019	56.12	33.67	30.00	18.00
Semantic-Adv [9]	2020	67.69	48.21	36.50	19.50
RSTAM$_2$ [17]	2022	71.18	–	–	–
EHSRA [18]	2023	–	70.61	–	–
DFPP [19]	2024	64.80	–	28.30	–
Adv-Diffusion [20]	2024	61.70	–	28.30	–
Ours	**2024**	**76.81**	**55.90**	**40.09**	**31.82**

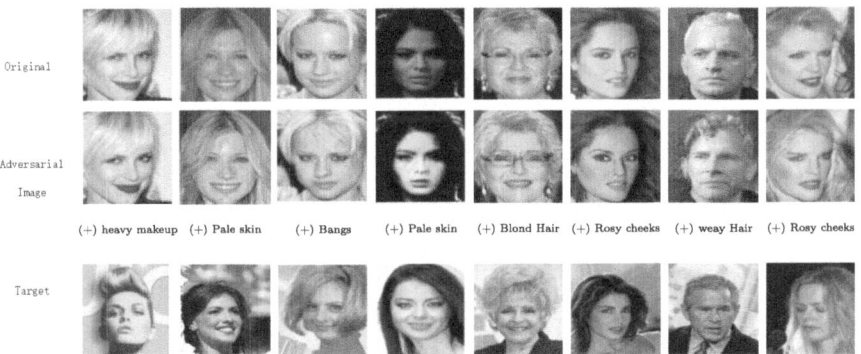

Fig. 2. The Experimental Results Show Successful Attacks Using Adversarial Face Images Generated by StarAdv. The four columns on the left represent the Face++ online API, while the four columns on the right represent the AliYun online API. This demonstrates the effectiveness of StarAdv in generating realistic images.

3.3 Comparison Experiments

To comprehensively compare the differences in adversarial facial images generated by various methods, a visual image comparison experiment was conducted, as shown in Fig. 3. FGSM and PGD attacks produce visually blurry and low-quality adversarial facial images. Patch-based methods generate adversarial patches cover critical areas of facial images, but these patches are easily noticeable, making the adversarial examples more detectable. Our method, which uses multi-feature map fusion, generates high-quality adversarial facial images that are visually realistic and well-concealed, making them harder to detect. In summary, application of this model in facial adversarial attacks significantly enhances the performance of the generative network and provides an effective approach for generating highly concealed and high-fidelity adversarial facial images.

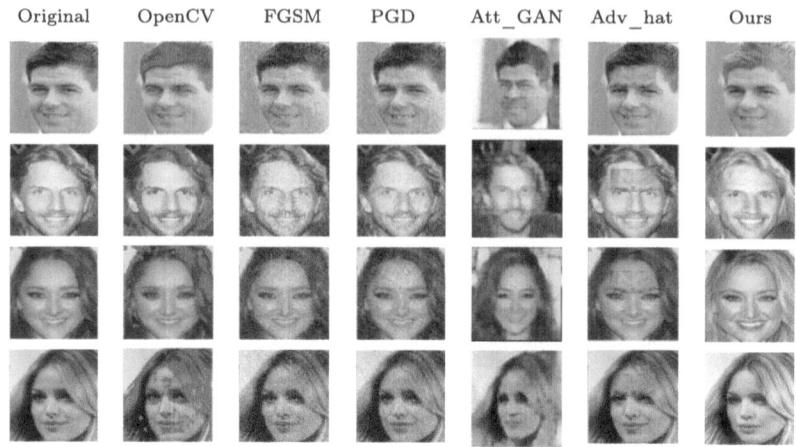

Fig. 3. Adversarial Facial Images Generated by Different Methods. Compared to these methods, StarAdv can produce high-quality adversarial facial images.

4 Conclusion

This paper presents StarAdv, a subtle and transferable adversarial attack method that offers improved controllability compared to existing approaches. We introduce a multi-layer feature map fusion strategy during the optimization process, using an adaptive fusion method to balance the contributions of feature maps from different layers, thus generating more deceptive adversarial facial images, achieves a high attack success rate and strong transferability in both white-box and black-box settings. We believe this work will open up new research opportunities and challenges in the field of adversarial learning.

Acknowledgments. This work is supported by the Natural Science Foundation for the Higher Education Institutions of Anhui Province (No. 2022AH050091, 2022AH040019, 2022AH05008637).

References

1. Yang, L., Song, Q., Wu, Y.: Attacks on state-of-the-art face recognition using attentional adversarial attack generative network. Multimed. Tools Appl. **80**, 855–875 (2021)
2. Yin, B., et al.: Adv-makeup: a new imperceptible and transferable attack on face recognition. arXiv preprint arXiv:2105.03162 (2021)
3. Nguyen, D.L., Arora, S.S., Wu, Y., Yang, H.: Adversarial light projection attacks on face recognition systems: a feasibility study. In: Proceedings of the IEEE/CVF Conference on Computer Vision and Pattern Recognition Workshops, pp. 814–815 (2020)

4. Jia, S., et al.: Adv-attribute: inconspicuous and transferable adversarial attack on face recognition. Adv. Neural. Inf. Process. Syst. **35**, 34136–34147 (2022)
5. Karras, T., Laine, S., Aila, T.: A style-based generator architecture for generative adversarial networks. In: Proceedings of the IEEE/CVF Conference on Computer Vision and Pattern Recognition, pp. 4401–4410 (2019)
6. Goodfellow, I.J., Shlens, J., Szegedy, C.: Explaining and harnessing adversarial examples. arXiv preprint arXiv:1412.6572 (2014)
7. Kurakin, A., Goodfellow, I., Bengio, S.: Adversarial machine learning at scale. arXiv preprint arXiv:1611.01236 (2016)
8. Komkov, S., Petiushko, A.: AdvHat: real-world adversarial attack on arcface face id system. In: 2020 25th International Conference on Pattern Recognition (ICPR), pp. 819–826. IEEE (2021)
9. Qiu, H., Xiao, C., Yang, L., Yan, X., Lee, H., Li, B.: SemanticAdv: generating adversarial examples via attribute-conditioned image editing. In: Vedaldi, A., Bischof, H., Brox, T., Frahm, J.-M. (eds.) ECCV 2020. LNCS, vol. 12359, pp. 19–37. Springer, Cham (2020). https://doi.org/10.1007/978-3-030-58568-6_2
10. Gao, L., Zhang, Q., Song, J., Liu, X., Shen, H.T.: Patch-wise attack for fooling deep neural network. In: Vedaldi, A., Bischof, H., Brox, T., Frahm, J.-M. (eds.) ECCV 2020. LNCS, vol. 12373, pp. 307–322. Springer, Cham (2020). https://doi.org/10.1007/978-3-030-58604-1_19
11. Hu, S., et al.: Protecting facial privacy: generating adversarial identity masks via style-robust makeup transfer. In: Proceedings of the IEEE/CVF Conference on Computer Vision and Pattern Recognition, pp. 15014–15023 (2022)
12. Khedr, Y.M., Xiong, Y., He, K.: Semantic adversarial attacks on face recognition through significant attributes. Int. J. Comput. Intell. Syst. **16**(1), 196 (2023)
13. Wang, H., Wang, S., Chen, C., Tistarelli, M., Jin, Z.: A multi-task adversarial attack against face authentication. arXiv preprint arXiv:2408.08205 (2024)
14. Carlini, N., Wagner, D.: Towards evaluating the robustness of neural networks. In: 2017 IEEE Symposium on Security and Privacy (SP), pp. 39–57. IEEE (2017)
15. Choi, Y., Choi, M., Kim, M., Ha, J.W., Kim, S., Choo, J.: StarGAN: unified generative adversarial networks for multi-domain image-to-image translation. In: Proceedings of the IEEE Conference on Computer Vision and Pattern Recognition, pp. 8789–8797 (2018)
16. Dong, Y., Pang, T., Su, H., Zhu, J.: Evading defenses to transferable adversarial examples by translation-invariant attacks. In: Proceedings of the IEEE/CVF Conference on Computer Vision and Pattern Recognition, pp. 4312–4321 (2019)
17. Liu, X., Shen, F., Zhao, J., Nie, C.: RSTAM: an effective black-box impersonation attack on face recognition using a mobile and compact printer. arXiv preprint arXiv:2206.12590 (2022)
18. Liang, Y., Wang, H., Cao, S., Wang, J., Tao, L., Zhang, J.: Efficient and high-quality black-box face reconstruction attack. In: Proceedings of the 15th International Conference on Digital Image Processing, pp. 1–7 (2023)
19. Sun, Y., Yu, L., Xie, H., Li, J., Zhang, Y.: DiffAM: diffusion-based adversarial makeup transfer for facial privacy protection. In: Proceedings of the IEEE/CVF Conference on Computer Vision and Pattern Recognition, pp. 24584–24594 (2024)
20. Liu, D., Wang, X., Peng, C., Wang, N., Hu, R., Gao, X.: Adv-diffusion: imperceptible adversarial face identity attack via latent diffusion model. In: Proceedings of the AAAI Conference on Artificial Intelligence, vol. 38, pp. 3585–3593 (2024)

A High Accuracy Text CAPTCHA Recognition Approach Through Opertimized Vision Transformer

Wei Hao, Shoulai Shang(✉), and Yepeng Zhang

Anhui University of Science and Technology, Huainan, Anhui, China
{whao,2023201191,2024201314}@aust.edu.cn
https://www.aust.edu.cn

Abstract. CAPTCHA (Completely Automated Public Turing test to tell Computers and Humans Apart) is an automated verification mechanism designed to distinguish between human visitors and automated systems as a vital component for cyber security. However, current CAPTCHA recognition technologies still struggle to keep pace with the sophistication CAPTCHA generation algorithms. This paper introduces a novel approach to CAPTCHA recognition, utilizing the Vision Transformer (ViT) and enhancing it with two key optimizations: the Permutation Visual Model, which improves the model's spatial understanding, and Transfer Learning for the Vision Transformer, which accelerates the model's adaptation to new tasks. The method was rigorously tested against a diverse set of practical data, simulating real-world scenarios. The results are promising, with an accuracy rate of over 95% across multiple websites, indicating a significant advancement in CAPTCHA recognition technology.

Keywords: CAPTCHA · image recognition · Vision Transformer · Permutation Visual Model · cyber security

1 Introduction

CAPTCHA also called HIPs (Human Interaction Proofs), is a computer test to distinguish between human users and computer algorithms and automated systems, in order to prevent various forms of online computer attaches, such as spam, unauthorized database access, online voting manipulation, and distributed denial of service (DDoS) attacks [1]. By implementing CAPTCHA systems, online services can filter out malicious traffic and avoid unwanted disruptions. Therefore, CAPTCHA is important for both automated tests and online attacks.

Since the commercialization of the internet in 1995, CAPTCHA technology has evolved significantly, with its concept introduced by Luis von Ahn in 2002 [2]. The rise of deep learning, particularly convolutional neural networks (CNNs), has shifted CAPTCHA recognition from simple character identification to complex models. In 2017, Le et al. [3] developed an efficient CAPTCHA-breaking model using CNNs and RNNs. More recent advancements, like Generative Adversarial Networks (GANs), have further improved success rates, with Zhang et al. [4] using GANs to crack complex CAPTCHAs. Despite progress, CAPTCHA segmentation remains challenging due to noise, distortions, and diverse designs, which hinder current models' ability to extract meaningful features [5]. Future methods, including GANs and transfer learning, are expected to enhance both CAPTCHA systems and breaking techniques.

This study enhances CAPTCHA recognition by integrating the [6]Vision Transformer (ViT) model with the Permutation Visual Model (PVM), combining Attention Mask techniques with pre-training and fine-tuning strategies to improve handling of complex visual information. Pre-training the ViT on large image datasets reduces the need for extensive annotated data, providing a robust foundation for fine-tuning. The optimized Attention Mask enables precise control over attention allocation, reducing noise and improving accuracy, while PVM allows flexible adjustment of image patch sequences to enhance adaptability. Experimental results show that the fine-tuned ViT model significantly improves CAPTCHA recognition accuracy, while reducing computation time.

The main contributions of this paper are as follows:

- **Application of ViT Model to CAPTCHA Recognition:** This work is the first to apply the ViT model to CAPTCHA recognition. By leveraging ViT's strengths in image feature extraction and contextual understanding, it provides a more accurate and faster solution for breaking CAPTCHAs.
- **Model Optimization and Recognition Improvement:**
  - **Attention Mask Optimization.** By refining the attention distribution of the ViT model, background noise interference is reduced, significantly improving CAPTCHA recognition accuracy.
  - **Transfer Learning Implementation.** The pre-trained ViT model is applied to CAPTCHA recognition, reducing training time while enhancing recognition performance.
- **Validation of Method Effectiveness:** Extensive experiments were conducted on public datasets and real-world website data to demonstrate the method's effectiveness and robustness in CAPTCHA recognition tasks.

Through this research, we propose a more efficient solution for CAPTCHA breaking and lay an important foundation for the advancement of cybersecurity.

The structure of this paper is as follows: Sect. 1 introduces the significance of CAPTCHA technology and its associated security challenges. Section 2 reviews related work, covering traditional CAPTCHA recognition methods and advanced deep learning approaches. Section 3 presents the methodology, detailing the ViT architecture and optimizations, including the PVM and pre-training techniques. Section 4 evaluates the PVIT model's performance through comparative experiments, and Sect. 5 concludes by discussing the advantages of the proposed approach and potential directions for future research.

2 Related Work

Common CAPTCHA recognition techniques include template matching, support vector machines (SVM) [7], and neural network methods. Template matching is efficient but struggles with image rotation and scaling, limiting its accuracy. Traditional machine learning methods, such as SVM with preprocessing techniques [8], have been used to recognize simple CAPTCHAs, but they struggle with overlapping characters. Advanced methods like Hopfield neural networks [9] and the droplet algorithm can handle more complex CAPTCHAs, though preprocessing remains a limitation.

Deep learning techniques, particularly CNNs [10], have rapidly gained prominence in CAPTCHA recognition. CNNs excel at recognizing patterns and invariant features in images, surpassing manually designed features. For instance, CNN-based methods with enhanced VGG structures [11] have achieved high recognition rates across multiple datasets. These networks extract local features like edges and textures, improving robustness to noise. RNNs and their variants, like LSTM and GRU, are used for processing sequences, providing context to character recognition.

Advanced models like Deep CNNs (DCNNs) and Residual Networks [12] (ResNets) can extract more abstract features, improving performance in complex CAPTCHA tasks. For example, CNN-RNN architectures with attention mechanisms, such as Wan's Adaptive CAPTCHA model [13], reduce parameter count and speed up training while maintaining high accuracy. Additionally, GANs help by generating training samples to improve model generalization. Despite not being convolutional, Transformers offer advantages in capturing relationships between characters with their self-attention mechanisms.

3 Methodology

The following sections describe the ViT architecture, optimization techniques, and the specific adjustments made for improved performance in CAPTCHA tasks.

3.1 ViT Transformer

The ViT model is a state-of-the-art architecture for image recognition tasks, and in this work, we enhance its capability by incorporating the PVM to improve CAPTCHA recognition.

Embedding (Linear Projection of Flattened Patches): Initially, the input image is divided into patches of fixed size. Each patch is flattened and passed through a linear layer to convert it into a feature vector. Additionally, a positional embedding is added to each feature vector to indicate its relative position in the original image.

Transformer Encoder: These pre-processed feature vectors are fed into a series of Transformer layers for encoding. Each Transformer layer consists of a multi-head attention mechanism and a feed-forward network. Between each Transformer layer, layer normalization is applied to ensure consistency in the distribution of features.

MLP Head and Classification: Finally, the output from the last Transformer encoder layer is passed through a multi-layer perceptron (MLP) head, which is typically a simple fully connected layer, to perform the final task, such as image classification. At this stage, the model maps the learned features into the class probability space.

- **Patch Embedding:** To handle CAPTCHA images, we define the image x as a 3D array $x[h, w, c]$, where h is the height, w is the width, and c is the number of color channels (typically 3 for RGB). The image is then divided into patches (e.g., 8×8 or 4×4). Each patch undergoes a linear transformation, mapping pixel values to lower-dimensional vectors through a learnable fully connected layer. These vectors form the patch embeddings, capturing local features. Finally, the embeddings are concatenated based on their positions and fed into a Transformer encoder for further processing, converting the CAPTCHA recognition task into a sequence-to-sequence problem.
- **Multi Head Attention:** The attention mechanism is the core of Transformer operations. In scaled dot-product attention, the similarity score between two d_k-dimensional vectors, q (query) and k (key), is computed using their dot product and is used to transform a d_v-dimensional vector v (value). Formally, scaled dot-product attention is defined as:

$$Attention(\mathbf{q}, \mathbf{k}, \mathbf{v}) = softmax(\frac{\mathbf{q}\mathbf{k}^T}{\sqrt{d_k}})\mathbf{v}$$

where T denotes the transpose operation. It accepts an optional attention mask that restricts which keys the queries can attend to. In Transformers, where the dimension of tokens is $d_k = d_v = d_{model}$Multi-head Attention (MHA) is an extension of scaled dot-product attention to multiple representation subspaces or heads. To keep the computational cost of MHA unchanged with respect to the number of heads, the dimensionality of vectors is reduced to $d_{head} = \frac{d_{model}}{h}$, where h is the number of heads.

$$head_i = Attention(\mathbf{q}\mathbf{W}_i^q, \mathbf{k}\mathbf{W}_i^k, \mathbf{v}\mathbf{W}_i^v)$$

$$MHA(\mathbf{q}, \mathbf{k}, \mathbf{v}) = Concat(head_1, ..., head_h)\mathbf{W}^o$$

The dimensions of W^q W^k and W^v are all in $R^{d_{model}*d_{head}}$,while W^orepresents the output projection matrix,with dimensions in $R^{d_{model}*d_{head}}$.
- **Encoder:** The ViT extends Transformers to images, using Multi-Head Attention (MHA) for self-attention where $q = k = v$. The 12-layer encoder excludes the classification head and [CLS] token. An image $x \in \mathbb{R}^{W \times H \times C}$ is divided into $p_w \times p_h$ patches. Each patch is flattened and projected into tokens of

dimension d_{model} via a patch embedding matrix $W_p \in \mathbb{R}^{(p_w p_h C) \times d_{\text{model}}}$, resulting in $\frac{WH}{p_w p_h}$ tokens. Learned positional embeddings are added before processing in the first ViT layer.

$$z = Enc(x) \in R^{\frac{WH}{p_w p_h}} \times d_{model}$$

- **Decoder:** The decoder follows the same architecture as the preLayerNorm Transformer decoder, but uses twice the number of attention heads, i.e., $n_{head} = \frac{d_{model}}{32}$. It has three required inputs: position, context, and image tokens, as well as an optional attention mask.

$$h_c = p + MHA(p, c, c, m) \in R^{(T+1) \times d_{model}}$$

Where T is the context length, $p \in \mathbb{R}^{(T+1) \times d_{model}}$ represents the positional tokens, $c \in \mathbb{R}^{(T+1) \times d_{model}}$ denotes the context embeddings with positional information, and $m \in \mathbb{R}^{(T+1) \times (T+1)}$ is the optional attention mask. Note that the total sequence length is increased to T+1 using special delimiter tokens ([B] or [E]).

Positional tokens encode target positions, allowing the model to effectively learn from the Permutation Visual Model (PVM). Without these tokens, using context tokens as queries, as in standard transformers, would prevent meaningful learning. During training, the attention [14] mask is generated by random permutation, while during inference, a standard left-to-right forward mask is used.

The model also uses a second Multi-Head Attention (MHA) mechanism for image position attention, refining the representation by considering the relative positions of image patches. The final decoder state is computed using a Multi-Layer Perceptron (MLP), and the output logits predict the characters in the CAPTCHA sequence, with each token representing a specific position in the output sequence.

3.2 Optimization1: Permutation Visual Model

To enhance the ViT's capacity for modeling the sequential arrangement of image patches, we introduce a method called PVM, as shown in Fig. 1. The key idea behind PVM is to enable the model to learn more nuanced spatial information by permuting the patch order while maintaining the correspondence between image inputs and outputs. In our implementation of ViT, we use a straightforward yet effective approach based on Attention Masks. The Attention Mask is a vector of the same length as the image patch sequence, where a value of 0 indicates that the corresponding patch is attended to, and a value of 1 signifies that it is masked. By designing different Attention Masks, we can control which patches each patch can see according to a specific permutation order. For instance, if the image consists of [Patch1, Patch2, Patch3, Patch4], a reverse-order Attention Mask like [1, 1, 1, 0] allows Patch4 to see all patches, Patch3 to see only Patch4, and so forth, thus establishing a reverse spatial dependency.

This approach offers the following advantages:

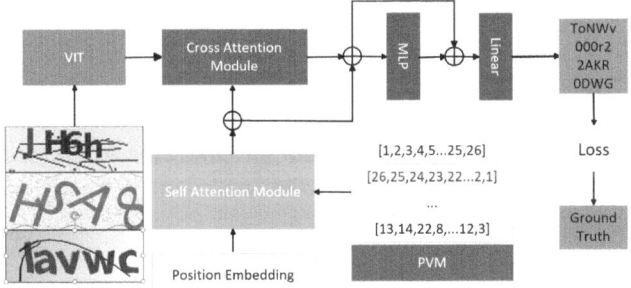

Fig. 1. Modified architecture of vit Transformer

1. No need to actually change the image input-output correspondence, reducing training complexity;
2. Ability to learn multiple spatial order representations, enhancing robustness to varying orders;
3. Provides richer spatial context modeling.

PVM provides ViT with a mechanism to control patch ordering through Attention Masks, effectively enhancing its capability to model and utilize complex spatial information, potentially improving image understanding and analysis.

4 Experiments

4.1 Experiment Configuration

Hardware. All models were trained using a single GPU setup with an RTX 3090 graphics card. The training configuration included 12 workers, a batch size of 384, and a learning rate of 7e-4. The optimizer used was Adam [15], in conjunction with the 1 cycle learning rate scheduler. At 75% of the total number of iterations, stochastic weight averaging (SWA) was applied, and the SWA scheduler replaced the 1cycle scheduler. Validation was performed every 1000 training steps. Due to SWA averaging weights at the end of each epoch, the final checkpoint was selected at the end of training.

Training Settings for PVIT. A permutation of K=6 was used with patch sizes of 8×4 During training, the maximum label length was set to T=25, and the character set size was S=94, which includes a mix of uppercase and lowercase alphanumeric characters and punctuation. The character set for the test set included only digits and lowercase letters, as real-world testing typically does not distinguish between uppercase and lowercase. CAPTCHA images were first augmented, resized, and then normalized to the range of $[-1, 1]$. Data augmentation operations were primarily performed using RandAugment [16], excluding sharpening operations. Gaussian blur and Poisson noise were also applied for data augmentation, with a RandAugment strategy comprising 3 layers and a magnitude of 5. Images were uniformly resized to 128 × 32 pixels.

Evaluation Metrics. Character Accuracy is a primary metric for CAPTCHA recognition. A prediction is considered correct only if all characters at all positions match the ground truth. The formula for character accuracy is:

$$ACC = \frac{C_{predict}}{C_{label}}$$

This metric provides a stringent measure of the model's performance, as it requires exact matches across all character positions.

Datasets. CAPTCHA images are often affected by complex distortions, deformations, and overlaps, and contain multiple categories such as letters, digits, and symbols. Pre-trained models excel at initializing feature representations, providing a better starting point for fine-tuning. We performed pre-training on large-scale real-world datasets (TextOCR and OpenVINO), allowing the model to capture general text features and establish prior knowledge for subsequent tasks. Through fine-tuning, the pre-trained model quickly adapts to different CAPTCHA styles and achieves satisfactory recognition performance with a small number of labeled samples. The use of pre-trained weights significantly reduces training costs and accelerates performance, enabling the model to effectively transfer the benefits of large-scale datasets to smaller datasets.

Open Data. In CAPTCHA recognition tasks, the Python ImageCaptcha Library [17] is commonly used to randomly generate CAPTCHAs for training networks. Additionally, public CAPTCHA datasets can be obtained from sources such as Kaggle and the official Baidu PaddlePaddle website. Examples of publicly available datasets are shown in Fig. 2.

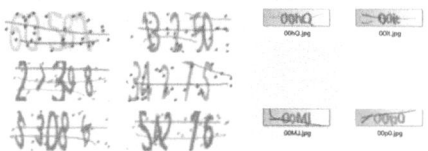

Fig. 2. Samples from Training Dataset

4.2 Ablation Experiment

To validate the effectiveness of the ViT+PVM model in complex CAPTCHA recognition, we conducted a series of experiments using the Adam optimizer with an initial learning rate of 7e−4, a 1-cycle scheduler with SWA applied at 75% of the training, a batch size of 384, 12 workers, patch size of 8×4, a maximum label length of 25, and a character set containing 94 characters, while the test set included only lowercase letters and digits, and data augmentation was applied using RandAugment (3 layers, strength 5), along with Gaussian blur and Poisson noise, and through ablation experiments where PVM was removed while keeping other parameters constant, we found that removing PVM significantly degraded performance, particularly on the D003 dataset, which contains complex interference like distortion, noise, and severe interference lines, with PVM introducing diversified structures that enhanced the model's robustness, improved its ability to handle complex noise and language variants, achieving high accuracy after 1000 training iterations, and further improving training efficiency with a varying learning rate, as shown in Fig. 3.

Test Dataset	Sample size	Model Accuracy	
		VIT	VIT + PVM
D001	5280	89.62%	96.27%
D002	4832	99.40%	100.00%
D003	11307	14.79%	79.76%
D004	1000	94.20%	100.00%
D006	1000	95.60%	99.20%
D007	3396	84.16%	99.23%
D008	1135	98.77%	100.00%
D009	2400	81.21%	92.25%
D010	582	99.83%	100.00%
D011	2100	92.14%	95.86%
D012	2421	99.79%	100.00%
D013	4734	98.75%	99.68%
D014	1500	100.00%	100.00%
D015	1000	97.60%	99.50%
D016	1501	94.20%	98.53%
D017	533	94.37%	98.50%
D018	996	99.90%	100.00%
D022	2706	95.40%	99.86%
D024	1500	94.27%	95.33%
D026	653	94.49%	98.93%
average		90.92%	97.65%

Fig. 3. Ablation Study Data Comparison

4.3 Results and Comparison

We compared the PVIT method with two of the most commonly used CAPTCHA recognition methods. The first is ddddocr, which has garnered 6.1k stars on GitHub, highlighting its importance as a benchmark in CAPTCHA recognition. The ddddocr (beat) version utilizes the latest built-in models. The second method is PaddleOCR [18], an outstanding open-source Chinese OCR tool known for its high accuracy, speed, and ease of use. PaddleOCR enables the rapid integration of Chinese OCR capabilities into various systems, improving recognition performance. It has been widely used in fields such as document digitization and license plate recognition, and has received 32.7k stars on GitHub.

We used the latest version of PaddleOCR for our experiments. For fairness, we did not alter any parameters; the PVIT model was evaluated directly on the public dataset without further fine-tuning. Similarly, we evaluated ddddocr, ddddocr (beat), and PaddleOCR (latest) under the same conditions.

Test Dataset	Sample Size	Model Accuracy			
		ddddocr	ddddocr(beat)	PaddleOCR	VIT(0803version)
D013	4734	77.40%	79.68%	20.89%	99.68%
D014	1500	96.00%	97.73%	72.40%	100.00%
D015	1000	50.70%	56.90%	0.30%	99.50%
D016	1501	53.96%	64.56%	2.07%	98.53%
D017	533	51.41%	54.03%	59.66%	98.50%
D018	996	99.80%	100.00%	35.74%	100.00%
D022	2706	1.52%	2.66%	12.53%	99.86%
D024	1500	82.00%	80.47%	31.33%	95.33%
D026	653	22.05%	12.71%	6.59%	98.93%
D027	248	87.10%	95.56%	92.74%	100.00%
D028	247	100.00%	100.00%	99.60%	100.00%
D029	250	93.20%	90.80%	61.60%	96.80%
D030	250	93.60%	92.00%	74.80%	99.20%
D031	251	99.20%	99.20%	94.42%	100.00%
D037	4913	99.29%	98.84%	82.90%	99.62%
D039	10028	89.85%	99.42%	55.34%	93.90%
D040	4893	75.78%	22.93%	36.93%	99.52%
D101	17700	71.45%	75.53%	6.31%	99.18%
average		74.68%	73.50%	47.01%	98.81%

Fig. 4. Comparison of Results with Traditional Models on Open Datasets

Open Data Comparison Experiment. To validate the effectiveness of the proposed algorithm, we compared it with several advanced algorithms, including ddddocr, ddddocr (beat version), and PaddleOCR (latest), using the Open Data dataset. As shown in Fig. 4, the first column represents the dataset identifiers (ranging from d013 to d040, etc.), the second column indicates the sample size of each dataset, and the subsequent columns display the recognition accuracy of each model on the corresponding dataset.

Compared to the proposed improved algorithm, both PaddleOCR (latest) and ddddocr show lower accuracy rates across all test datasets, resulting in a lower average accuracy. Although the ddddocr (beat version) algorithm outperforms the proposed algorithm on the D039 test dataset, its accuracy is lower on the other test datasets. As a result, the average accuracy of the proposed algorithm is significantly higher than that of ddddocr (beat version). These results demonstrate the superior performance of the proposed algorithm.

5 Conclusion

In this paper, we apply the PVM to the ViT model and, for the first time, use the ViT model for CAPTCHA recognition tasks. By jointly tuning both image and text, we achieve state-of-the-art (SOTA) results on publicly available datasets, and our model also significantly outperforms current mainstream and traditional models on real-world website datasets. The end-to-end model effectively addresses the challenges of low recognition efficiency for different characters in CAPTCHA images and low accuracy in recognizing complex interference CAPTCHAs. Through extensive validation on a large number of CAPTCHA datasets, we have demonstrated the feasibility of the model and the superiority of our approach. In future work, due to the varying types of interference in real-world website CAPTCHAs, we will continue to collect data from different real-world websites for training and validation analysis to improve the model's generalization capability.

References

1. Zhu, T., Qiu, X., Rao, Y., Yan, H., Zhou, Y., Shi, G.: HiAtGang: how to mine the gangs hidden behind DDoS attacks. Chin. J. Electron. **31**(2), 293–303 (2022)
2. Von Ahn, L., Blum, M., Hopper, N. J., Langford, J.: CAPTCHA: using hard AI problems for security. In: Theory and Application of Cryptographic Techniques, pp. 294–311 (2003)
3. Le, T., Baydin, A., Zinkov, R., Wood, F.: Using synthetic data to train neural networks is model-based reasoning. IEEE (2017)
4. Zhang, N., Ebrahimi, M., Li, W., Chen, H.: Counteracting dark web text-based captcha with generative adversarial learning for proactive cyber threat intelligence. ACM Trans. Manage. Inform. Syst. (TMIS) **13**(2), 1–21 (2022)
5. Lin, X., Li, L., Ren, Y.: Deep learning captcha recognition for mobile based on tensorflow. In: Third International Seminar on Artificial Intelligence, Networking, and Information Technology (AINIT 2022), vol. 12587, pp. 381–387 (2023)
6. Asifullah, K., et al.: A survey of the vision transformers and their CNN-transformer based variants. Artif. Intell. Rev. **56**(Suppl 3), 2917–2970 (2023)
7. Huang, W., et al.: Railway dangerous goods transportation system risk identification: comparisons among SVM, PSO-SVM, GA-SVM and GS-SVM. Appl. Soft Comput. **109**, 107541 (2021)
8. Guo, Z., Qiu, Y.: Research and implementation of character and graphical captcha recognition algorithm based on SVM. Comput. Program. Skills Maint. **12**, 163–165 (2021)
9. Bao, Q., Li, W., Zhang, Q.: Research on captcha recognition on the android platform. Sci. Technol. Bull. **33**(09), 73–75+219 (2017)
10. Pan, X., Wang, H.: License plate similar character recognition based on deep learning. Comput. Sci. **44**(S1), 229–231 (2017)
11. Wang, Z., Shi, P.: Captcha recognition method based on CNN with focal loss. Complexity **2021**(1), 6641329 (2021)
12. He, K., Zhang, X., Ren, S., Sun, J.: Deep residual learning for image recognition. In: Proceedings of the IEEE Conference on Computer Vision and Pattern Recognition, pp. 770–778 (2016)
13. Wan, X., Johari, J., Ruslan, F.: Adaptive captcha: a CRNN-based text captcha solver with adaptive fusion filter networks. Appl. Sci. **14**(12), 5016 (2024)
14. Yang, Z.: XLNet: generalized autoregressive pretraining for language understanding. arXiv preprint arXiv:1906.08237 (2019)
15. Manju, S., Selvam, S.: Optimizing hyperspectral image classification through advanced deep learning models enhanced by Adam. Traitement du Signal **41**(2) (2024)

16. Xiao, A., Shen, B., Tian, J., Hu, Z.: Differentiable randaugment: learning selecting weights and magnitude distributions of image transformations. IEEE Trans. Image Process. **32**, 2413–2427 (2023)
17. Nouri, Z., Rezaei, M.: Deep-CAPTCHA: a deep learning based captcha solver for vulnerability assessment. arxiv 2020. CoRR abs/2006.08296 (2020)
18. Gabriella, M., et al.: A comprehensive framework for industrial sticker information recognition using advanced OCR and object detection techniques. Appl. Sci. **13**(12), 7320 (2023)

LightMamba-UNet: Lightweight Mamba with U-Net for Efficient Skin Lesion Segmentation

Wanzhen Hou[1], Shiwei Zhou[2], and Haifeng Zhao[1]($\boxtimes$)

[1] School of Computer Science and Technology, Anhui University, Hefei, China
senith@163.com
[2] School of Artificial Intelligence, Anhui University, Hefei, China

Abstract. Recently, advancements in Vision Mamba Models (VMMs) have achieved notable success in dense prediction. However, existing Vision-Mamba-like approaches often improve performance by adopting more complex modules, making them poor for medical applications where computational resources are constrained. Meanwhile, focusing on segmentation tasks, these methods prefer to redesign the internal architecture of the encoder and decoder, potentially overlooking the thorough exploration of skip connections, which in turn affects the accuracy of segmentation. To minimize the parameters of the model derived from Vision Mamba, we introduce a novel Lightweight Mamba with U-Net (LightMamba-UNet) for processing the deep features in parallel. Furthermore, to address the demand for more accurate segmentation, we introduce a full-scale lightweight skip connection to bridge the semantic gap between the encoder and decoder. Extensive experiments demonstrate that our model outperforms the state-of-the-art (SOTA) on the public ISIC17, ISIC18, and PH^2 datasets, concurrently the proposed parameter efficient framework can reduce the total model size by 36.73%.

Keywords: Skin Lesion Segmentation · Mamba · U-Net

1 Introduction

The continuous development of computer vision in the medical domain has led to the extensive application of computer-aided diagnosis in clinical settings, with medical image segmentation serving as a pivotal component. Existing medical image segmentation is typically realized through deep learning networks, including Convolutional Neural Networks (CNNs) [2,7,18,19,28,30] and Vision Transformers (ViTs) [1,3,4,9,11,26]. CNNs have exceptional capabilities in extracting localized features, yet it lacks efficiency in establishing the correlation of remote information. As for ViTs, while the self-attention mechanism addresses the challenge of extracting remote information through sequences of contiguous patches, it also concurrently increases computational complexity.

Recently, State Space Models (SSMs) [10,14,23,29,31,33] have demonstrated efficiency in capturing long-range dependencies and exhibit linear complexity with respect to input size and memory occupation, offering a great balance between computational efficiency and model parameters. Nonetheless, for enhancing the predictive accuracy of computer-aided diagnosis, methods based on Vision Mamba Models (VMMs) often require a large number of parameters. For instance, U-Mamba [14] has a parameter count of 173.53M, which significantly impacts its practical feasibility in clinical applications. Therefore, achieving excellent performance while maintaining a low computational load has become one of the pressing issues in medical image segmentation tasks.

To address this challenge, VMMs fuel the development of lightweight medical image segmentation algorithms with its strengths in capturing long-range dependencies and streamlining information processing, particularly for skin lesion segmentation tasks. For instance, UltraLight VM-UNet [29] achieves competitive segmentation performance while reducing the number of parameters. However, many existing methods tend to focus on redesigning the internal architecture of the encoder and decoder, often neglecting the thorough exploration of skip connections. This oversight results in insufficient fusion of low-level details with high-level semantics, ultimately degrading the model's performance in fine-grained segmentation tasks, especially for complex structures or small targets. Although LightM-UNet [8] enhances feature fusion by straightforward integrating skip connections, which mitigates the issue to some extent, yet it is still incapable of exploring sufficient information from fulls cales.

In this paper, we construct a slimmed-down version of the model based on Vision Mamba, named Lightweight Mamba with U-Net (LightMamba-UNet). To address the challenge of high computational load in clinical applications, we introduce Parallel Channel Mamba layer (PCM) for processing of deep features in parallel. Where inputs are split into four equal groups along the channel axis, each processed by the SSM-based Mamba block operation, yielding outstanding performance with a negligible computational burden. Moreover, due to inconsistent target sizes and shapes in medical image segmentation, acquiring multi-scale information is vital. Hence, LiteSkip Full-scale (LSF) is utilized to extract low-level details with high-level semantics from full scales, mitigating the semantic gap's harmful effects on segmentation with a minimal parameter footprint. By means of the aforementioned design, LightMamba-UNet achieves excellent performance with surprisingly low parameters in the model based on Vision Mamba. Our contributions and findings are threefold:

- Proposing an innovative PCM for processing features in parallel, it achieves exceptional performance with a minimal computational load.
- Devising a novel LSF to incorporate fine-grained details and coarse-grained semantics from feature maps in full scales, which improves segmentation performance with a minimal parameter footprint.
- Conducting extensive experiments on three public skin lesion datasets, the proposed LightMamba-UNet, with only 0.031M parameters, consistently exhibits improvements over a range of baseline models.

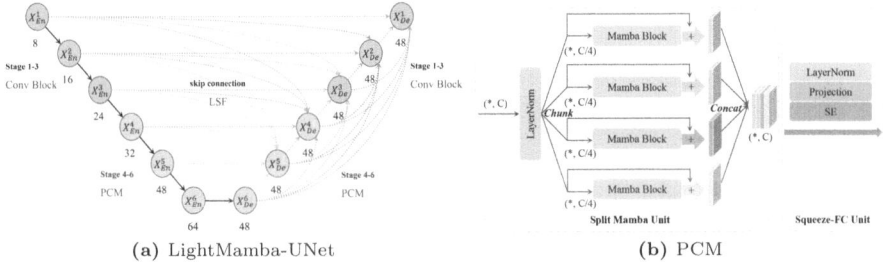

Fig. 1. (a) The proposed LightMamba-UNet architecture. (b) PCM architecture.

2 Method

As shown in Fig. 1(a), LightMamba-UNet is built upon a 6-layer U-shaped architecture with the encoder having channel dimensions of [8, 16, 24, 32, 48, 64] and the decoder having a uniform channel count of 48. While the first three stages employ a Conv Block (a plain convolution with a kernel size of 3) to extract shallow features respectively, the last three stages utilize the proposed PCM to extract the deep features in parallel. Furthermore, LSF is utilized for the information fusion of low-level details and high-level semantics in full scales.

2.1 Parallel Channel Mamba

Medical image segmentation is constrained by computational resources in clinical settings, so achieving excellent performance with a low computational load is vital for the model's design. Therefore, PCM is proposed in this paper to address the aforementioned challenge. As shown in Fig. 1(b), this module is composed of two sub-units: Split Mamba Unit (SM) and Squeeze-FC Unit (SFC). SM first passes the feature map through a LayerNorm layer, then splits the feature map along the channel dimension into four parts, and obtains both global (Selection Mechanism and Fully Connected layers) and local (1D Convolutional layers) feature information through Mamba block. Concurrently, to capture contextual information more effectively, a residual connection operation is applied, and then a concatenation operation is performed along the channel dimension to restore the size of the feature map. Next, for SFC, the feature map is processed by LayerNorm and Projection operation independently to interact with the global and local information. Finally, the attention map with the same shape as the input feature is generated using the Squeeze-and-Excitation block to inhibit the irrelevant regions of feature information transmitted by SM, thereby enabling the model to focus more on the paramount information. The specific operation can be briefly expressed by formulas (1) to (4):

$$X_1, X_2, X_3, X_4 = \text{Chunk}_4\left(\text{LN}(X)\right), \tag{1}$$

$$X'_i = \text{Mamba}(X_i) + X_i, \quad i \in \{1,2,3,4\}, \tag{2}$$

$$X' = \text{Concat}(X'_1, X'_2, X'_3, X'_4), \tag{3}$$

$$\text{OUT} = \text{SE}\left(\text{Proj}\left(\text{LN}(X')\right)\right), \tag{4}$$

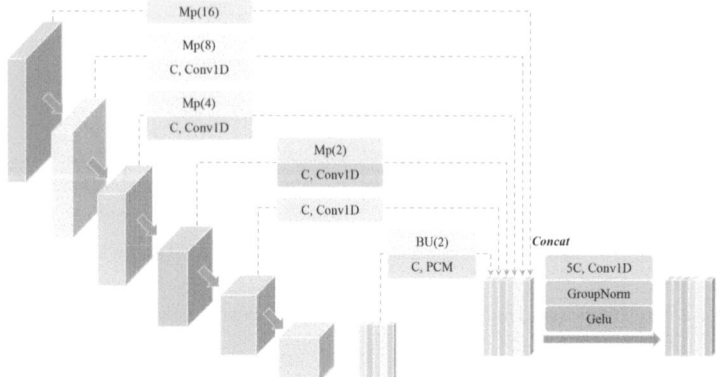

Fig. 2. Overview of the proposed LSF module architecture.

where LN(·) refers to the LayerNorm operation, $\text{Chunk}_4(\cdot)$ indicates that the input feature map is split into four parts along the channel axis, Mamba(·) denotes the Mamba block operation, Concat(·) is used for the concatenation operation along the channel dimension, Proj(·) stands for the Projection operation and SE(·) represents the Squeeze-and-Excitation block operation.

2.2 LiteSkip Full-Scale

The fusion of low-level details with high-level semantics has been proven to contribute to the segmentation of objects of different sizes, which is deemed pivotal for improving performance. Therefore, we propose LiteSkip Full-scale, called LSF, which takes six inputs to make full use of the multi-scale features. As shown in Fig. 2, this module is employed to synthesize a channel attention map via the concatenation of features from disparate stages at the channel axis to better facilitate the integration of information. Specifically, pointwise convolution, bilinear interpolation, and max-pooling are used to resize the multi-level features, so as to match the size of the target features. Then, LSF passes the feature map through a pointwise convolution to extract information at full scales. Finally, a GroupNorm operation and a Gelu operation are applied to obtain the output. The proposed module can be expressed by formulas (5) to (7):

$$\begin{cases} Y_{En}^1 = \text{Mp}\left(X_{En}^1\right), \\ Y_{En}^2 = \text{Conv1D}\left(\text{Mp}\left(X_{En}^2\right)\right), \\ Y_{En}^3 = \text{Conv1D}\left(\text{Mp}\left(X_{En}^3\right)\right), \\ Y_{En}^4 = \text{Conv1D}\left(X_{En}^4\right), \\ Y_{De}^6 = \text{Conv1D}\left(\text{BU}\left(X_{De}^6\right)\right), \\ Y_{De}^5 = \text{PCM}\left(\text{BU}\left(X_{De}^5\right)\right), \end{cases} \quad (5)$$

$$Y' = \text{Cat}\left(Y_{En}^1, Y_{En}^2, Y_{En}^3, Y_{En}^4, Y_{De}^6, Y_{De}^5\right), \quad (6)$$

$$OUT = \text{Gelu}\left(\text{GN}\left(\text{Conv1D}\left(Y'\right)\right)\right), \quad (7)$$

Table 1. Performance comparison of the proposed method against the state-of-the-art approaches on skin lesion segmentation task.

Dataset	Model	Params↓	GFLOPs↓	DSC↑	Acc↑	Spe↑	Sen↑
ISIC2017	U-Net [20]	7.77	13.78	86.99	95.65	97.43	86.82
	UTNetV2 [5]	12.80	15.50	87.23	95.84	98.05	84.85
	TransFuse [32]	26.27	11.53	88.40	96.17	97.98	87.14
	VM-UNet [21]	27.43	4.11	89.03	96.29	97.58	**89.90**
	UltraLight VM-UNet [29]	0.049	**0.060**	87.06	96.31	98.37	87.44
	MALUNet [22]	0.175	0.083	88.13	96.18	98.47	84.78
	UNeXt-S [24]	0.300	0.100	89.80	95.95	97.74	87.04
	LightMamba-UNet (Ours)	**0.031**	0.421	**90.97**	**96.41**	**98.55**	89.14
ISIC2018	U-Net [20]	7.77	13.78	87.55	94.69	96.69	85.86
	UTNetV2 [5]	12.80	15.50	88.25	94.32	96.48	87.6
	SANet [25]	23.90	5.99	88.59	94.39	95.97	89.46
	TransFuse [32]	26.27	11.53	89.27	94.66	95.74	**91.28**
	VM-UNet [21]	27.43	4.11	89.71	94.91	96.13	91.12
	UltraLight VM-UNet [29]	0.049	**0.060**	89.28	95.75	97.16	90.74
	MALUNet [22]	0.175	0.083	89.04	94.62	96.19	89.74
	UNeXt-S [24]	0.300	0.100	88.33	94.39	96.72	87.15
	LightMamba-UNet (Ours)	**0.031**	0.421	**90.31**	**95.86**	**98.09**	89.83
PH^2	U-Net [20]	7.77	13.78	89.36	92.33	95.88	91.25
	Attention U-Net [16]	8.73	16.74	90.03	92.76	96.40	92.05
	C^{2}SDG [6]	22.00	7.97	90.30	91.37	94.76	**93.67**
	VM-UNet [21]	27.43	4.11	90.33	93.69	94.83	91.31
	UltraLightVM-UNet [29]	0.049	**0.060**	90.06	95.31	96.77	92.45
	MALUNet [22]	0.175	0.083	88.65	92.63	94.25	89.22
	LightM-UNet [8]	0.403	0.391	91.56	94.57	96.13	91.29
	SCR-Net [27]	0.801	1.567	89.89	93.39	94.46	91.14
	LightMamba-UNet (Ours)	**0.031**	0.421	**92.23**	**95.72**	**96.89**	92.83

where Mp(·) stands for the Maxpooling operation, Conv1D(·) indicates the Pointwise Convolution operation, BU(·) denotes the Bilinear Upsample operation, PCM(·) represents the PCM operation, GN(·) is used for GroupNorm operation, and Gelu(·) indicates the Gelu activation function.

3 Experiments

3.1 Datasets and Evaluation Metrics

To validate the effectiveness of the proposed method at the parameter of 0.031M, extensive comparison experiments were conducted on three publicly available dermatological lesion datasets, namely ISIC2017 [5], ISIC2018 [22], and PH^2 [15], containing 2150, 2694, and 200 dermatoscope images with segmentation mask labels, respectively. Among them, consistent with prior research [29], the datasets were allocated as follows: 1,250 and 1,815 images for training, 150 and 259 for validation, and 600 for testing in the ISIC2017 and ISIC2018 datasets, respectively, whereas all 200 images in the PH^2 dataset were used for external validation.

3.2 Implementation Details

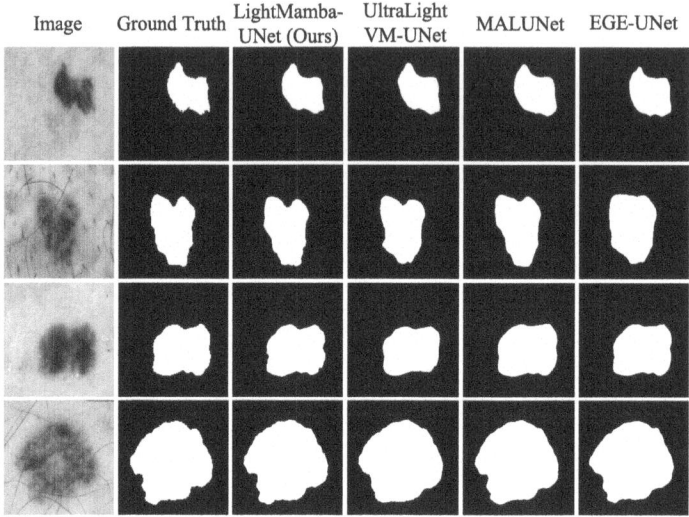

Fig. 3. Visualization of segmentation graphs for comparison experiments on the ISIC2017 dataset.

All experiments were implemented by Pytorch [17] framework, and were run on a single NVIDIA GeForce RTX 3090 GPU with 24 GB of memory. To more fairly assess the model's performance, we maintained consistency with [29] by applying horizontal and vertical flips, as well as random rotations. The loss function is the BceDice loss. AdamW [12] with an initial learning rate of 1e−4 is used as the optimizer, and CosineAnnealingLR [13] with a minimum learning rate of 1e−6 is applied for learning rate scheduling. The training epoch is set equal to 250, and the batch size is 8. All dermatoscope images from three datasets are standardized to 256 × 256 pixels when inputting the model.

3.3 Comparison Results

This section compared the proposed LightMamba-UNet with several state-of-the-art lightweight and classical medical image segmentation models. As shown in Table 1, the comparative experimental results indicated that LightMamba-UNet exhibited a state-of-the-art performance in terms of balancing parameters, computational complexity, and segmentation performance. Specifically, the parameters of our model were 36.73% lower than those of the current lightest Vision Mamba UNet model, namely UltraLight VM-UNet [29], yet it achieved excellent and highly competitive performance. For the ISIC2018 dataset, compared with TransFuse [32], although our model was 1.45% lower in Sen, it could not be ignored that LightMamba-UNet was 847× and 275× lower than TransFuse in terms of parameters and computational complexity, with all other three metrics surpassing TransFuse. For other datasets, the performance of our model also exceeded that of the best-performing model. Besides, some visualizations of segmentation graphs for comparison experiments on the ISIC2017 dataset were shown in Fig. 3. It is evident that the proposed LightMamba-UNet achieved competitive performance with more precise contours and without hollows.

Table 2. Ablation performances of the single module.

Model	Params↓	GFLOPs↓	DSC↑	Acc↑	Spe↑	Sen↑
Baseline	0.049	**0.060**	87.06	96.31	98.37	87.44
Baseline + LSF	**0.029**	0.420	88.30	96.26	98.15	88.53
Baseline + PCM	0.052	0.060	87.65	96.38	98.39	88.48
Baseline+ PCM + LSF	0.031	0.421	**90.97**	**96.41**	**98.55**	**89.14**

Table 3. Ablation experiments on the effect of PCM in LightMamba-UNet.

Model	Params↓	GFLOPs↓	DSC↑	Acc↑	Spe↑	Sen↑
Baseline	**0.031**	**0.421**	**90.97**	**96.41**	**98.55**	89.14
Encoder_Conv	0.068	0.431	90.71	96.23	98.12	**89.53**
Decoder_Conv	0.052	0.436	89.43	96.33	98.33	87.58
(En+De)_Conv	0.089	0.447	89.01	95.98	97.82	88.14

Table 4. Ablation performances of LightMamba-UNet under various parallel connection count settings.

Connections	Params↓	GFLOPs↓	DSC↑	Acc↑	Spe↑	Sen↑
1	0.109	**0.421**	90.87	96.32	98.01	**90.37**
2	0.049	**0.421**	85.95	95.58	**98.88**	80.91
4	0.031	**0.421**	**90.97**	**96.41**	98.55	89.14
8	**0.025**	**0.421**	90.83	96.38	98.46	89.47

3.4 Ablation Experiments

Ablation Study on the Single Module. We conducted the ablation experiment of the single module to prove the effectiveness of our proposed modules on ISIC2017 dataset. The baseline utilized in our work was referenced from UltraLight VM-UNet [29]. Detailed results of the ablation experiments were presented in Table 2.

Specifically, we conducted the macro ablation on PCM and LSF, respectively. Initially, we substituted the skip-connection operation in baseline with LSF. Due to the efficient integration of low-level details with high-level semantics of LSF, it not only enhanced performance beyond the baseline, but also significantly decreased both parameter count and computational load. Subsequently, PCM was integrated into the final three layers of the baseline, replacing Parallel Vision Mamba (PVM), resulting in performance gains. Ultimately, by seamlessly integrating LSF and PCM into the proposed LightMamba-UNet, we achieved exceptional performance with surprisingly low parameter count in the model.

Ablation Study on the Effect of PCM. Additionally, a suite of ablation experiments on the ISIC2017 dataset were conducted to validate the impact of the proposed PCM on LightMamba-UNet. As shown in Table 3, which depicts the substitution of PCM in the encoder and decoder, individually and simultaneously, with a plain convolution with a kernel size of 3. The table indicates that, when PCM was individually replaced in the encoder and decoder, the parameters increased by 119.35% and 67.74%, respectively, with GFLOPs rising for both but performance declining. Particularly, simultaneous replacement of PCM in both the encoder and decoder resulted in a 187% increase in the parameters and a 1.12% decline in Spe. In conclusion, PCM played a crucial role in reducing the parameters, GFLOPs, and improving segmentation performance.

Ablation Study on Various Parallel Connection Count Settings. Moreover, a series of ablation experiments were performed to verify the validity of the proposed method of Channel Mamba (CM) with different parallelism. Moreover, a set of ablation experiments on the ISIC2017 dataset was conducted to explore the effect of varying parallel connection counts in Channel Mamba (CM) on segmentation performance. As shown in Table 4, "Connections = 1" indicates CM with one parallel connection, with the number of channels unchanged; "Connections = 2" refers to CM with two parallel connections, splitting the channels into two parts, each with $C/2$, and so forth. The table reveals that at "Connections = 4", the model achieves the best trade-off between the parameters and segmentation performance. Hence, in this paper, we adopted CM with four parallel connections as the critical structure for the proposed PCM.

4 Conclusion

In this paper, we propose LightMamba-UNet with a parameter of only 0.031M based on Vision Mamba for skin lesion segmentation tasks. It is consisted of two advanced sub-modules. Initially, to tackle the high computational load in clinical applications, PCM method is proposed for processing the deep features in parallel. In addition, LSF is introduced to fully utilize multi-scale features and alleviate the adverse impact of semantic gaps on segmentation performance. Experimental results validate the effectiveness of the proposed method, demonstrating strong performance competitiveness with minimal parameters. Currently,

LightMamba-UNet is tailored only for skin lesion segmentation and we intend to extend our lightweight approach to other tasks.

Acknowledgments. This work was supported in part by the National Natural Science Foundation of China (No.62076005, No.62106006, No. U20A20398), the University Synergy Innovation Program of Anhui Province, China (No. GXXT-2021-030, GXXT-2021-065, GXXT-2022-029, GXXT-2022-031). We also thank the High-performance Computing Platform of Anhui University for providing computing resources.

References

1. Aghdam, E.K., Azad, R., Zarvani, M., Merhof, D.: Attention swin U-net: cross-contextual attention mechanism for skin lesion segmentation. In: 2023 IEEE 20th International Symposium on Biomedical Imaging (ISBI), pp. 1–5. IEEE (2023)
2. Cai, S., Tian, Y., Lui, H., Zeng, H., Wu, Y., Chen, G.: Dense-UNet: a novel multiphoton in vivo cellular image segmentation model based on a convolutional neural network. Quant. Imaging Med. Surg. **10**(6), 1275 (2020)
3. Cao, H., et al.: Swin-UNet: UNet-like pure transformer for medical image segmentation. In: European Conference on Computer Vision, pp. 205–218. Springer (2022)
4. Chen, J., et al.: TransUNet: transformers make strong encoders for medical image segmentation. arXiv preprint arXiv:2102.04306 (2021)
5. Gao, Y., Zhou, M., Metaxas, D.N.: UTNet: a hybrid transformer architecture for medical image segmentation. In: Medical Image Computing and Computer Assisted Intervention–MICCAI 2021: 24th International Conference, Strasbourg, France, 27 September–1 October 2021, Proceedings, Part III 24, pp. 61–71. Springer (2021)
6. Hu, S., Liao, Z., Xia, Y.: Devil is in channels: contrastive single domain generalization for medical image segmentation. In: International Conference on Medical Image Computing and Computer-Assisted Intervention, pp. 14–23. Springer (2023)
7. Khanna, A., Londhe, N.D., Gupta, S., Semwal, A.: A deep residual u-net convolutional neural network for automated lung segmentation in computed tomography images. Biocybern. Biomed. Eng. **40**(3), 1314–1327 (2020)
8. Liao, W., Zhu, Y., Wang, X., Pan, C., Wang, Y., Ma, L.: LightM-UNet: mamba assists in lightweight UNet for medical image segmentation. arXiv preprint arXiv:2403.05246 (2024)
9. Lin, A., Chen, B., Xu, J., Zhang, Z., Lu, G., Zhang, D.: Ds-transunet: dual swin transformer u-net for medical image segmentation. IEEE Trans. Instrum. Meas. **71**, 1–15 (2022)
10. Liu, Y., et al.: VMamba: visual state space model 2024. arXiv preprint arXiv:2401.10166 (2024)
11. Liu, Z., et al.: Swin transformer: hierarchical vision transformer using shifted windows. In: Proceedings of the IEEE/CVF International Conference on Computer Vision, pp. 10012–10022 (2021)
12. Loshchilov, I.: Decoupled weight decay regularization. arXiv preprint arXiv:1711.05101 (2017)
13. Loshchilov, I., Hutter, F.: SGDR: stochastic gradient descent with warm restarts. arXiv preprint arXiv:1608.03983 (2016)

14. Ma, J., Li, F., Wang, B.: U-mamba: enhancing long-range dependency for biomedical image segmentation. arXiv preprint arXiv:2401.04722 (2024)
15. Mendonça, T., Ferreira, P.M., Marques, J.S., Marcal, A.R., Rozeira, J.: PH 2-a dermoscopic image database for research and benchmarking. In: 2013 35th Annual International Conference of the IEEE Engineering in Medicine And Biology Society (EMBC), pp. 5437–5440. IEEE (2013)
16. Oktay, O., et al.: Attention U-net: learning where to look for the pancreas. arXiv preprint arXiv:1804.03999 (2018)
17. Paszke, A., et al.: PyTorch: an imperative style, high-performance deep learning library. Adv. Neural. Inf. Process. Syst. **32** (2019)
18. Peng, Y., Sonka, M., Chen, D.Z.: U-net v2: rethinking the skip connections of U-net for medical image segmentation. arXiv preprint arXiv:2311.17791 (2023)
19. Rao, Y., Zhao, W., Tang, Y., Zhou, J., Lim, S.N., Lu, J.: HorNet: efficient high-order spatial interactions with recursive gated convolutions. Adv. Neural. Inf. Process. Syst. **35**, 10353–10366 (2022)
20. Ronneberger, O., Fischer, P., Brox, T.: U-net: Convolutional networks for biomedical image segmentation. In: Medical Image Computing and Computer-Assisted Intervention–MICCAI 2015: 18th International Conference, Munich, Germany, 5–9 October 2015, Proceedings, Part III 18, pp. 234–241. Springer (2015)
21. Ruan, J., Xiang, S.: VM-UNet: vision mamba UNet for medical image segmentation. arXiv preprint arXiv:2402.02491 (2024)
22. Ruan, J., Xiang, S., Xie, M., Liu, T., Fu, Y.: MALUNet: a multi-attention and lightweight UNet for skin lesion segmentation. In: 2022 IEEE International Conference on Bioinformatics and Biomedicine (BIBM), pp. 1150–1156. IEEE (2022)
23. Tang, H., Cheng, L., Huang, G., Tan, Z., Lu, J., Wu, K.: Rotate to scan: UNet-like mamba with triplet SSM module for medical image segmentation. arXiv preprint arXiv:2403.17701 (2024)
24. Valanarasu, J.M.J., Patel, V.M.: UNext: MLP-based rapid medical image segmentation network. In: International Conference on Medical Image Computing and Computer-Assisted Intervention, pp. 23–33. Springer (2022)
25. Wei, J., Hu, Y., Zhang, R., Li, Z., Zhou, S.K., Cui, S.: Shallow attention network for polyp segmentation. In: Medical Image Computing and Computer Assisted Intervention–MICCAI 2021: 24th International Conference, Strasbourg, France, 27 September–1 October 1 2021, Proceedings, Part I, pp. 699–708. Springer (2021)
26. Wu, H., Chen, S., Chen, G., Wang, W., Lei, B., Wen, Z.: Fat-net: feature adaptive transformers for automated skin lesion segmentation. Med. Image Anal. **76**, 102327 (2022)
27. Wu, H., Zhong, J., Wang, W., Wen, Z., Qin, J.: Precise yet efficient semantic calibration and refinement in convnets for real-time polyp segmentation from colonoscopy videos. In: Proceedings of the AAAI Conference on Artificial Intelligence, pp. 2916–2924 (2021)
28. Wu, R., et al.: MHorUNet: High-order spatial interaction UNet for skin lesion segmentation. Biomed. Signal Process. Control **88**, 105517 (2024)
29. Wu, R., Liu, Y., Liang, P., Chang, Q.: Ultralight VM-UNet: parallel vision mamba significantly reduces parameters for skin lesion segmentation. arXiv preprint arXiv:2403.20035 (2024)
30. Wu, R., Lv, H., Liang, P., Cui, X., Chang, Q., Huang, X.: HSH-UNet: hybrid selective high order interactive U-shaped model for automated skin lesion segmentation. Comput. Biol. Med. **168**, 107798 (2024)

31. Zhang, M., Yu, Y., Jin, S., Gu, L., Ling, T., Tao, X.: VM-UNet-v2: rethinking vision mamba UNet for medical image segmentation. In: International Symposium on Bioinformatics Research and Applications, pp. 335–346. Springer (2024)
32. Zhang, Y., Liu, H., Hu, Q.: TransFuse: fusing transformers and CNNs for medical image segmentation. In: Medical image computing and computer assisted intervention–MICCAI 2021: 24th international conference, Strasbourg, France, 27 September–1 October 2021, Proceedings, Part I, pp. 14–24. Springer (2021)
33. Zhou, Z., Siddiquee, M.M.R., Tajbakhsh, N., Liang, J.: UNet++: redesigning skip connections to exploit multiscale features in image segmentation. IEEE Trans. Med. Imaging **39**(6), 1856–1867 (2019)

Exploiting Memory-Aware Q-Distribution Prediction for Nuclear Fusion via Modern Hopfield Network

Qingchuan Ma[1], Shiao Wang[1], Tong Zheng[1], Xiaodong Dai[1], Yifeng Wang[2], Qingquan Yang[2], and Xiao Wang[1(✉)]

[1] School of Computer Science and Technology, Anhui University, Hefei, China
xiaowang@ahu.edu.cn
[2] Institute of Plasma Physics, Chinese Academy of Sciences, Hefei, China

Abstract. This study addresses the critical challenge of predicting the Q-distribution in long-term stable nuclear fusion task, a key component for advancing clean energy solutions. We introduce an innovative deep learning framework that employs Modern Hopfield Networks to incorporate associative memory from historical shots. Utilizing a newly compiled dataset, we demonstrate the effectiveness of our approach in enhancing Q-distribution prediction. The proposed method represents a significant advancement by leveraging historical memory information for the first time in this context, showcasing improved prediction accuracy and contributing to the optimization of nuclear fusion research.

Keywords: Q-distribution Prediction · Controllable Nuclear Fusion · Modern Hopfield Network · Associative Memory

1 Introduction

Controlled nuclear fusion is one of the key approaches expected to solve humanity's challenge of clean energy. Currently, the main international and domestic nuclear fusion devices include ITER, EAST, HL-2M, HT-7, etc. Research in the field of nuclear fusion, powered by artificial intelligence, is beginning to attract the attention of an increasing number of researchers. This includes areas such as disruption prediction [1,4], deep reinforcement learning based control system optimization [9], magnetic control method [2], and more.

In long-term stable nuclear fusion missions, the prediction of the Q-distribution is one of the key points, and this article explores this issue. As it is a new research problem for Q-distribution prediction using artificial intelligence, the collection of training and testing data is the first thing we need to do. Therefore, we used a simulation program to generate input signals containing various physical quantities, along with the corresponding Q-distribution ground truth. Finally, we obtain a dataset containing 5753 paired input-output samples and split them into a training and testing subset which contains 5166 and 587 samples, respectively.

Based on the newly collected database, in this paper, we propose to predict the Q-distribution using deep neural networks. Considering the raw data of each physical variable are all 1-D vectors in a sample, it is intuitive to encode them using an MLP (Multi-Layer Perceptron) network and regress the Q-distribution using a dense layer. It will be a simple baseline but not a good approach for high-performance Q-distribution prediction. Because the samples collected from nearby shots are highly correlated to each other, the performance will be better once the history memory information is aggregated. As shown in Fig. 1, in this work, we propose to encode associative memory of context shots using the Modern Hopfield Networks (MHN). Specifically, we take the historical example as another input and extract their features using the shared MLP. The two input features will be concatenated and added with position encoding features, then, we feed them into the MHN for associative memory augmented feature representation learning. Meanwhile, we also concatenate the learnable parameters with features of raw signal and feed them into the MHN. These two features will be concatenated and fed into the dense layer for Q-distribution prediction. Extensive experiments demonstrate the effectiveness of our proposed memory-augmented network for Q-distribution prediction.

To sum up, we draw the contributions of this work as the following two aspects:

1) We propose a novel associative memory-augmented Q-distribution prediction framework using modern Hopfield networks. It is the first time to exploit the historical memory information for this research task.
2) We conducted extensive experiments on the newly collected dataset, and the results demonstrate the effectiveness of historical memory.

2 Our Proposed Approach

2.1 Problem Formulation

A key challenge in nuclear fusion research is the accurate prediction of the Q-distribution, which plays a pivotal role in determining plasma stability and behavior within a tokamak. However, the highly nonlinear and complex dynamics of plasma physics make this task exceedingly difficult for traditional empirical models, which are typically based on physics-driven insights. We formulate this problem as a regression task [5], where the model takes time-series data as input and outputs the Q-distribution. To address the limitations of conventional methods, we propose a novel deep learning approach leveraging the Modern Hopfield Network for history memory-guided learning. By utilizing the memory mechanisms of the Hopfield Network, we provide a more reliable and effective solution for Q-distribution prediction.

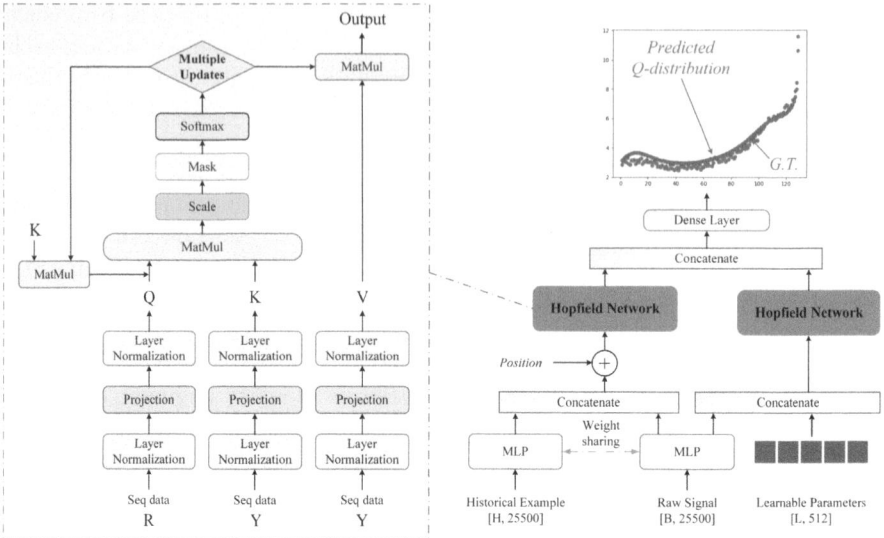

Fig. 1. An overview of our proposed associative memory augmented Q-distribution using the Modern Hopfield Networks.

2.2 Overview

In this paper, we delve into the practical significance of utilizing the memory storage capabilities of the Hopfield Network in predicting nuclear fusion sequence information within the realm of physics applications. To maximize the accuracy of predictive performance, our focus lies in treating nuclear fusion data as sequential data with temporal information and utilizing the Hopfield Network to extract temporal sequence features. Specifically, within each shot, there are multiple temporal samples. We first identify historical samples for each sample and utilize the Hopfield Network to assist in extracting historical information for the current sample. It has been proven that this approach of historical feature extraction enhances the predictive capability of Q-distribution. By incorporating historical information extracted by the Hopfield Network and global information into the features of current samples, we strengthen the overall predictive capability. Our network framework is illustrated in Fig. 1.

2.3 Input Processing

In this study, our original nuclear fusion data comprises a total of 141 indicators, with each indicator consisting of multiple floating-point numbers. The $npsip - q$ and q indicators represent the Q-distribution that we aim to predict (i.e., the model output labels), while all other indicators represent information regarding the state of plasma disruption during the nuclear fusion process. We concatenate the other 139 indicators together as a vector, which serves as the input. For each sample, the input consists of a total of 25,500 dimensions. For a batch, the input shape is (batch size, 25,500), and the label shape is (batch size, 129). Through this approach, we ensure that each batch has both input and label data [11].

Additionally, the nuclear fusion data consists of 22 shots, with each shot comprising over 200 different samples at different times. Therefore, the nuclear fusion data can be viewed as sequential data, where the data within each shot is sorted according to time, resulting in a sequence of data for each shot. Consequently, we treat each shot as a sequence. For instance, considering a particular sample as the current sample, the samples preceding it in time constitute the historical samples of this sample. This enables us to obtain the historical samples for each sample, which are then inputted into the model. If the time gap between samples is too large, the historical information is disregarded. Furthermore, for testing samples to extract historical features effectively, they also require historical samples. Hence, during the process of extracting the test set, continuous samples are extracted to ensure that the vast majority of samples have historical information from their adjacent samples.

2.4 Network Architecture

Based on the concept of deep learning for sequential data and global features, we have designed a network model capable of predicting Q-distribution in nuclear fusion tasks, as depicted in Fig. 1. The main components of the model include two Hopfield Network modules, which are utilized for extracting features from historical data and global features, respectively, aiding us in prediction. The detailed process overview is outlined as follows.

• **Raw Data Processing.** By processing the data labeled with 139 indicators of nuclear fusion, we can transform the raw data into tensor form to be used as input, with the input size expressed as $\mathbb{R}^{B \times D}$. Here, B represents the batch size, indicating the number of samples included in each batch during training or testing, while D denotes the number of features for each individual data point, also referred to as the data dimension, with a specific value of 25500. We feed a batch of data into the neural network, as depicted in Fig. 1. We utilize an MLP as the basic feature extractor, extracting original features from the raw data. MLP, or Multi-Layer Perceptron, comprises multiple linear layers and multiple activation functions [3], where the activation functions introduce nonlinear features. This can be summarized as follows:

$$feature = Linear((\sigma(Linear(V_i))) * N) \qquad (1)$$

where N represents the number of such combinations of linear layers and activation functions. The features extracted here only represent the characteristics of individual samples and require further processing to enable learning of historical and global features. The historical samples can also be inputted into the MLP to obtain features for the historical samples.

• **Incorporating Temporal Information.** Historical information and current sample features are concatenated along a new dimension, commonly referred to as the sequence length in deep learning. This yields new features represented as $\mathbb{R}^{B \times S \times D}$, where S denotes the quantity of historical information plus the number of current samples. After processing with the position embedding module, positional information, or in other words, temporal information, can be incorporated. The position embedding module is used to add positional encoding to sequential data. Here, we employ sine and cosine functions to compute the positional encoding [12], which can be expressed by the following equation:

$$PE_{(pos,2i)} = \sin\left(\frac{pos}{10000^{\frac{2i}{hidden\ size}}}\right) \tag{2}$$

$$PE_{(pos,2i+1)} = \cos\left(\frac{pos}{10000^{\frac{2i}{hidden\ size}}}\right) \tag{3}$$

$$x_{(b,s,h)} = x_{(b,s,h)} + PE_{(1,s,h)} \tag{4}$$

At this point, the features represent true sequential data, enriched with temporal information.

- **Extraction of Historical and Global Features.** The Hopfield Network is employed to further process features, serving as a module equipped with memory storage capabilities. When sequential features are inputted into the Hopfield Network, it can extract historical information features based on the similarity between the current sample and historical sample features. The output shape remains the same as $\mathbb{R}^{B \times S \times D}$, and at this point, the features of the last sequence (the most recent features in the temporal dimension), which represent the features of the current sample, are extracted. In comparison to features extracted by MLP, these features contain historical information, thereby completing the feature extraction from historical information. The process of extracting global features is similar. The detailed architecture is depicted in Fig. 1. The processing formula of the Hopfield Network can be summarized as follows:

$$Hopfield(R,Y,Y) = softmax(\beta \cdot RW_Q W_K^t Y^t) Y W_K W_v \tag{5}$$

Here is a detailed explanation of the calculation process of the Hopfield Network. β is a hyperparameter. All W matrices represent projections of dimensions, implemented using linear layers. For instance, W_R is used to transform the feature dimension (the last dimension) to the Hopfield space of dimension D. Similarly, W_K and W_V serve the same purpose. With these projections, we can ignore the dimension mapping, thus simplifying the formula to the following:

$$Hopfield(R,Y,Y) = softmax(\beta \cdot RY^t)Y \tag{6}$$

where R represents the state (query) patterns to be retrieved, while Y represents the stored key patterns. Assuming both R and Y have shapes of (B, S, D), first, matrix operations are performed between R and Y along the D dimension, enabling the calculation of the similarity between a certain pattern and other patterns. The resulting shape is (B, S, S). For example, $r_{b,i,j}$ represents the similarity between the ith state pattern of R and the jth stored pattern of Y in the bth batch.

Then, the obtained similarity matrix is multiplied with Y along the S dimension, allowing for pattern retrieval from Y based on similarity. The resulting shape is (B, S, D). It represents the patterns retrieved from Y using R as the query. If both R and Y are regarded as sequence features obtained through position embedding, this process can be viewed as using the Hopfield Network to extract information from different positions in the sequence features, thereby extracting historical information. The features of the last pattern outputted by the Hopfield Network correspond to the features of the current sample with historical information incorporated. The method for extracting global features is similar to that for extracting historical features. Finally, the output features of the two Hopfield Networks are concatenated and then inputted into a linear layer to project the features into the output dimension, yielding the model's prediction results.

2.5 Loss Function

Predictions can be obtained by passing the data through the final linear layer of the model. We utilize the Mean Squared Error (MSE) loss function [7,13] to gauge the disparity between the predicted results and the ground truth (the actual Q-distribution). The specific computation process of the loss function can be expressed by Eq. 7.

$$MSE = \frac{1}{n}\sum_{i=1}^{n}(GT_i - Q_i)^2 \qquad (7)$$

In the equation, Q_i represents the distribution prediction result for the x_i sample generated by the model, while GT_i denotes the Q-distribution of the x_i sample.

3 Experiments

3.1 Dataset and Evaluation Metric

In the nuclear fusion dataset, comprising 22 shots, a total of 5753 samples were recorded. The temporal resolution between adjacent samples is notably fine, approximately 10 milliseconds. To construct our test set, we adopted a consistent ratio (1/10) for extracting continuous samples from each shot. It is worth noting that the vast majority of samples have adjacent historical samples in time. Ultimately, we partition the nuclear fusion dataset into training and testing subsets, comprising 5166 and 587 raw data points, respectively. To evaluate our model's performance, we utilize the MSE (Mean Squared Error) metric, which quantifies the disparity between our predicted Q-distribution and the ground truth.

3.2 Implementation Details

The models were trained end-to-end. All experimental results were obtained after 140 epochs of training. A learning rate of 0.001 and a batch size of 16 were utilized, and training was performed using the SGD optimizer [10]. Subsequently, the models were evaluated on the test set using the MSE loss function to obtain the error results. The code was implemented in Python using the PyTorch [8] framework. The computations were carried out on a server equipped with a CPU Intel(R) Xeon(R) CPU E5-2620 v4 @ 2.10 GHz and a GPU TITAN XP with 12 GB of memory.

Table 1. Comparison with Other Models.

Algorithm	MLP	RNN	GRU	LSTM	Transformer	Ours
MSE	0.0666	0.0656	0.0670	0.0700	0.0659	0.0584

3.3 Comparison with Other Algorithms

To effectively extract visual features from sequential data, we compared various mainstream sequence models [6] to determine the optimal approach. Table 1 summarizes our findings. We evaluated several models, including RNNs, GRUs, LSTMs, Transformers, and our Hopfield Network. Through comprehensive experimental comparisons, we observed that employing the Hopfield Net as our visual backbone network yielded the best performance. This can be attributed to the intrinsic memory storage capability and powerful feature extraction ability of the Hopfield Network, which helps extract historical information from sequential data. Surprisingly, the main computational formulas of the Transformer and Hopfield Network are quite similar, but there is a significant performance gap. We speculate that this is because the Transformer (specifically, the encoder layer of the Transformer model we used) is more complex, and our dataset is not large enough, which can lead to overfitting. In contrast, our Hopfield Network is more flexible to use and has been optimized specifically for the nuclear fusion task, allowing it to demonstrate the best results.

3.4 Ablation Study

• **Component Analysis.** This paper explores the application of deep learning in predicting nuclear fusion results. Our experiments demonstrate that treating nuclear fusion data as sequential information and utilizing the memory storage capabilities of the Hopfield Network can effectively reduce the fitting error in the Q-distribution prediction task. Initially, when the data undergoes processing through several linear layers, such as Multi-Layer Perceptron (MLP), although achieving fitting, the results are unsatisfactory, with an MSE loss of

Table 2. Component analysis results. LParam is short for Learnable Parameter.

No.	MLP	Hopfield	Position	LParam	MSE
1	✓				0.0666
2	✓	✓			0.0620
3	✓	✓	✓		0.0593
4	✓			✓	0.0658
5	✓	✓	✓	✓	0.0584

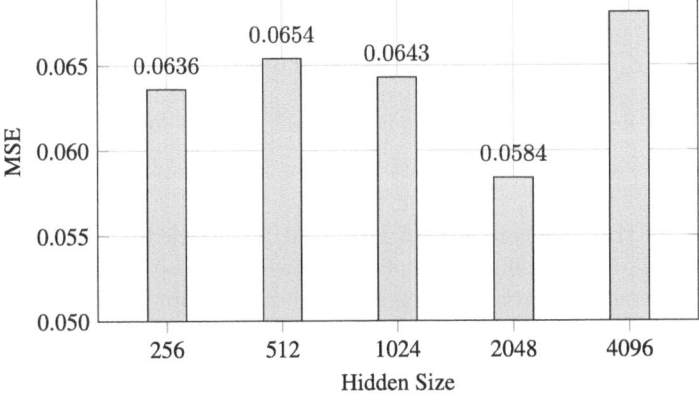

Fig. 2. Impact of Hidden Size on Hopfield Network for Historical Samples

0.0666, as indicated in Table 2. However, when we leverage the Hopfield Network effectively functions as an attention mechanism, improving feature extraction capability, and reducing the MSE loss to 0.0620. Remarkably, introducing temporal information via the position embedding module, which provides positional information for samples at different times, significantly enhances fitting performance, resulting in an MSE loss of 0.0593. Furthermore, the incorporation of a module with learnable parameters aids in extracting global information, further enhancing the predictive performance of Q-distribution, leading to a decrease in the MSE loss to 0.0584. For comparison, experiments using only MLP and the module with learnable parameters were conducted. It is evident that compared to the effectiveness of the Hopfield Network and position embedding module, the module with learnable parameters does not exhibit such prominent performance. These experimental results emphasize the effectiveness of treating nuclear fusion data as sequential data and utilizing the Hopfield Network with position embedding, both of which significantly enhance the predictive performance of the Q-distribution. Moreover, incorporating a module with learnable parameters allows for learning global information, further improving the effectiveness of the model.

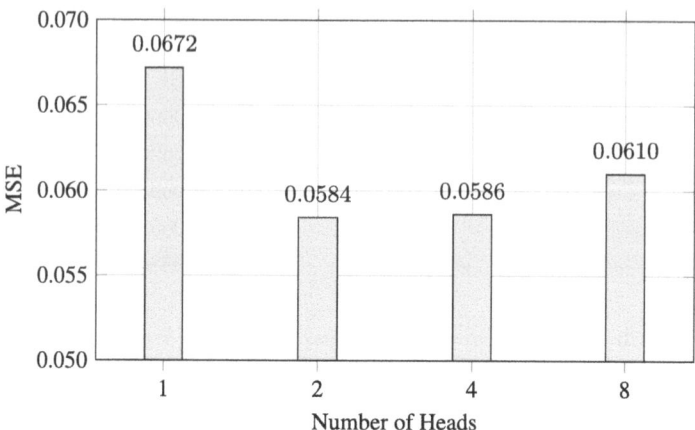

Fig. 3. Ablation studies on the number of heads of MHN for historical samples.

- **Analysis of the Hidden Size of Hopfield Network for Historical Samples.** The hidden size of the Hopfield Network is also an important factor affecting the results. A larger hidden size introduces more learnable parameters, but it also poses a risk of overfitting. Thus, we set the Hopfield Network with different hidden sizes to find the most suitable hidden size, and the experimental results are shown in Fig. 2. We can observe that the best result is achieved when we choose a hidden size of 2048 for input, with an error of 0.0584.

- **Analysis of the Number of Heads of Hopfield Network for Historical Samples.** Like the multi-head attention mechanism in Transformers, increasing the number of heads in the Hopfield Network allows it to extract different features. These features can then be mixed through an MLP, resulting in more powerful feature extraction. Therefore, the number of heads is also a significant parameter for enhancing the effectiveness of the Hopfield Network model. As shown in Fig. 3, we obtained different MSE loss results by changing the hyperparameter, the number of heads. Here, the minimum MSE loss is achieved when the number of heads is set to 2.

- **Analysis of the Number of Layers of Hopfield Network for Historical Samples.** A Hopfield Network with more layers tends to have stronger fitting capabilities but also comes with fitting risks. Conversely, a Hopfield Network with fewer layers exhibits weaker fitting capabilities and may lead to underfitting issues. Therefore, finding the appropriate number of layers is also an important hyperparameter issue. We experimented with different numbers of layers and recorded the results in Fig. 4. We found that setting the number of layers to 1 achieved the best performance.

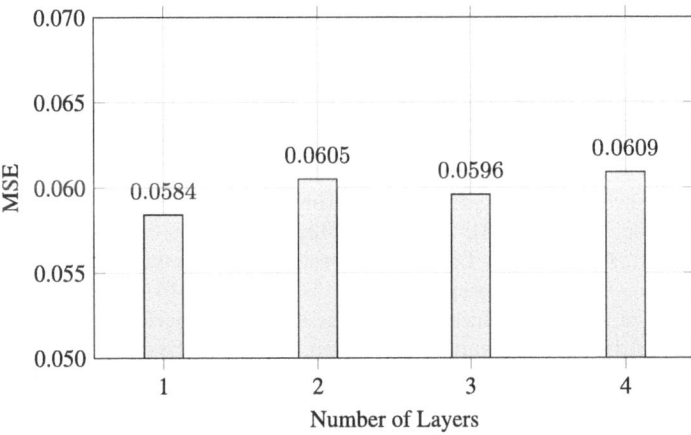

Fig. 4. Ablation studies on the number of layers of Hopfield Network for historical samples.

4 Conclusion

This study has successfully developed a novel approach to predicting the Q-distribution in long-term stable nuclear fusion missions by leveraging deep neural networks and associative memory augmentation. The collection and utilization of a comprehensive dataset comprising 5753 paired input-output samples have provided a solid foundation for our research. By introducing Modern Hopfield Networks (MHN) to encode associative memory of context shots, we have enhanced the feature representation learning for Q-distribution prediction. The proposed framework not only incorporates historical memory information for the first time in this context but also demonstrates significant improvements in prediction accuracy through extensive experimentation.

Acknowledgement. This work is supported by the National Natural Science Foundation of China under Grant U24A20342, 62102205, the Anhui Provincial Natural Science Foundation under Grant 2408085Y032. The authors acknowledge the High-performance Computing Platform of Anhui University for providing computing resources.

References

1. Churchill, R.: Deep convolutional neural networks for multi-scale time-series classification and application to disruption prediction in fusion devices. Technical report, Princeton Plasma Physics Lab. (PPPL), Princeton, NJ, USA (2020)
2. Degrave, J., et al.: Magnetic control of tokamak plasmas through deep reinforcement learning. Nature **602**(7897), 414–419 (2022)
3. Dubey, S.R., Singh, S.K., Chaudhuri, B.B.: Activation functions in deep learning: a comprehensive survey and benchmark. Neurocomputing **503**, 92–108 (2022)

4. Kates-Harbeck, J., Svyatkovskiy, A., Tang, W.: Predicting disruptive instabilities in controlled fusion plasmas through deep learning. Nature **568**(7753), 526–531 (2019)
5. Lathuilière, S., Mesejo, P., Alameda-Pineda, X., Horaud, R.: A comprehensive analysis of deep regression. IEEE Trans. Pattern Anal. Mach. Intell. **42**(9), 2065–2081 (2019)
6. Lim, B., Zohren, S.: Time-series forecasting with deep learning: a survey. Phil. Trans. R. Soc. A **379**(2194), 20200209 (2021)
7. Mathieu, M., Couprie, C., LeCun, Y.: Deep multi-scale video prediction beyond mean square error. arXiv preprint arXiv:1511.05440 (2015)
8. Paszke, A., et al.: PyTorch: an imperative style, high-performance deep learning library. In: Advances in Neural Information Processing Systems, vol. 32 (2019)
9. Seo, J., et al.: Feedforward beta control in the KSTAR tokamak by deep reinforcement learning. Nucl. Fusion **61**(10), 106010 (2021)
10. Sun, S., Cao, Z., Zhu, H., Zhao, J.: A survey of optimization methods from a machine learning perspective. IEEE Trans. Cybern. **50**(8), 3668–3681 (2019)
11. Tawakuli, A., Havers, B., Gulisano, V., Kaiser, D., Engel, T.: Survey: time-series data preprocessing: a survey and an empirical analysis. J. Eng. Res. (2024). https://doi.org/10.1016/j.jer.2024.02.018
12. Wolf, T., et al.: Transformers: state-of-the-art natural language processing. In: Proceedings of the 2020 Conference on Empirical Methods in Natural Language Processing: System Demonstrations, pp. 38–45 (2020)
13. Yessou, H., Sumbul, G., Demir, B.: A comparative study of deep learning loss functions for multi-label remote sensing image classification. In: IGARSS 2020-2020 IEEE International Geoscience and Remote Sensing Symposium, pp. 1349–1352. IEEE (2020)

Multi-modal Fusion Based Q-Distribution Prediction for Controlled Nuclear Fusion

Shiao Wang[1], Yifeng Wang[2(✉)], Qingchuan Ma[1], Xiao Wang[1], Ning Yan[2], Qingquan Yang[2], Guosheng Xu[2], and Jin Tang[1]

[1] School of Computer Science and Technology, Anhui University, Hefei, China
[2] Institute of Plasma Physics, Chinese Academy of Sciences, Hefei, China
yfwang@ipp.ac.cn

Abstract. Q-distribution prediction is a crucial research direction in controlled nuclear fusion, with deep learning emerging as a key approach to solving prediction challenges. In this paper, we leverage deep learning techniques to tackle the complexities of Q-distribution prediction. Specifically, we explore multimodal fusion methods in computer vision, integrating 2D line image data with the original 1D data to form a bimodal input. Additionally, we employ the Transformer's attention mechanism for feature extraction and the interactive fusion of bimodal information. Extensive experiments validate the effectiveness of our approach, significantly reducing prediction errors in Q-distribution.

Keywords: Q-distribution Prediction · Multi-modal Fusion · Controlled Nuclear Fusion

1 Introduction

The problem of nuclear fusion has always been a major challenge that humanity needs to solve. In recent years, many researchers have made outstanding contributions to the development of controllable nuclear fusion [1–3]. Research on international magnetic confinement controlled nuclear fusion began in the 1950s. Over the years, various approaches have been explored, including magnetic confinement [4,5], magnetic mirrors [6,7], stellarators [8–10], and Tokamaks [11–13], all aimed at improving key plasma parameters to achieve the conditions necessary for controlled nuclear fusion reactions eventually. Since the 1970s, the Tokamak method has gradually demonstrated unique advantages, becoming the mainstream approach in magnetic confinement fusion research. With the global advancement of Tokamak experiments, the comprehensive plasma parameters have been continuously improved, and significant progress has been made in fusion engineering technologies. However, many key technologies still face considerable challenges before practical applications can be realized.

Among the many sub-tasks in controlled nuclear fusion research, Q-distribution prediction is both a critical and highly challenging problem. In this paper, we leverage artificial intelligence techniques to address the Q-distribution prediction challenge within the context of controlled nuclear fusion. To begin, we create a dataset that contains 5,753 data samples using a simulation toolkit. For efficient model training and evaluation, these data are divided into 5,166 samples for training and 587 for testing. Based on this dataset, we developed a deep learning model specifically designed for Q-distribution prediction. Our approach focuses on exploring multimodal fusion techniques within the field of computer vision. To capture temporal trends in the data, we transformed the original 1D data into 2D line charts, visually representing how key indicators change over time. We then employed the Vision Transformer (ViT) [14], a state-of-the-art model for image analysis, as the core of our multimodal fusion network. This network efficiently extracts and fuses features from the multimodal inputs. Finally, the model performs regression to predict the Q-distribution and assess its degree of alignment with the ground truth. The results of our approach demonstrate the efficacy of using deep learning for this task, offering a significant improvement in prediction accuracy over previous methods.

Overall, we have made the following contributions to this work:

1) We propose a multimodal fusion framework for Q-distribution prediction using convolutional neural networks, multi-layer perceptron, and Transformer networks. It is the first multimodal framework for the Q-distribution prediction.
2) We simulate a new dataset that contains 5753 pieces of data for the Q-distribution prediction and divide them into training and testing sets to train our network.
3) We conduct extensive experiments on this dataset, which fully demonstrate the effectiveness of our proposed multimodal fusion method for Q-distribution prediction.

2 Method

2.1 Problem Formulation

In the task of predicting plasma ruptures, accurately forecasting the Q-distribution is of critical importance. This paper focuses on applying deep learning methods to minimize errors in Q-distribution prediction. Selecting an appropriate deep learning approach for modeling this unique type of nuclear fusion data is crucial. Moreover, from a multimodal artificial intelligence perspective, it is also essential to explore how to integrate information from multiple viewpoints to enhance model accuracy. Consequently, we conducted comprehensive experiments with various mainstream deep learning models and identified the optimal data fusion method to achieve effective Q-distribution prediction. We design a simple and effective multimodal fusion framework for Q-distribution prediction, as shown in Fig. 1.

2.2 Overview

Deep learning has garnered significant attention in recent years due to its remarkable capabilities in various domains. In this paper, we delve into the practical implications of utilizing deep learning methodologies for nuclear fusion prediction in physical applications. To maximize the accuracy of prediction performance, our emphasis lies in leveraging multimodal deep learning network frameworks such as Transformer [15] and convolutional neural networks (CNNs) [16]. Specifically, we begin by generating line charts for each nuclear fusion indicator, illustrating their respective changes over time as visual information. This bimodal data approach proves effective in capturing nuanced patterns. Subsequently, we employ linear network layers and convolutional neural networks to extract feature information of the raw numerical data and visual representations, respectively. Finally, we fuse the extracted features from both modalities, enhancing the overall predictive capability. Our network framework diagram is shown in Fig. 1. To be specific, we leverage multiple linear network layers and Attention mechanisms [15] to extract original features from the raw fusion index data. Concurrently, for the newly generated image data, Convolutional Neural Networks (CNNs) are employed alongside Attention mechanisms to extract visual features. The utilization of Attention mechanisms significantly enriches the expression of features, owing to its outstanding performance. Subsequently, we harness the formidable capabilities of multimodal Transformer blocks to effectively fuse the features extracted from both modalities, culminating in the final prediction. Further elaboration on the network architecture and the application of deep learning-based methods in nuclear fusion prediction will be provided in the subsequent subsections.

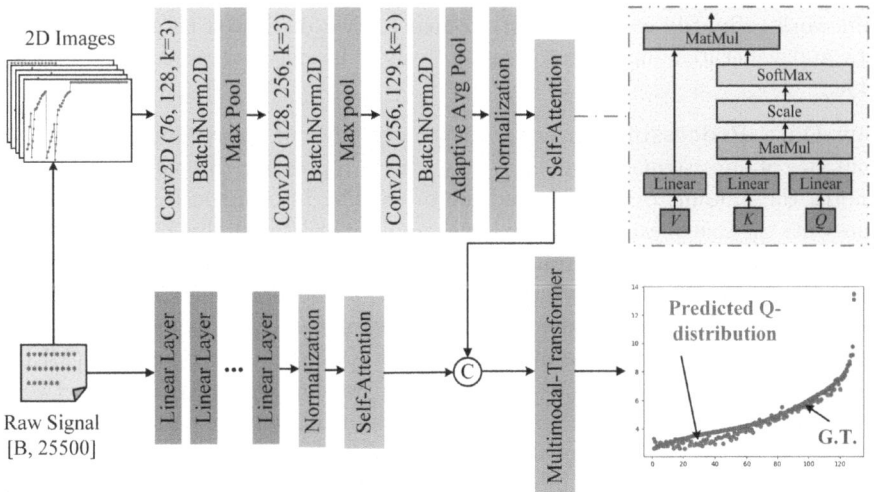

Fig. 1. An overview of our proposed model for the Q-distribution.

2.3 Input Processing

In this study, our nuclear fusion raw data comprises 141 physical indicators, denoted as $\mathcal{R}$, which encompass variables such as $npsip - q$ and q distribution. Notably, with the exception of $npsip - q$ and q, all other indicators signify the plasma rupture status during the nuclear fusion process (The $npsip - q$ and q distribution represent the final predicted ground truth). To facilitate the network's interpretation and processing, we transform the content of each indicator in the raw data into tensor representations. Subsequently, we concatenate these tensors along the first dimension and feed them into the network.

In addition to the raw data for nuclear fusion prediction, we have also generated 2D image data illustrating changes in translation data indicators. Specifically, to address the varying values and scales of the raw data, we selected 76 representative predictive indicators and transformed them into 2D visual line charts. For the convenience of drawing and calculation, we established a standard sampling approach: for each indicator of the raw data, if the number of values contained in the indicator N is greater than M, we sample M values at intervals; otherwise, we retain the original N values. Based on this standard, we have drawn line charts for each representative indicator. Subsequently, we input both the raw data and the generated 2D visual data into multiple linear network layers and convolutional neural networks (CNNs) respectively for feature extraction. The detailed network structures will be introduced in the next section.

2.4 Network Architecture

We have devised a network framework for nuclear fusion prediction tasks based on the multimodal fusion concept of deep learning, as illustrated in Fig. 1. The framework primarily comprises two branches, each dedicated to processing raw data and generated image data, respectively. The detailed process is outlined below.

Raw Data Processing. In order to achieve efficient expression of data, the input size of our original data is $\mathbb{R}^{B \times n}$, where B represents the batch size, which denotes the amount of data loaded for processing in a single iteration, and n represents the size of each individual data point, which is usually 25500. We input a batch of data into the neural network, as shown in the lower part of Fig. 1. We begin by feeding the raw signal data into a backbone network comprising several consecutive linear network layers interspersed with sigmoid activation functions, forming what is commonly known as a Multi-Layer Perceptron (MLP), to extract the raw data features. Subsequently, we employ an Attention network to establish global dependencies among the data points, the network structure diagram of the Attention network is shown in the top right corner of Fig. 1. We transform the raw data into Query (Q), Key (K), and Value (V) vectors, which are then fed into the linear layer within the Attention mechanism for weighting. Following this, we perform matrix multiplication between the Q and K matrices, which are subsequently normalized using a scaling factor. Finally, applying the softmax activation function to these normalized values, we obtain attention weights, which are then used to dot multiply with the Value vectors (V), yielding the final output denoted as $output_1$. The processing formula for the raw data can be summarized as follows,

$$output1 = Attention(Norm((MLP(R_i))*N)) \tag{1}$$

$$Attention(Q,K,V) = softmax(\frac{QK^T}{\sqrt{d_k}})V \tag{2}$$

where R_i represents the input of the raw data, MLP represents the multi-layer perception, Norm represents the linear normalization layer, N represents The number of MLP layers, Q, K, and V are query, key, and value matrices separately, d_k in the $Attention$ formula represents the scaling factor, and $softmax$ represents the normalization operation.

The Generated 2D Images Processing. Multi-modal fusion represents a pivotal research avenue in deep learning. This paper proposes a fusion prediction approach leveraging multimodal fusion technology. To integrate multimodal data input, we generated 2D images derived from the raw data to encompass visual information. After obtaining the 2D images that reflect the changes in indicators, we input the corresponding images into the upper part shown in Fig. 1. Firstly, we resize the input 2D images to a predetermined fixed resolution size, which is 224 * 224 used in this paper. Following the resizing step, we overlay a batch of data images along the first dimension to construct a $4-dimensional$ visual image dataset, which is $\mathbb{R}^{B \times K \times W \times H}$. Here B refers to batch size, K refers to the number of images generated by a single data point, H and W refer to the height and width of the images. We proceed by feeding the preprocessed visual data into the visual backbone network. Our visual backbone network primarily comprises convolutional neural networks (CNNs). More precisely, we employ 2D CNNs integrated with BatchNorm layers, followed by a max pooling layer aimed at gradually reducing the resolution size of the images. It's noteworthy that the pooling layer utilized in the final stage is the global pooling layer, which considers the global information of the images. Following the extraction of visual features, we further utilize the Attention network to encapsulate the global information of visual images, thereby enhancing the feature representation capabilities. Ultimately, the output obtained from the Attention mechanism serves as the final output of the visual branch, which we call $output_2$. The processing formula for vision data is as follows,

$$output2 = Attention(AvgPool((MaxPool(BN(Conv2D(V_i))))*N)) \tag{3}$$

where V_i is the generated 2D images from raw data, N is the number of convolutional layers, and BN denotes the BatchNorm layer.

The Multimodal Fusion Network. To leverage multimodal fusion methods for enhancing prediction accuracy and reducing prediction errors, we combine the features extracted from the raw data with those obtained from the generated 2D images. Drawing from the remarkable success of the Transformer architecture, this paper employs a multimodal Transformer approach to accomplish effective multimodal fusion. Specifically, based on the $output_1$ and $output_2$ we obtained above, we first concatenate the two outputs, and then input it to the multimodal Transformer for feature interaction and fusion. Firstly, we feed the fused features into the multi-head self-attention network to facilitate feature interaction and enhancement. Subsequently, the output from this attention mechanism is passed through a linear network layer and a feedforward network. Notably, at each step, the output is augmented by adding it to the input, thereby maximizing the preservation of the original data features. After traversing through N layers of Transformer blocks, we obtain the final fused interaction output. The formula is as follows,

$$output = Transformer([output1, output2]) \quad (4)$$

where $[\cdot]$ represents the concatenation operation, and $Transformer$ is the multimodal Transformer. These outputs are then utilized to calculate the loss, which is computed based on the Q-distribution of the raw groundtruth. The detailed calculation process of the loss is delineated in the subsequent section.

2.5 Loss Function

Through multimodal Transformers, we obtained the final output of the network. To reflect the error between our predicted results and the groundtruth, we use the MSE (mean squared error) loss function to measure it. The specific loss function can be represented as,

$$MSE = \frac{1}{n}\sum_{i=1}^{n}(GT_i - Q_i)^2 \quad (5)$$

where GT_i represents the groundtruth of the i-th sample, and Q_i represents the distribution predicted result of the i-th sample.

3 Experiments

3.1 Dataset and Evaluation Metric

The data for nuclear fusion is split into training/testing subsets which contain 5177 and 575 raw data, respectively. The MSE (Mean Squared Error) metric is adopted to measure the distance between our predicted Q-distribution and the groundtruth.

3.2 Implementation Details

The training of our network is conducted end-to-end. We train our network with multimodal inputs for 130 epochs. The learning rate is 0.001, and the batch size is 16. We opt for Stochastic Gradient Descent (SGD) [17] as the optimizer. Our code is implemented using Python based on PyTorch [18] framework and the experiments are conducted on a server with CPU Intel(R) Xeon(R) Gold 5318Y CPU @2.10 GHz and GPU RTX3090 with 24 GB memory.

3.3 Comparison with Other Models

In order to effectively extract visual features from the generated 2D images, we compared various mainstream visual backbones to identify the optimal method. Table 1 summarizes our findings. We evaluated several architectures including Residual Networks (ResNet18, ResNet34, ResNet50) [19], Multi-Layer Perception networks (MLP), Vision Transformer networks (ViT-B, ViT-L, ViT-H) [14], VGGNet [20] (VGG16, VGG19), and Ours Convolutional Neural Networks (CNNs). Through comprehensive experimental comparisons, we observed that utilizing convolutional neural networks as our visual backbone network yielded the best performance. This can be attributed to the local perception ability and strong generalization capability inherent in convolutional neural networks.

Table 1. Comparison with Other Models.

Algorithm	ResNet18	ResNet34	ResNet50	MLP	ViT-B
MSE	0.1037	0.1239	0.0768	0.0748	0.0759
Algorithm	ViT-L	ViT-H	VGG16	VGG19	Ours
MSE	0.0743	0.0697	0.0746	0.0759	0.0696

3.4 Ablation Study

Component Analysis. This paper is about the application of deep learning in the field of nuclear fusion prediction. Our experiments have shown that incorporating deep learning into network models can indeed reduce the prediction error of Q-distribution. As shown in Table 2, what we can see is that using only the raw data through some linear network layers yields an unsatisfactory result, with an MSE loss value of 0.1125. However, when we leverage the powerful modeling ability of the Attention network, the prediction loss is significantly reduced to 0.0884. Even more surprisingly, inspired by the 2D image data from computer vision, when we generated 2D image data from the raw data and added it for appropriate fusion with the raw data, the result further improved to 0.0696. These results fully demonstrate that the prediction results of nuclear fusion have been improved after the addition of the multimodal fusion network for deep learning.

Table 2. Component Analysis results.

No.	MLP	Attention	Multi-modal fusion	MSE
1	✓			0.1125
2	✓	✓		0.0884
3	✓	✓	✓	0.0696

Analysis of the Maximum Number of Spaced Sampling Points of Images. According to the above, when we added visual information, the prediction results were further improved. However, due to the fact that some indicators in the raw data are composed of thousands of values, it is difficult to generate 2D images representing the numerical changes of the indicators. Based on past experience, we conducted interval sampling on indicators with excessive values and specified the collection of only M points when the numerical value N of the indicator is much greater than M. Thus, what is the most suitable value for M? We conducted experiments on the maximum interval point sampling data volume as shown in Table 3. It is obvious that when we set the maximum number of points for interval sampling to 100, the MSE error reaches its minimum. When the number of sampling points is too small, the loss of unsampled data can significantly increase the prediction error. On the other hand, when there are too many sampling points, the dense clustering can cause indistinguishable points in the generated image, also reducing prediction accuracy. Therefore, selecting an appropriate number of samples is critically important.

Table 3. Ablation studies on the maximum number of spaced sampling points of images.

Sample Points	50	100	200	500
MSE	0.0812	0.0696	0.0785	0.0772

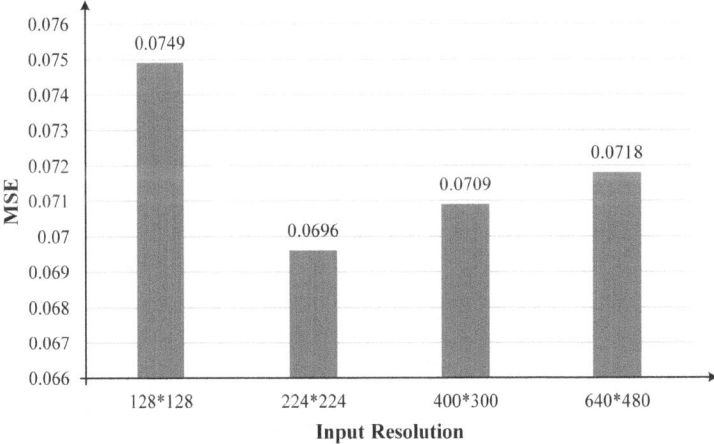

Fig. 2. Visual comparison of different resolutions of input 2D images.

Analysis of the Scale of 2D Images. We have analyzed the maximum number of spaced sampling points of the images above. In addition, the resolution of the input images is also an important factor affecting the results. So, we set the input of images with different resolutions to find the most suitable resolution input, and the experimental results are shown in Fig. 2. We can observe that the best result is achieved when we choose an image resolution of 224 * 224 for input, with an error reduction of 0.0696.

Analysis of the Number of Multi-modal Transformer Blocks. Inspired by the idea of multimodal fusion in the field of computer vision, we also use multimodal Transformers to achieve multimodal fusion between raw data and generated image data. The network structure of the Transformer mainly consists of multiple Transformer blocks, each of which contains modules such as multi-head attention. In order to investigate the impact of the number of Transformer blocks on the final result, we also conduct research on the number of multi-modal transformer blocks, The experimental results are shown in Fig. 3. Through this set of experiments, We can clearly see that the number of blocks does not directly correlate with the quality of the effect. Using only two blocks yields the optimal effect.

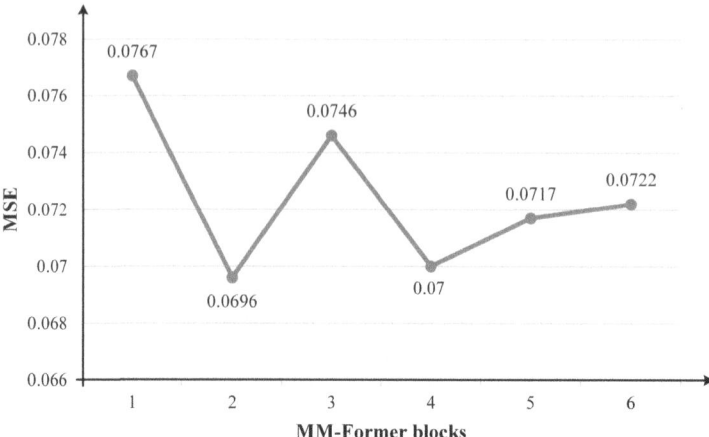

Fig. 3. Visual comparison of the number of different multimodal Transformer blocks.

4 Conclusion

In this paper, we primarily investigate how to use deep learning models to improve the predictive ability of the Q-distribution task in nuclear fusion from the perspective of multimodal fusion. Our initial attempts using simple neural networks (MLPs) to process raw data yielded suboptimal results. Consequently, we transformed the raw data into 2D visual images and leveraged Convolutional Neural Networks (CNNs) along with Attention mechanisms to effectively model the 2D image information. We then employed a multimodal Transformer network to seamlessly integrate the 1D raw information with the 2D image data, significantly reducing prediction errors in Q-distribution. Extensive analytical experiments have corroborated the efficacy of our proposed multimodal fusion framework, demonstrating its capability to effectively control Q-distribution errors in nuclear fusion. Looking ahead, we aim to further explore the potential of deep learning models in addressing various challenges in nuclear fusion tasks, including Q-distribution prediction, thereby contributing to sustainable energy development.

Acknowledgement. This work is supported by the National Natural Science Foundation of China under Grant U24A20342, 62102205, the Anhui Provincial Natural Science Foundation under Grant 2408085Y032. The authors acknowledge the High-performance Computing Platform of Anhui University for providing computing resources.

References

1. Lennholm, M., et al.: Plasma control for the step prototype power plant. Nucl. Fusion **64**(9), 096036 (2024)

2. Seo, J., et al.: Avoiding fusion plasma tearing instability with deep reinforcement learning. Nature **626**(8000), 746–751 (2024)
3. Lennholm, M., et al.: Controlling a new plasma regime. Philos. Trans. A **382**(2280), 20230403 (2024)
4. Dal Molin, A., et al.: Measurement of the gamma-ray-to-neutron branching ratio for the deuterium-tritium reaction in magnetic confinement fusion plasmas. Phys. Rev. Lett. **133**(5), 055102 (2024)
5. Wang, Z.: Current status of research on magnetic confinement fusion and superconducting tokamak devices. Procedia Comput. Sci. **228**, 163–170 (2023)
6. van Efferen, C., Fischer, J., Costi, T.A., Rosch, A., Michely, T., Jolie, W.: Modulated kondo screening along magnetic mirror twin boundaries in monolayer mos2. Nat. Phys. **20**(1), 82–87 (2024)
7. Velasco, J.L., Calvo, I., Sánchez, E., Parra, F.: Robust stellarator optimization via flat mirror magnetic fields. Nucl. Fusion **63**(12), 126038 (2023)
8. Thienpondt, H., et al.: Prevention of core particle depletion in stellarators by turbulence. Phys. Rev. Res. **5**(2), L022053 (2023)
9. Goodman, A.G., et al.: Quasi-isodynamic stellarators with low turbulence as fusion reactor candidates. PRX Energy **3**(2), 023010 (2024)
10. Rodríguez, E., Mackenbach, R.: Trapped-particle precession and modes in quasisymmetric stellarators and tokamaks: a near-axis perspective. J. Plasma Phys. **89**(5), 905890521 (2023)
11. Zheng, W., et al.: Disruption prediction for future tokamaks using parameter-based transfer learning. Commun. Phys. **6**(1), 181 (2023)
12. Schwander, F., Serre, E., Bufferand, H., Ciraolo, G., Ghendrih, P., Tamain, P.: Global fluid simulations of edge plasma turbulence in tokamaks: a review. Comput. Fluids 106141 (2023)
13. Berkery, J.W., et al.: NSTX-U research advancing the physics of spherical tokamaks. Nuclear Fusion (2024)
14. Dosovitskiy, A., et al.: An image is worth 16×16 words: transformers for image recognition at scale. In: ICLR (2021)
15. Vaswani, A., et al.: Attention is all you need. In: Advances in Neural Information Processing Systems, vol. 30 (2017)
16. LeCun, Y., Bottou, L., Bengio, Y., Haffner, P.: Gradient-based learning applied to document recognition. Proc. IEEE **86**, 2278–2324 (1998)
17. Bottou, L.: Stochastic gradient descent tricks. Neural Netw. (2012)
18. Paszke, A., et al.: PyTorch: an imperative style, high-performance deep learning library. In: Advances in Neural Information Processing Systems, vol. 32 (2019)
19. He, K., Zhang, X., Ren, S., Sun, J.: Deep residual learning for image recognition. In: IEEE Conference on Computer Vision and Pattern Recognition, pp. 770–778 (2015)
20. Simonyan, K., Zisserman, A.: Very deep convolutional networks for large-scale image recognition. CoRR, vol. abs/1409.1556 (2014)

Deformable Transformer for 3D Medical Image Segmentation

Haifeng Zhao, Tianxia Yang, Minghui Xu, and Yanping Fu(✉)

Anhui Provincial Key Laboratory of Multi-modal Cognitive Computation,
School of Computer Science and Technology, Anhui University, Hefei, China
ypfu@ahu.edu.cn

Abstract. In recent years, CNNs have demonstrated remarkable success in medical image processing due to their powerful feature extraction capabilities. However, their inherent limitation hampers their ability to learn global features. In contrast, Transformers excel at modeling long-range dependencies, offering a solution to this limitation. Yet, the computational demands of self-attention make high-resolution medical image processing-particularly for 3D data—extremely resource-intensive. To address these challenges, we introduce DeTransNet, a hybrid segmentation method that harnesses the strengths of both CNNs and deformable Transformer. First, we introduce a multi-channel feature fusion module (MCFF) that efficiently integrates multi-scale features from CNNs to enhance the input for the Transformer network. Additionally, we propose a deformable attention mechanism to improve feature extraction, enabling the model to capture global information while reducing self-attention complexity. Extensive experiments on the public BCV dataset demonstrate that the proposed method outperforms existing methods.

Keywords: Multi-channel feature fusion · Deformable Transformer · 3D medical image segmentation

1 Introduction

Medical image segmentation refers to the process of separating regions of interest from the background in medical images. With the development of medical imaging technology, the application of 3D medical images such as CT and MRI has become increasingly widespread. Therefore, there is an urgent need for an efficient method to segment 3D medical images to meet application demands. Convolutional Neural Networks (CNNs) are designed for data with structured patterns, particularly images. A prominent model is U-Net [1], featuring a symmetric encoder-decoder with skip connections for better feature retention during upsampling. Following U-Net, various variants [2,3] have emerged in 3D medical image segmentation. Traditional methods processed volumetric slices independently, often losing essential 3D context. The introduction of 3D U-Net [4] overcame these limitations by leveraging entire volumetric data with 3D operations, setting a new benchmark in the field of 3D image segmentation.

CNNs excel at feature learning and generalization but struggle with global information due to convolutional kernel sizes. In contrast, Transformers are effective at capturing long-range dependencies, particularly in natural language processing. Vision Transformer (ViT) [5] applies this to images, leading to various competitive visual Transformer models. UNETR [6] employs a U-shaped encoder-decoder structure with Transformers as encoders to extract features from volumetric sequences, enabling global information capture. While this approach significantly enhances the ability to capture global context in medical imaging, the quadratic complexity of Transformer-based attention mechanisms presents a major computational challenge, especially when processing 3D data.

To leverage the strengths of both CNNs and Transformers, we introduce the Deformable Transformer Network (DeTransNet), a groundbreaking hybrid architecture that fuses CNNs with parallel deformable Transformers. The network begins with convolutional layers to extract multi-scale features from the input image. These features are then processed by effective parallel deformable Transformer (EDETrans) blocks for adaptive long-range dependency learning. The EDETrans block creates reference point coordinates via grid sampling, utilizing learnable offsets from a bias network. This innovative mechanism enables adaptive multi-head self-attention on dynamically adjusted reference points, efficiently capturing complex, long-range dependencies. Feature fusion occurs between branches, and the decoder upscales the features, utilizing skip connections to link corresponding resolutions, resulting in highly precise segmentation outputs. To sum up, our main contributions are as follows:

(1) We introduce a novel network architecture that combines the strengths of CNNs and deformable Transformers for 3D medical image segmentation.
(2) We design a multi-channel feature fuse block to enhance the details of the medium-resolution features extracted by the CNN encoder, which helps with further feature extraction by the Transformer.
(3) We propose an effective parallel deformable Transformer module for adaptive long-range dependency learning, which can effectively enhance the segmentation performance.

2 Related Work

2.1 Convolutional Neural Networks

CNNs primarily learn local information due to the limited size of the convolutional kernel. While increasing kernel size [7,8] can expand the receptive field, it significantly elevates computational costs. Dilated convolution [9] enhances the receptive field by inserting zeros between kernel elements, potentially causing early computations of distant pixels, which is not optimal for long-range feature extraction. Deformable convolution [10] alters sampling positions with learned offsets, enabling adaptability to input features. Deformable convolution methods [11,12] outperform traditional segmentation techniques and have been adapted for Transformer models.

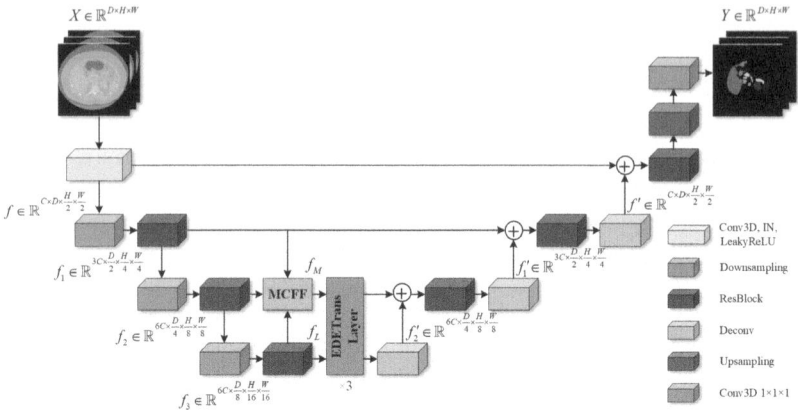

Fig. 1. The overall architecture of the proposed DeTransNet.

2.2 Transformers

Transformers model long-range dependencies but incur high computational costs. To address this, variants like Swin Transformer [13] use window attention with a sliding mechanism, achieving linear time complexity relative to input size. Deformable Transformer [14–16] focuses on sampled positions as pre-filters to highlight important features. Our Deformable Efficient Transformer adopts this approach for more flexible attention to key areas through bias addition.

2.3 Hybrid CNN and Transformer

Researchers have developed hybrid models to merge CNN and Transformer benefits, like TransUNet [17], which uses Transformers in the encoder to maintain high-resolution details for medical image segmentation. However, direct use of ViT increases the complexity of 3D images. CoTr [18] combines 3D convolution with deformable Transformers, focusing the attention on critical small details and handling high-resolution feature maps efficiently. TransHRNet [19] employs parallel Transformer blocks to process multi-scale features from CNNs, enhancing overall model performance with a fusion step before each Transformer layer.

3 Method

The architecture of the proposed DeTransNet is illustrated in Fig. 1. The preprocessed 3D image $X \in \mathbb{R}^{D \times H \times W}$ is first passed through a CNN-based encoder equipped with residual and downsampling blocks, efficiently capturing multi-scale features. To maximize the feature representation, High-resolution and low-resolution features are seamlessly integrated into medium-resolution features using a multi-channel feature fusion module (MCFF). These fusion features and low-resolution features are then processed through three EDETrans blocks

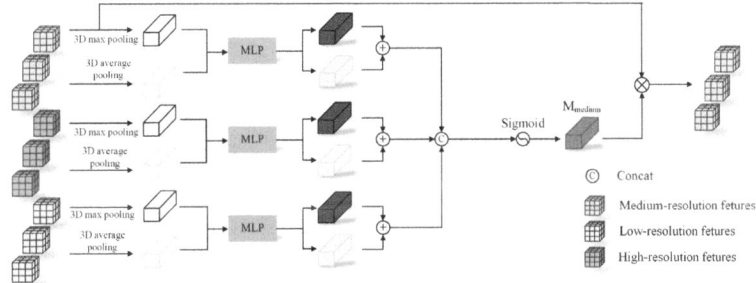

Fig. 2. The architecture of the multi-channel feature fusion block (MCFF).

to capture long-range dependencies. The resulting enriched features are finally passed to the decoder, generating precise segmentation results that align perfectly with the input dimensions.

3.1 CNN-Based Encoder

To effectively capture multi-scale features, we employ a CNN-based encoder with the same structure as [18,19] for feature extraction. We first employ 3D convolution followed by instance normalization (IN) and Leaky Rectified Linear Unit (LeakyReLU) activation for downsampling, yielding a feature representation of f. This representation is then processed through three consecutive convolutional layers, extracting features at three distinct scales: high-resolution f_1, medium-resolution f_2, and low-resolution f_3.

3.2 Transformer Encoder

Multi-channel Feature Fusion Block (MCFF). To fully leverage the information extracted at various scales, we design a multi-channel attention module that facilitates channel-level fusion of diverse scale information, as shown in Fig. 2. Think of max pooling retaining essential information from the feature map, while average pooling preserves more overall information. We first send various scale features $f_i(i=1,2,3)$ into parallel 3D max pooling and 3D average pooling layers to obtain two complementary feature vectors Fi_{max} and $Fi_{avg}(i=1,2,3)$. This dual extraction not only enhances the richness of the feature set but also enables the model to effectively integrate fine-grained details with overall context, leading to improved performance in subsequent tasks.

$$Fi_{max}(c) = \max_{d,h,w}\left(f_i(c,d,h,w)\right) \tag{1}$$

$$Fi_{avg}(c) = \frac{1}{D_i \times H_i \times W_i} \sum_{d=1}^{d_i}\sum_{h=1}^{h_i}\sum_{w=1}^{W_i} f_i(c,d,h,w), \tag{2}$$

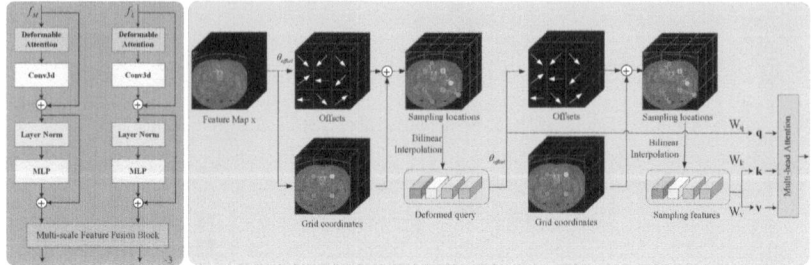

Fig. 3. Architecture of EDETrans block (left) and Deformable Attention (right).

where $Fi(c)$ refers to the $c-th$ channel of Fi. The obtained feature vectors are then passed through Multi-Layer Perceptrons (MLPs) to generate corresponding weight vectors, which is then combined through element-wise addition to enhance feature interaction. These weights are subsequently ranked by relevance, concatenated, and refined using a sigmoid function to produce the final channel attention weights. These optimized weights are then multiplied by their respective feature maps to emphasize the most relevant information. Finally, the weighted features are input into the Transformer encoder.

Effective Deformable Transformer Block (EDETrans). The self-attention computation in Transformers for 3D medical image processing tasks requires a huge amount of computational power. To address this, we draw inspiration from deformable convolutional networks and design a Transformer with deformable attention, as shown in Fig. 3. This approach uses learnable positional offsets to dynamically sample feature maps, dramatically reducing the computational burden while directing focus toward the most critical regions of the image.

For the input feature map f_M (with similar processing applied to f_L), we first feed it into the deformable attention module to calculate the sampling features. Firstly, we use an offset network $\theta_{\text{offset}}(\cdot)$ to learn the bias from f_M. The offset network consists of two 3D convolutions, followed by normalization and activation, designed to effectively learn local feature shifts. Then, the feature map is fed into a 3D convolution with a stride of r followed by normalization and activation to capture local features. This produces an output $\Delta p \in \mathbb{R}^{D_G \times H_G \times W_G \times 3}(D_G = D_i/r, H_G = H_i/r, W_G = W_i/r(i=1,2))$, representing the learned offsets in three spatial directions. The offsets $\Delta p = \theta_{\text{offset}}(f_M)$ are then used to adjust reference point coordinates $\{(p_x, p_y, p_z)|(0,0,0), ..., (D_G - 1, H_G - 1, W_G - 1)\}$, which are evenly distributed on a grid. By adding the learned bias to these reference coordinates, we accurately determine the sampling locations for queries. Finally, we use a sampling function $\phi(\cdot, \cdot)$ with bilinear interpolation to selectively extract features from critical regions.

$$\phi(f_M, p + \Delta p) = \sum_{(r_x, r_y, r_z)} g(p_x, r_x)g(p_y, r_y)g(p_z, r_z)x[r_z, r_y, r_x, :], \quad (3)$$

Table 1. Dice scores of different approaches on the BCV dataset.

Methods	Organs											Avg.
	Sp	Ki	Gb	Es	Li	St	Ao	IVC	PSV	Pa	AG	
SETR(ViT-B/16-rand) [20]	95.2	92.3	55.6	71.3	96.2	80.2	89.7	83.9	68.9	68.7	60.5	78.4
SETR(ViT-B/16-pre) [20]	94.8	91.7	55.2	70.9	96.2	76.9	89.3	82.4	69.6	70.7	58.7	77.8
TransUNet [17]	95.9	93.7	63.1	77.8	97.0	86.2	91.0	87.8	77.8	81.6	**73.9**	84.2
CoTr w/o CNN-encoder [18]	95.2	92.8	59.2	72.2	96.3	81.2	89.9	85.1	71.9	73.3	61.0	79.8
CoTr [18]	**96.4**	93.9	67.8	78.0	**97.1**	87.0	90.8	87.9	77.5	**83.5**	73.7	84.9
TransBTS [21]	95.9	90.9	**75.6**	77.8	96.7	**88.6**	89.0	**88.4**	**78.4**	81.2	70.0	84.8
TransHRNet_S [19]	96.3	93.9	66.5	78.0	97.0	88.1	91.1	**88.4**	77.5	83.1	73.7	84.9
DeTransNet	96.0	**94.7**	71.1	**78.2**	97.0	86.1	**91.2**	88.1	77.8	82.8	72.9	**85.1**

where $g(a, b)$ is the bilinear sampling weight function and (r_x, r_y, r_z) indexes all locations on f_M. We note the output of Eq. 3 as q. After getting the deformed query, we use $\theta_{\text{offset}}(\cdot)$ to learn offset according to q to sample features for keys and values. Then, the deformed keys and values are obtained by the projections W_k and W_v, respectively. We summarize the above components as follows:

$$k = \tilde{x}W_k, v = \tilde{x}W_v, \text{with} \Delta p^{'} = \theta_{\text{offset}}(q), \tilde{x} = \phi(f_M, p + \Delta p^{'}) \quad (4)$$

Finally, the deformable multi-head attention receives the queries, keys, and values as input and outputs a refined feature.

The acquired feature representation is restored to its original size through the convolutional layer. Then, the features corresponding to the scale are outputted through the Layer Normalization (LN) layer and MLP with residual connections to ensure smooth gradient flow. The medium- and low-resolution features are fused together using a multi-scale feature fusion module [19], maximizing the synergy between different resolutions. These fused features are then passed through successive EDETrans blocks, with three consecutive blocks used to process and refine the features. A key innovation lies in the cross-fusion of features from two parallel EDETrans layers, facilitating an efficient exchange of information between different resolutions.

3.3 Decoder

We use the transpose convolution to upsample the feature maps to the original resolution progressively. We incorporate skip connections after each transpose convolution layer to connect features of the same resolution from the encoder and decoder, enhancing segmentation quality. Our loss function is the sum of the Dice loss [22] and Cross-Entropy loss.

4 Experiments

Dataset. The BCV [23] dataset comprises 50 abdominal CT scans. We use 21 cases for training and reserve the remaining cases for testing. We evaluate the

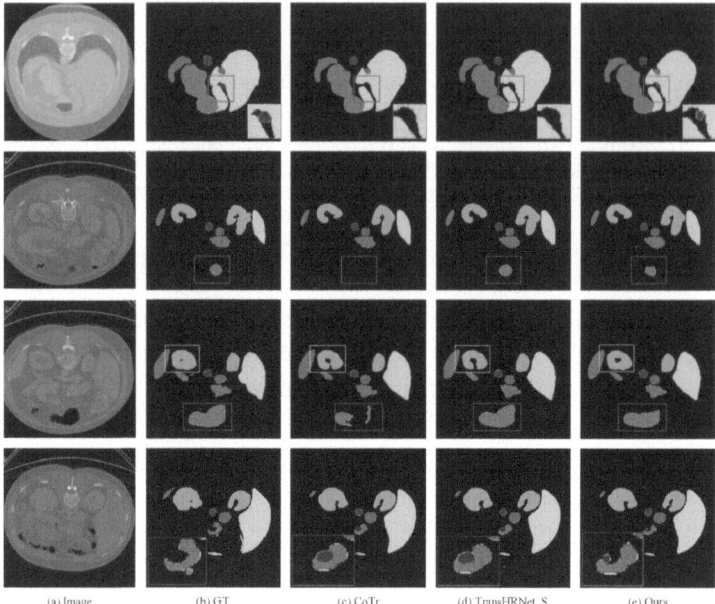

Fig. 4. Qualitative comparison results for the segmentation task of BCV dataset.

model's performance by reporting the average Dice scores across eleven critical abdominal organs: spleen (Sp), kidney (Ki), gallbladder (Gb), Esophagus (Es), liver (Li), stomach (St), aorta (Ao), inferior vena cava (IVC), portal vein and splenic vein (PSV), pancreas(Pa) and adrenal gland (AG).

Implementation Details. We implement the proposed method using PyTorch on an NVIDIA 3090 GPU. The model input consists of randomly cropped subvolumes of size $48 \times 192 \times 192$ from the CT scans, enhanced through data augmentation techniques such as random rotation, scaling, flipping, and adding Gaussian white noise. DeTransNet is trained for 1000 epochs with a batch size of 2, momentum of 0.99, and an initial learning rate of 0.01.

4.1 Quantitative and Qualitative Results

Quantitative Results. We conduct a comprehensive comparison of our proposed method against various CNN-based, Transformer-based, and hybrid CNN-Transformer methods using the same dataset to ensure a fair and consistent evaluation. Table 1 shows the numerical results of different methods on different organs. We can see that our method significantly improves the performance of multi-organ segmentation, achieving optimal results across various organs. Notably, kidney segmentation accuracy increases by 0.8% over the next best method, while esophagus segmentation improves by 0.2%, contributing to a 0.2%

Table 2. Ablation study of the effect on the functional blocks.

EDETrans	MCFF	Sp	Ki	Gb	Es	Li	St	Ao	IVC	PSV	Pa	AG	Avg.
		96.0	92.6	63.8	77.9	97.0	83.6	90.8	87.8	76.7	81.2	72.6	83.6
✓		96.6	94.2	61.9	77.3	97.0	86.0	90.9	87.7	77.5	83.0	73.9	84.2
✓	✓	96.0	94.7	71.1	78.2	97.0	86.1	91.2	88.1	77.8	82.8	72.9	85.1

increase in the overall average Dice score. These experimental results demonstrate the effectiveness of our method as a robust solution for 3D medical image segmentation, excelling in both accuracy and consistency across diverse anatomical structures.

Qualitative Results. Figure 4 shows the segmentation results of different methods. The zoomed-in images in the lower left corner of the first row and the lower right corner of the fourth row highlight critical details. These close-ups reveal that our segmentation method delivers superior visual quality, particularly in complex regions. In challenging areas like the portal and splenic veins (PSV, depicted in dark blue), competing methods suffer from noticeable regional segmentation errors. In contrast, our approach, utilizing deformable self-attention, effectively captures and integrates multi-scale feature relationships, significantly improving segmentation accuracy and consistency.

4.2 Ablation Studies

To demonstrate the effectiveness of the proposed modules, we conduct several experiments to validate the EDETrans block and MCFF block. The experimental data results are shown in the Table 2. The EDETrans block is an important component for extracting global features, addressing the shortcomings of pure CNN architectures, resulting in a 0.6% increase in the average Dice score for organs. Meanwhile, the MCFF module, which enhances medium-resolution features through effective multi-channel interaction with both high- and low-resolution features, results in an even more substantial improvement, boosting the average Dice score by 0.9%. These results demonstrate the critical role of both modules in driving superior segmentation performance.

5 Conclusion

This paper introduces DeTransNet, a 3D medical image segmentation model that integrates CNNs and deformable Transformers. The MCFF module enhances feature integration across resolutions, optimizing the Transformer's learning. The deformable attention-based Transformer utilizes learnable biases to prioritize significant regions, reducing computational load. Extensive BCV experiments show our method outperforms other advanced networks. We aim to offer fresh insights into deformable attention-based architectures.

Acknowledgment. This study is funded by the National Natural Science Foundation of China (62472004, 62076005, U20A20398), the Natural Science Foundation of Anhui Province (2308085MF214), the University Synergy Innovation Program of Anhui Province, China (GXXT-2021-002, GXXT-2021-030, GXXT-2022-029), and the Key Natural Science Project of Anhui Provincial Education Department (No.2023AH050065).

References

1. Ronneberger, O., Fischer, P., Brox, T.: U-Net: convolutional networks for biomedical image segmentation. In: Navab, N., Hornegger, J., Wells, W.M., Frangi, A.F. (eds.) MICCAI 2015. LNCS, vol. 9351, pp. 234–241. Springer, Cham (2015). https://doi.org/10.1007/978-3-319-24574-4_28
2. Meyer, A., et al.: Anisotropic 3d multi-stream cnn for accurate prostate segmentation from multi-planar mri. Comput. Methods Prog. Biomed. **200**, 105821 (2021)
3. Qamar, S., Jin, H., Zheng, R., Ahmad, P., Usama, M.: A variant form of 3d-unet for infant brain segmentation. Futur. Gener. Comput. Syst. **108**, 613–623 (2020)
4. Çiçek, Ö., Abdulkadir, A., Lienkamp, S.S., Brox, T., Ronneberger, O.: 3D U-Net: learning dense volumetric segmentation from sparse annotation. In: Ourselin, S., Joskowicz, L., Sabuncu, M.R., Unal, G., Wells, W. (eds.) MICCAI 2016. LNCS, vol. 9901, pp. 424–432. Springer, Cham (2016). https://doi.org/10.1007/978-3-319-46723-8_49
5. Dosovitskiy, A., et al.: An image is worth 16x16 words: transformers for image recognition at scale. arXiv preprint arXiv:2010.11929 (2020)
6. Hatamizadeh, A., et al.: Unetr: transformers for 3d medical image segmentation. In: Proceedings of the IEEE/CVF Winter Conference on Applications of Computer Vision, pp. 574–584 (2022)
7. Wang, W., Li, S., Shao, J., Jumahong, H.: Lkc-net: large kernel convolution object detection network. Sci. Rep. **13**(1), 9535 (2023)
8. Huang, T., Chen, J., Jiang, L.: Ds-unext: depthwise separable convolution network with large convolutional kernel for medical image segmentation. SIViP **17**(5), 1775–1783 (2023)
9. Yu, F., Koltun, V.: Multi-scale context aggregation by dilated convolutions. arXiv preprint arXiv:1511.07122 (2015)
10. Dai, J., et al.: Deformable convolutional networks. In: Proceedings of the IEEE International Conference on Computer Vision, pp. 764–773 (2017)
11. Shen, N., et al.: Multi-organ segmentation network for abdominal ct images based on spatial attention and deformable convolution. Expert Syst. Appl. **211**, 118625 (2023)
12. Jin, Q., Meng, Z., Pham, T.D., Chen, Q., Wei, L., Su, R.: DUNet: a deformable network for retinal vessel segmentation. Knowl.-Based Syst. **178**, 149–162 (2019)
13. Liu, Z., et al.: Swin transformer: hierarchical vision transformer using shifted windows. In: Proceedings of the IEEE/CVF International Conference on Computer Vision, pp. 10012–10022 (2021)
14. Zhu, X., Su, W., Lu, L., Li, B., Wang, X., Dai, J.: Deformable detr: deformable transformers for end-to-end object detection. arXiv preprint arXiv:2010.04159 (2020)

15. Xia, Z., Pan, X., Song, S., Li, L.E., Huang, G.: Vision transformer with deformable attention. In: Proceedings of the IEEE/CVF Conference on Computer Vision and Pattern Recognition, pp. 4794–4803 (2022)
16. Azad, R., et al.: Beyond self-attention: deformable large kernel attention for medical image segmentation. In: Proceedings of the IEEE/CVF Winter Conference on Applications of Computer Vision, pp. 1287–1297 (2024)
17. Chen, J., et al.: Transunet: transformers make strong encoders for medical image segmentation. arXiv preprint arXiv:2102.04306 (2021)
18. Xie, Y., Zhang, J., Shen, C., Xia, Y.: CoTr: efficiently bridging CNN and transformer for 3D medical image segmentation. In: de Bruijne, M., et al. (eds.) MICCAI 2021. LNCS, vol. 12903, pp. 171–180. Springer, Cham (2021). https://doi.org/10.1007/978-3-030-87199-4_16
19. Yan, Q., et al.: 3d medical image segmentation using parallel transformers. Pattern Recogn. **138**, 109432 (2023)
20. Zheng, S., et al.: Rethinking semantic segmentation from a sequence-to-sequence perspective with transformers. In: Proceedings of the IEEE/CVF Conference on Computer Vision and Pattern Recognition, pp. 6881–6890 (2021)
21. Wang, W., Chen, C., Ding, M., Yu, H., Zha, S., Li, J.: TransBTS: multimodal brain tumor segmentation using transformer. In: de Bruijne, M., Cattin, P.C., Cotin, S., Padoy, N., Speidel, S., Zheng, Y., Essert, C. (eds.) MICCAI 2021. LNCS, vol. 12901, pp. 109–119. Springer, Cham (2021). https://doi.org/10.1007/978-3-030-87193-2_11
22. Milletari, F., Navab, N., Ahmadi, S.A.: V-net: Fully convolutional neural networks for volumetric medical image segmentation. In: 2016 Fourth International Conference on 3D Vision (3DV), pp. 565–571. IEEE (2016)
23. Landman, B., Xu, Z., Igelsias, J., Styner, M., Langerak, T., Klein, A.: Miccai multi-atlas labeling beyond the cranial vault–workshop and challenge. In: Proceedings of MICCAI Multi-Atlas Labeling Beyond Cranial Vault—Workshop Challenge, vol. 5, p. 12 (2015)

On the Gap Between AI-Generated and Human-Written Patent Texts

Zhanhao Xiao[1,2], Wei Hu[1], Yanqiang Wu[1], Weiqi Chen[1], Huihui Li[1,2], and Xiaoyong Liu[3(✉)]

[1] School of Computer Science, Guangdong Polytechnic Normal University, Guangzhou 510665, China
[2] Guangdong Provincial Key Laboratory of Intellectual Property and Big Data, Guangdong Polytechnic Normal University, Guangzhou 510665, China
[3] School of Data Science and Engineering, Guangdong Polytechnic Normal University, Guangzhou 510665, China
liuxy@gpnu.edu.cn

Abstract. Since the GPT-X models have made progress in generative tasks, a large number of large language models (LLMs) have sprung up. When the powerful features of LLMs have attracted the interest of numerous researchers, their misuse has also become a source of growing concern for human beings. In fact, LLMs have been used to generate fake news, fake academic papers, and fake patent application documents. Detecting whether content is generated by artificial intelligence (AI) has been a significant problem. Unfortunately, to our knowledge, there is currently no existing research focused on AI-generated patent text detection, nor are there any datasets tailored for patents publicly available. In this paper, to explore the differences between AI-generated and human-written patent texts, we generate a set of patent abstract texts by ChatGPT, in Chinese and English, from granted patent claims. Each generated patent abstract text corresponds to its original patent abstract. We analyze the linguistic characteristics of two types of patent texts by various comparison experiments. We anticipate that our work can assist people in identifying the patents generated by AI from the ocean of patents.

Keywords: Artificial Intelligence Generation · Large Language Model · Text Generation Detection · Patent Texts

1 Introduction

Recently, large language models (LLMs) have revolutionized the field of natural language processing research. Artificial intelligence (AI)-generated content technology continues to advance, starting with the progress made with the GPT-X

This work was partially supported by Guangdong Provincial Key Laboratory Project of Intellectual Property and Big Data (2018B030322016).

models developed by OpenAI. LLMs can generate texts or pictures based on prompt words or sentences provided by the users. However, the potential misuse of LLMs raises human concerns. For example, they are used to generate fake news headlines and content, which are used to manipulate public opinion or mislead the public to undermine social stability or promote specific political agendas. In addition, what has attracted particular attention is the abuse of LLMs to generate patent content to deceive patent examination offices to obtain patent grants. This will lead to a consequence: patents may be generated by one sentence without deliberation and the meaningless and low-quality patents may become ubiquitous. Furthermore, the improper use of AI technology will result in intellectual property infringement. In many intellectual property offices, patent applications generated by AI are either required to be disclosed or are prohibited entirely. Thus, such a detector assists examiners in identifying them and streamlines the examination process. Therefore, exploring the gap between AI-generated and human-written patent texts is a necessary research task.

To the best of our knowledge, there is currently no existing research focused on AI-generated patent text detection, nor are there any datasets tailored for patents publicly available. Different from general text content, patent content exhibits unique professional and normative characteristics, which may make patent text detection distinct from generic detection. In this paper, we construct a dataset for patent text generation and detection for the first time. This dataset includes patents from four domains: artificial intelligence, biomedicine, electrical engineering, and machinery manufacturing. Then we analyze the linguistic differences between patent texts generated by LLMs and written by humans. Specifically, we attempt to explore the following questions by experiments:

1. What are the differences in vocabulary features between AI-generated and human-written patent texts?
2. Do AI-generated texts have different features on the part of speech?
3. What differences in dependency distributions exist between AI-generated and human-written patent texts?
4. Do AI-generated and human-written patent texts differ in sentiment polarity?
5. What are the disparities in language model perplexity performance between AI-generated and human-written patent texts?

2 Related Work

2.1 Text Generation Detection Datasets

Recently, researchers have proposed various datasets to study identifying AI-generated texts.

- **CHEAT** [1]: The CHEAT dataset is the most comprehensive publicly available resource for detecting academic content generated by ChatGPT. It consists of 15,000 human-authored abstracts and 35,000 ChatGPT-generated

abstracts. The human-authored abstracts are sourced from the IEEE Xplore database and span a broad range of topics in computer science. The ChatGPT-generated abstracts are organized into three categories:
  i) Generating a 200-word abstract by inputting a paper title and some keywords;
  ii) Polishing a human-written abstract using ChatGPT through specific prompt templates;
  iii) Mixing refined abstracts with human-written ones by using randomly constructed masks to determine the sentences to be replaced.
- **HC3** [2] is one of the most known datasets, which includes both human-written and ChatGPT-generated texts. It gathers the responses from humans and ChatGPT to the same questions. The dataset was constructed using a prompt template to input questions into ChatGPT, adjusting the temperature parameters to ensure the generated content aligns with the intended answer, which comprises Chinese and English branches. Specifically, the English branch, HC3-en, includes 58K human answers and 26K ChatGPT answers across 24K questions, primarily sourced from the ELI5 dataset, WikiQA dataset, etc. The Chinese branch, HC3-zh, includes 22K human answers and 17K ChatGPT answers covering more domains, such as medicine, finance, psychology, law, etc. The human answers in HC3-zh are sourced from WebTextQA, BaikeQA, etc.

There are other related datasets including Turing Bench [3], GROVER [4], TweepFake [5], MGTBench [6], ArguGPT [7], DeepfakeTextDetect [8], M4 [9], Scientific-articles Benchmark [10]. Unfortunately, the above datasets do not involve patent texts. However, the majority of existing detection datasets are limited to a specific domain, such as academic papers, financial news, Wikipedia, or certain question-and-answer platforms. Different from the above content, patent content contains more detailed and professional technical descriptions, often uses repetitive terminology, adheres to a specific format, and generally features complex sentence structures. These features may make patent content distinct from general text content. This paper proposes a dataset about patent abstracts to explore the gap between AI-generated and human-written patent texts.

2.2 Text Generation Detection Methods

Existing methods for detecting AI-generated texts can be broadly divided into two categories: metric-based and model-based methods.

Generally, metric-based methods utilize a pre-trained model to extract the distinguishing features of the input texts. For instance, GLTR [11] is a classical tool for detecting text generation based on probability ranking. DetectGPT [12] detects AI-generated texts by examining the curvature of the probability function for a given text. It originates from their finding that AI-generated texts often fall within the negative curvature region of the log probability function, while human-written texts hardly do. Metric-based detectors also includes Log-Likelihood [13], Log-Rank [12], Entropy [11], LRR and NPR [14], etc.

Model-based approaches are typically trained on a corpus containing both human-written and AI-generated texts, enabling the classification model to distinguish the texts generated by AI. For example, Guo et al. [2] proposed a RoBERTa-based detector to distinguish human-written texts from ChatGPT-generated texts. The detector was fine-tuned using the HC3 dataset. The authors provide two ways to train the detector. The first one only leverages the pure answer texts, and the second one leverages the question-answer text pairs to train the model jointly. The experimental results indicate that training with question-and-answer pairs leads to superior performance, potentially due to the richer semantic information encapsulated within these pairs. Additionally, Amrita et al. [15] explored the effectiveness of ChatGPT as a detector, trained on publicly available datasets TuringBench [3], NeuralNews [16], and TweepFake [5]. Experimental results indicate that ChatGPT frequently misclassifies AI-generated texts as human-written texts. They indicated that AI-generated texts typically possess greater fluency and consistency, which makes them superficially resemble human-written texts. Other model-based methods also include ConDA [17]. However, these methods may suffer from the issue of generalization.

3 A Novel Dataset for Patent Texts

To explore the difference between AI-generated and human-written patent texts, we constructed a novel dataset called Patent Abstract for Detection (PAD), which contains Chinese and English patent texts covering four domains: artificial intelligence, biomedicine, machinery manufacturing, and electrical engineering.

Our original patent data comes from Google Patents[1]. We chose the granted patents from 2019 to 2021 when LLMs, in particular ChatGPT, are not used universally. We put each patent claim to an LLM, ChatGPT3.5-Turbo, with appropriate temperature parameters to generate an abstract. We then store the abstract text produced by the model alongside its corresponding human-authored patent abstract as paired entries in the dataset. Table 1 shows the quantitative features of the dataset.

Table 1. Details of the PAD dataset

Domain	*English	*Chinese
arti	9118	17827
bio	7613	12508
elec	10581	18694
mech	19746	18873
ALL	47058	67902

---

[1] https://patents.google.com (Accessible on 10 Sep 2024).

4 Experiments

Next, we give the experimental analysis performed on our dataset to explore the differences between patent abstracts generated by ChatGPT and by humans.

4.1 Vocabulary Feature Analysis

Next, we analyze the quantitative characteristics of the collected corpus, including the average length (AvgLen), the total number of words (T-words), the number of unique words (U-words), and word density. Word density is defined as the ratio of the unique word number to the total number of words. As shown in Table 2, the average and total scales of patent abstracts generated by ChatGPT are greater than those of human-written abstracts. However, the number of unique words in human-written patent abstracts is slightly lower than that of ChatGPT-generated abstracts. Consequently, the word density of human-written patent texts is higher, indicating that humans tend to employ a more diverse vocabulary in their expressions.

Table 2. Comparison of quantitative characteristics in PAD

	*English				*Chinese			
	AvgLen	U-words	T-words	Density	Avg.len	U-words	T-words	Density
human	259.1	104421	12197275	**0.0086**	221.3	98298	8531004	**0.0115**
ChatGPT	574.0	112300	270133111	0.0042	335.4	113475	12979049	0.0087

4.2 Part-of-Speech Analysis

We utilize the open-source Python library SpaCy[2] to analyze patent texts generated by ChatGPT and written by humans. The occurrence characteristics of different part-of-speech (POS) tags are compared, as presented in Fig. 1.

---

[2] We used the en_core_web_sm model for English and the zh_core_web_sm model for Chinese.

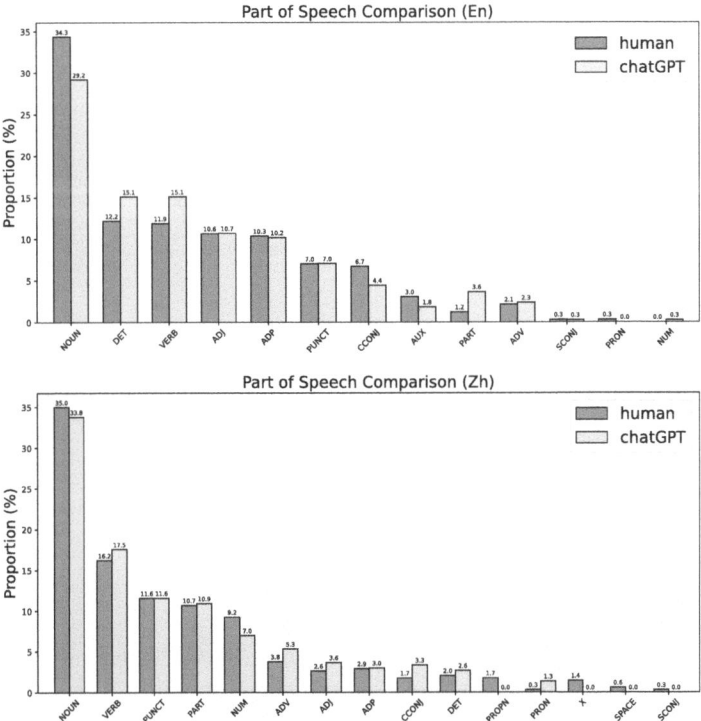

Fig. 1. Comparison of part-of-speech distribution between ChatGPT-generated and human-written patent abstracts (above: English; below: Chinese). The results are sorted descendingly by the ratios of human-written patent abstracts. "X" represents the part of speech not to be specifically classified.

Both the Chinese and English patent abstracts have high proportions of nouns (NOUN) and verbs (VERB). The dominance of nouns indicates that proper nouns are used to describe technologies. The high proportion of verbs demonstrates the dynamic characteristics of inventions in patents. In the English texts, humans use nouns 17.5% higher than ChatGPT, while in Chinese use 3.6% higher. However, in the English texts, the verb usage proportion of humans is 26.9% lower than ChatGPT, and in Chinese, it is 8% lower than ChatGPT. This is because ChatGPT often describes technical processes in detail, utilizing more verbs to maintain its clarity.

4.3 Dependency Relation Analysis

We utilize dependency relation analysis to reveal the relationships between words within patent text sentences. The experiments were also conducted using the open-source library SpaCy. Figure 2 shows the results of dependency relations in the English and Chinese patent texts.

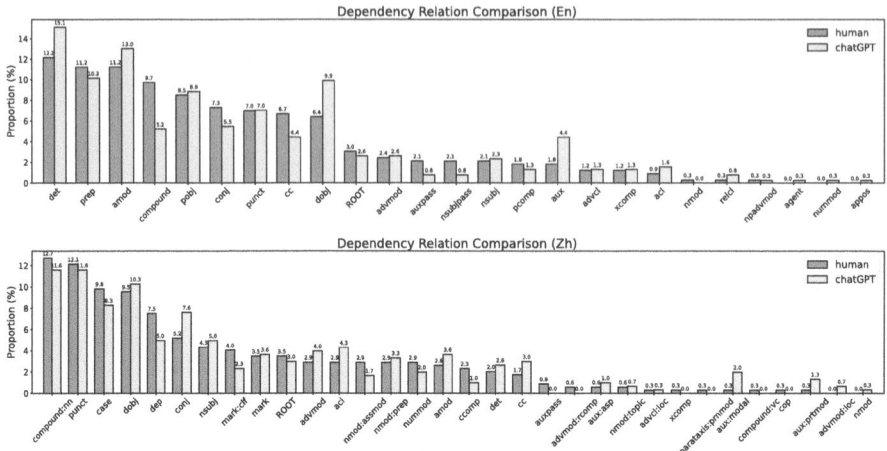

Fig. 2. Comparison of dependency relation ratios between patent abstract texts generated by humans and ChatGPT (above: English; below: Chinese). The results are sorted descendingly by the ratios of human-written texts.

In the English texts, both human-written and ChatGPT-generated texts have high percentages of determiner (det), prepositional modifier (prep), and adjectival modifier (amod) relationships. Each relationship accounts for more than 10%. This is because patent texts must describe technical details and features, which inherently require using more adjectives to modify nouns, using prepositions to modify noun phrases and provide supplementary explanations, and using qualifiers to define technical terms and contexts within the patent text precisely. In particular, humans use 86.5% more compound relationship structures, which indicates that human experts focus more on expression diversity. In contrast, for the direct object (dobj) and auxiliary (aux) relationships, the usage of ChatGPT is 54.7% and 144.4% higher than that of human-written texts. This is because ChatGPT often uses more straightforward sentence structures for clarity.

In the Chinese texts, both abstracts have high percentages of compound noun (compound:nn) and punctuation (punct) relationships, each of which accounts for more than 10%. Notably, the proportion of the compound noun relationship used by humans is 9.5% higher than ChatGPT. Additionally, humans use case marking (case) relationships 18.1% more frequently and dependent (dep) relationships 50% more frequently than ChatGPT. This indicates that human-written patent texts exhibit more complex sentence structures. However, ChatGPT uses the conjunct (conj) relationship 46.2% higher, and the adjectival clause (acl) relationship 48.3% higher than humans. This is because ChatGPT prefers concise and smooth sentence structures, leading to more conjunctions for parallel sentences. This contrasts with human writing, which frequently employs compound sentences, which results in a stronger inter-sentential correlation and makes the texts more complicated.

4.4 Sentiment Analysis

Generally, human emotions are inherently reflected in natural language. Since ChatGPT-generated texts are trained on human-written data, we aim to explore the emotional differences between human-written and model-generated texts. For sentiment analysis of patent texts, we employed a fine-tuned multilingual model, XLM-RoBERTa-base[3] in the experiments. The comparison results of the sentiment distributions are shown in Fig. 3.

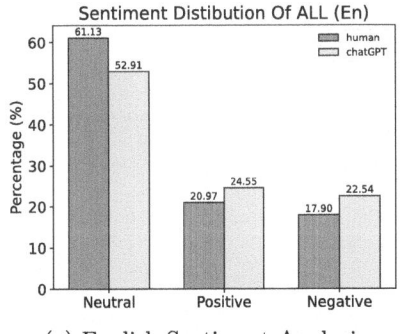
(a) English Sentiment Analysis

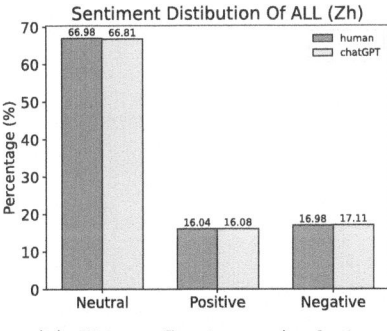
(b) Chinese Sentiment Analysis

Fig. 3. Proportions of neutral, positive, and negative emotional words

In Fig. 3, we find that the proportion of neutral emotions is the largest for both humans and ChatGPT, which account for more than 50% in both the Chinese and English texts. The results are in line with the objectivity of patent texts. More specifically, patent texts are descriptive which aim to accurately and objectively detail technological inventions. Also, these texts adhere to strict formatting and linguistic standards, which demand neutrality and professionalism in language. Both human-written and ChatGPT-generated patent texts conform to these conventions.

Notably, in the English texts, the proportion of neutral emotions in human-written content is 15.5% higher. However, the proportion of positive emotions is 17.1% lower, and that of negative emotions is 25.9% lower. The reason may be that humans strictly follow the patent writing specifications, while ChatGPT prefers the overall fluency of expression when generating patent texts.

4.5 Language Model Perplexity Analysis

The perplexity (PPL) is often used as a metric to evaluate the performance of a language model. A lower PPL indicates that the language model is more confident in its predictions and is considered a better model. The training of the

---

[3] https://huggingface.co/cardiffnlp/twitter-xlm-roberta-base-sentiment (Accessible on 10 September 2024).

language model is carried out on a large-scale text corpus, which can be considered that it has learned some common language patterns and text structures. Therefore, PPL can be used to measure how well the text conforms to common characteristics. We use the open source GPT-2 small[4] (Wenzhong-GPT 2-110M[5] in Chinese) model to calculate the PPLs of the constructed PAD dataset. The perplexity of a sentence is defined as the exponent of the negative average log-likelihood of its words under a language model. Formally,

$$\text{Perplexity}(S) = 2^{-\frac{1}{N}\sum_{i=1}^{N}\log_2 P(w_i)} \quad (1)$$

where N is the total number of words in the sequence S, $P(w_i)$ is the probability of the i-th word output by a given language model.

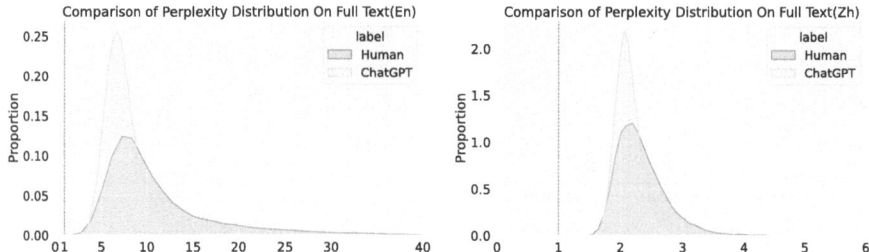

(a) PPL Distributions of the English Texts (b) PPL Distributions of the China Texts

Fig. 4. Comparison of perplexity in the English and Chinese texts. The horizontal axis represents the perplexity value, while the vertical axis represents the proportion of different perplexity values.

Figure 4 illustrates the results of the PPL distributions of human-written and ChatGPT-generated texts. It can be observed that the ChatGPT-generated texts have a relatively lower PPL compared to human-written texts. Also, the PPL values of ChatGPT-generated texts are more concentrated in the lower regions. This is because ChatGPT is good at capturing common patterns and structures in the texts, which leads to ChatGPT generating patent abstracts based on several patterns learned from the training corpora. Against, the writing of humans is uncertain in determining what words and what sentences to follow, which makes the human-written texts individual. Therefore, the human-written texts have higher PPL values and exhibit a long-tailed distribution.

5 Discussion

In this paper, we conducted an experimental exploration for several research questions, aiming to understand the differences between patent abstracts gener-

---

[4] https://huggingface.co/gpt2 (Accessible on 10 September 2024).
[5] https://huggingface.co/IDEA-CCNL/Wenzhong-GPT2-110M (Accessible on 10 September 2024).

ated by an LLM and by humans. From the vocabulary perspective, human-written patent abstracts typically have a larger vocabulary than ChatGPT-generated patent abstracts. This reflects that humans are more creative and utilize a greater variety of synonyms and diverse expressions when writing patent abstracts. Furthermore, humans possess more expert knowledge of the patent domain, while ChatGPT depends on its excellent summarization ability obtained from its training on the vast of corpora.

From the perspectives of parts of speech and dependent relations, humans use more nouns (NOUN), fewer verbs (VERB), fewer determiners (DET), and have more compound relationships (compound, compound:nn), fewer direct object relationships (dobj), and fewer determiner (det) relationships. This is because humans use more complex noun phrases and compound relationships to increase information density. Additionally, the writing specifications of patent texts require humans not to use demonstrative pronouns. On the contrary, ChatGPT focuses more on the fluency and readability of the generated texts and uses more demonstrative pronouns to make the texts shorter. Furthermore, ChatGPT-generated texts contain more verbs and determiners to describe technology details.

Regarding emotional polarity, for English patent texts, humans show more neutral, less positive, and fewer negative emotions, due to the objectivity requirements of patent writing. Whereas, ChatGPT uses different generation strategies, instead of pursuing a neutrality generation strategy.

Finally, from the perspective of perplexity, human-written abstracts have relatively higher perplexity values with a decentralized distribution, which indicates the creativity and individuality of human beings. On the other hand, ChatGPT may generate patent abstracts based on a latent common framework and avoid highly confusing statements to guarantee the smoothness and readability of the generated texts. It is easier for another language model to predict the words given sentence prefixes.

6 Conclusion

In this paper, we explore the differences between LLM-generated and human-written patent abstracts from five perspectives in natural language processing. The experimental results show their quantitative distinctions in these metrics. Through experimental analysis, it was found that the AI-generated and human-written patent abstracts differ in wording preferences, sentence structures, and perplexities, and have a similar distribution of emotional polarity.

Furthermore, the purpose of this exploration is to fill the research gap of lacking an AI generation detection dataset for patent texts. On the other hand, we hope our attempt contributes to building a better AI-generated patent text detector by further taking the found differences into account. Second, we only evaluate the texts generated by ChatGPT as it is a representative of LLMs. It would be interesting to explore the patent texts generated by other LLMs.

References

1. Yu, P., Chen, J., Feng, X., Xia, Z.: Cheat: a large-scale dataset for detecting ChatGPT-written abstracts. arXiv preprint arXiv:2304.12008 (2023)
2. Guo, B., Zhang, X., Wang, Z., et al.: How close is ChatGPT to human experts? comparison corpus, evaluation, and detection. arXiv preprint arXiv:2301.07597 (2023)
3. Uchendu, A., Ma, Z., Le, T., Zhang, R., Lee, D.: TURINGBENCH: a benchmark environment for turing test in the age of neural text generation. In: EMNLP, pp. 2001–2016 (2021)
4. Zellers, R., et al.: Defending against neural fake news, pp. 9051–9062 (2019)
5. Fagni, T., Falchi, F., Gambini, M., Martella, A., Tesconi, M.: TweepFake: about detecting deepfake tweets. PLoS ONE **16**(5), 1–16 (2021)
6. He, X., Shen, X., Chen, Z., Backes, M., Zhang, Y.: Mgtbench: benchmarking machine-generated text detection. arXiv preprint arXiv:2303.14822 (2023)
7. Liu, Y., Zhang, Z., Zhang, W., et al.: ArguGPT: evaluating, understanding and identifying argumentative essays generated by gpt models. arXiv preprint arXiv:2304.07666 (2023)
8. Li, Y., et al.: Mage: machine-generated text detection in the wild. In: ACL, pp. 36–53 (2024)
9. Wang, Y., et al.: M4: multi-generator, multi-domain, and multi-lingual black-box machine-generated text detection. In: EACL, pp. 1369–1407 (2024)
10. Mosca, E., Abdalla, M.H.I., Basso, P., Musumeci, M., Groh, G.: Distinguishing fact from fiction: a benchmark dataset for identifying machine-generated scientific papers in the llm era. In: TrustNLP 2023, pp. 190–207 (2023)
11. Gehrmann, S., Strobelt, H., Rush, A.M.: GLTR: statistical detection and visualization of generated text. In: ACL, pp. 111–116 (2019)
12. Mitchell, E., Lee, Y., Khazatsky, A., Manning, C.D., Finn, C.: Detectgpt: zero-shot machine-generated text detection using probability curvature. In: ICML, pp. 24950–24962. PMLR (2023)
13. Solaiman, I., et al.: Release strategies and the social impacts of language models. arXiv preprint arXiv:1908.09203 (2019)
14. Su, J., Zhuo, T.Y., Wang, D., Nakov, P.: Detectllm: leveraging log rank information for zero-shot detection of machine-generated text. In: EMNLP, pp. 12395–12412 (2023)
15. Bhattacharjee, A., Liu, H.: Fighting fire with fire: can chatgpt detect ai-generated text? ACM SIGKDD Explorat. Newsl. **25**(2), 14–21 (2024)
16. Tan, R., Plummer, B., Saenko, K.: Detecting cross-modal inconsistency to defend against neural fake news. In: EMNLP, pp. 2081–2106 (2020)
17. Bhattacharjee, A., Kumarage, T., Moraffah, R., Liu, H.: Conda: contrastive domain adaptation for ai-generated text detection. In: IJCNLP, pp. 598–610 (2023)

MRI-CT Brain Image Registration Based on SuperPCA and Block-Matching Algorithm

Wannan Zhang[✉]

School of Computer Science, Huainan Normal University, Huainan 232038, China
1508622762@qq.com

Abstract. Existing medical image registration algorithms generally have problems such as manual intervention and long registration time. In this paper, we propose a novel method based on superpixelwise principal component analysis (SuperPCA) and block-matching for registration between magnetic resonance imaging (MRI) and computed tomography (CT) brain images. Firstly, we apply SuperPCA to align MR and CT images roughly. We use SuperPCA to compute initial parameters which can prevent from falling into a local optimum. Next, registration is refined via a free-form deformation which is based on an block-matching algorithm. Here we adopt robust self-similarity descriptor as a similarity metric in both stages of registration, which can extract modality-invariant neighbourhood representations separately. Experimental results have shown that the performances of the obtained images are improved by comparing with the state-of-the-art methods, demonstrating potentials of the proposed method to be applied for brain tumor radiation therapy.

Keywords: Brain mage registration · SuperPCA · MR · CT

1 Introduction

In medical diagnosis and treatment, images are used more and more widely. The current medical imaging technology is developing rapidly, and there are more and more types of medical images. Each imaging method obtains certain information from the human body, and different imaging methods can provide different aspects of the same patient's information [1]. For example, CT uses X-rays to pass through the human body to be attenuated. Based on a large number of X-ray projection data, a mathematical reconstruction method is used to obtain a two-dimensional or three-dimensional X-ray density (tissue attenuation coefficient) distribution image [2,3]. Its characteristic is that the spatial resolution is extremely high, which can clearly reflect the bone characteristics, but the effect of reflecting the soft tissue information is not good.

MRI utilizes the relaxation of hydrogen ions magnetized by human tissue in a static magnetic field after being excited by an external radio frequency magnetic field for imaging, and can obtain tomographic images of various angles and orientations. Because MRI selectively images the distribution of protons, it is an ideal method for soft tissue imaging [4]. It can provide detailed structural images with high spatial resolution. The disadvantage is that the imaging time is long, it is not sensitive to calcification points, and geometric distortion will occur when interfered by external magnetic fields [5]. CT can provide clear bone tissue information, and MRI can provide clear soft tissue information. Especially for organs with complex structures such as the brain, comprehensive diagnostic information cannot be obtained from a certain image alone [6]. Images of different modalities often need to be registered and fused to obtain more information.

In the past, clinical diagnosis largely relied on the doctor's experience. When reading the film, the doctor would artificially analyze and synthesize various information in the mind to judge the location and severity of the lesion. This analysis method is subjectively affected, and its accuracy is questionable. If a variety of images can be registered and fused with the help of computer image processing technology, it will undoubtedly increase the accuracy and reliability of diagnosis and assist doctors in treatment [7]. Medical image registration refers to seeking one spatial transformation, so that the corresponding point on the other medical image is consistent in space [8]. This consistency means that the same anatomical point on the human body has the same spatial position on the two matching images. Medical image registration has very important clinical application value.

In this paper, we propose a novel method based on SuperPCA and block-matching for registration between MR and CT brain images. Firstly, we apply SuperPCA to align MR and CT images roughly. We use SuperPCA to compute initial parameters which can prevent from falling into a local optimum. Next, registration is refined via a free-form deformation which is based on an improved block-matching algorithm. Here we adopt robust self-similarity descriptor as a similarity metric in both stages of registration, which can extract modality-invariant neighbourhood representations separately.

2 Methodology

2.1 SuperPCA for Coarse Registration

Instead of the original PCA algorithm, SuperPCA is used for computing a linear transformation to map a high dimensional space into a lower dimensional space has been used for coarse registration between CT and MR images [9]. Firstly, PCA is used to obtain the first principal component of the image, and then the entropy rate superpixel (ERS) is used to obtain the superpixel segmentation. Set a different number of superpixels (Sc) to divide the image to obtain multiple segmentation maps of different scales. Specifically, the first principal component (If) of the image is divided into 2C+1 scales, and the number of pixels in the c-th scale is:

$$S_c = (\sqrt{2})^2 S_f, c = 0, \pm 1, \pm 2, \ldots, \pm C \qquad (1)$$

Sf is the basic number of super pixels set based on experience. For the segmented multi-scale image, the final decision fusion strategy is adopted, that is, the category label of each pixel in different scales is fused [10]. In this article, the majority vote decision fusion strategy is adopted, as shown in the following formula:

$$I = \underset{i\in\{1,2,...,G\}}{\operatorname{argmax}} N(i), N(i) = \sum_{j=1}^{2C+1} \alpha_j I(I_j = i) \quad (2)$$

Here I is one of the G possible categories, $N(i)$ represents the number of times that category i is predicted, and I is the indicator function α_j representing the voting weight of the j-th classifier.

Compared with the traditional PCA model, SuperPCA has four main characteristics: (1) Unlike the traditional PCA method based on the entire image, SuperPCA considers the diversity of different homogeneous regions, that is, different regions should have different predictions. (2) Most traditional feature extraction models cannot directly use spatial information, while SuperPCA can incorporate spatial context information into the unsupervised dimensionality reduction of superpixel segmentation. (3) Due to the uniformity of the regions obtained by superpixel segmentation, SuperPCA can extract potential low-dimensional features even under noise. (4) Although SuperPCA is an unsupervised method, it can achieve competitive performance compared with supervised methods. The resulting characteristics are distinguishable, compact and noise resistant, thereby improving classification performance.

2.2 Registration Registration Using Block-Matching

Block matching algorithm [11] is a common method in motion estimation, image filtering, and video compression. This method divides an image into image blocks of equal size, selects one of the image blocks as a query block to match other image blocks [12], and finds out The query block has K matching blocks with similar characteristics to form a block matching array.

Generate an image block of size $k \times k$ (k is an odd number) in the neighborhood of the center pixel, and use it as the basic unit for extracting the center coordinates, and use block matching to search for structures with a certain degree of similarity [13]. The structural image I generates an image block at the center pixel, where P represents the query block, and Q represents the image block matching the query block. When the similarity between P and Q is less than a certain threshold, it is deemed that the two are similar to belong to the same group. In the search process, the distance of the image block is represented by $d(P,Q)$:

$$d(P,Q) = \frac{\parallel P - Q \parallel^2}{k \times k} \quad (3)$$

In the formula, d represents the distance between two image blocks, the numerator represents the modulus of the query block and the matching block, and the denominator represents the size of the image block. The matching blocks whose distance from the query block P is less than a certain threshold form a grouping SP:

$$S_p = \{Q \in I | d(P,Q) \leq \tau\} \tag{4}$$

τ represents the similarity threshold between the matching block and the query block. When the difference between the two images is less than τ, the two images are deemed to have high similarity and have the same center coordinates.

2.3 Registration Registration Using Block-Matching Similarity Metric : Robust Self-similarity Descriptor

In this work, we adopt robust self-similarity descriptor (RSSD) to measure the resemblance between MR and CT images. RSSD estimates the self-similarity based on the patch similarity [14]. It can be defined by Eq. 5 as follows.

$$RSSD = \sum_{n=1}^{N-1} s(|g_n - g_c|) \cdot 2^{mod(n-Dd,N)} \tag{5}$$

Each $s(|g_n - g_c|)$ value is encoded into binary code as follows:

$$s(|g_n - g_c|) = \begin{cases} 1, |g_n - g_c| > \sigma \\ 0, |g_n - g_c| \leq \sigma \end{cases} \tag{6}$$

where $mod(n - Dd, N)$ represent $n - Dd$ modulo N.

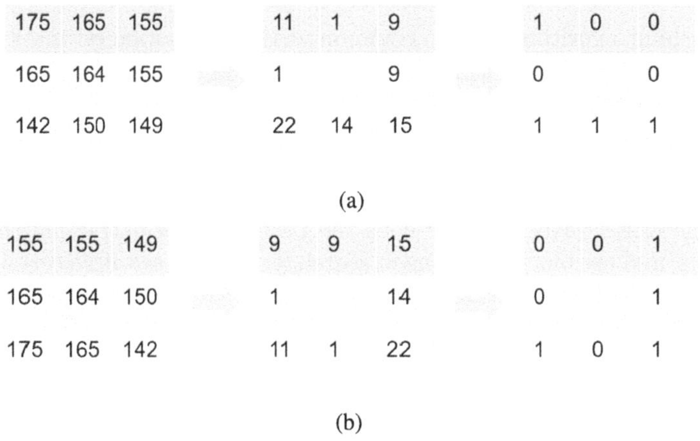

175	165	155		11	1	9		1	0	0
165	164	155			1	9		0		0
142	150	149		22	14	15		1	1	1

(a)

155	155	149		9	9	15		0	0	1
165	164	150			1	14		0		1
175	165	142		11	1	22		1	0	1

(b)

Fig. 1. Illustration of the RSSD with the 00 direction (a) and the 900 direction (b) in 2D image

In Fig. 1, we use a 3×3 patch in a 2D image to explain the RSSD method. The numbers [175 165 165; 165 164 155; 142 150 149] are the intensity values within the patch. Firstly, the distance between an individual neighboring pixel and the central pixel, $|g_n - g_c|$, is evaluated. Secondly, the standard deviation σ of the intensity values of the neighboring pixels is calculated and used as the threshold value. Here, in Fig. 1, the threshold value σ is 9.87. If the distance $|g_n - g_c|$ is greater than σ, the binary value 1 is assigned to the corresponding neighbor location. Otherwise, the binary value 0 is assigned instead. By repeating the process of encoding each neighboring pixel, we obtain the value of binary pattern 10100011. In determining the value of Dd, the position of the pixel with the maximum value of $|g_n - g_c|$ is returned, which is 7 in the case of the example in Fig. 1a. The intensity value of the pixel corresponding to that position is 156. When the image is rotated by 900 counter-clockwises, Dd is still the position of the pixel providing the maximum value of $|g_n - g_c|$, (shown as the gray cell in Fig. 1a. The weighted term of the binary result is assigned by $2mod(n - Dd, N)$. In Fig. 1a and 1b, it can be seen that the binary result is not varied when its corresponding patch is rotated.

3 Methodology Experimental Results and Analysis

To show the performance of our proposed algorithm, the experiments are carried out on brain images from 11 patients. All methods were carried out in accordance with relevant guidelines and regulations and all experimental protocols were approved by Xiangya Hospital. The x-rays were collected by Xiangya Hospital as part of routine patient treatment. We confirm that informed consent was obtained from all subjects. The sizes of MR and CT images are 128×128 and 256×256 pixels, respectively. We use anatomical landmarks to evaluate quantitatively the accuracy of the proposed method. Then, the mean Euclidean distance (±standard deviation) between these pairs is computed. We compare the results obtained by our method with those obtained by Andrei [15] and Mehrabian et al. [16]. The registration results are exhibited in Fig. 2(c), Fig. 2(d) and Fig. 2(e) respectively.

Average distance errors are reported in Table 1. The average distance error is 0.72±0.51 mm with the proposed method. It is less than 0.99±0.61 mm with Ji et al.and 1.32±0.66 mm with Mehrabian et al. [16]. The reason of high accuracy registration lies in that SuperPCA algorithm is performed to compute initial parameters which prevents from falling into a local optimum. Unlike the traditional PCA method based on a whole image, SuperPCA takes into account the diversity in different homogeneous regions. Meanwhile, the mean processing time of our method is also the least with 12.8 s, less than 14.3 s with Andrei and 13.1 s with Mehrabian et al.

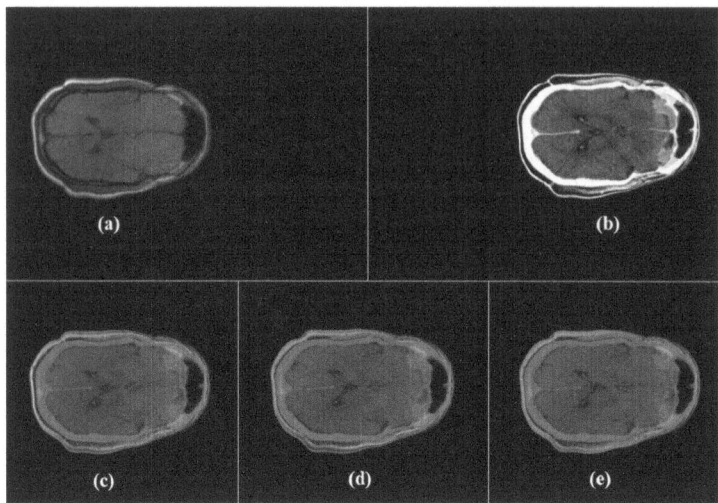

Fig. 2. MR/CT registration results on a slice of brain images based on different methods. (a) Original MR image, (b) original CT image, (c) Andrei [15], (d) Mehrabian et al. [16], (e) the proposed method.

Table 1. Mean errors. Values represent the mean and standard deviation of the spatial deviation and the distance error. We computed Euclidean distance between landmarks selected on the initial images. Values in square brackets represent the [min, max] range of error.

Modality	Algorithm	Δ x (mm)	Δ y (mm)	Distance Error (mm)	Time (s)
MR/CT	Andrei [15]	0.56±0.35	0.82±0.38	0.99±0.61	14.3
		[0.02,2.92]	[0.05,2.43]	[0.28,2.63]	
	Mehrabian et al. [16]	0.81±0.54	1.04±0.41	1.32±0.66	13.1
		[0.04,3.11]	[0.09,2.98]	[0.25,2.51]	
	Proposed	0.34±0.23	0.59±0.19	0.72±0.51	12.8
		[0.03,1.16]	[0.02,1.44]	[0.22,1.89]	

4 Conclusion

In this paper, we have proposed a new method for registration between CT and MR images, which is crucial for brain tumor radiation therapy. Firstly, we apply SuperPCA to align MR and CT images roughly. We use SuperPCA to compute initial parameters which can prevent from falling into a local optimum. Next, registration is refined via a free-form deformation which is based on an improved block-matching algorithm. Here we adopt robust self-similarity descriptor as a similarity metric in both stages of registration, which can extract modality-invariant neighbourhood representations separately. Experimental results show that the average distance error is reduced to 0.72±0.51 mm compared with the other state-of-the-art registration methods, and the mean processing time of our method is also the least with 12.8 s. The reason of high accuracy registration lies in that SuperPCA algorithm is performed to compute initial parameters which prevents from falling into a local optimum. Unlike the traditional PCA method based on a whole image, SuperPCA takes into account the diversity in different homogeneous regions. The experiments demonstrate the potentials of our method to be applied for brain tumor radiation therapy.

References

1. Reaungamornrat, S., et al.: MIND demons: symmetric diffeomorphic deformable registration of MR and CT for image-guided spine surgery. IEEE Trans. Med. Imaging **35**(11), 2413–2424 (2016). https://doi.org/10.1109/TMI.2016.2576360
2. Commandeur, F., et al.: MRI to CT prostate registration for improved targeting in cancer external beam radiotherapy. IEEE J. Biomed. Health Inf. **21**(4), 1015–1026 (2016). https://doi.org/10.1109/JBHI.2016.2581881
3. Swierczynski, P., Papiez, B.W., Schnabel, J.A., Macdonald, C.B.: A level-set approach to joint image segmentation and registration with application to CT lung imaging. Comput. Medical Imaging Graph. **65**, 58–68 (2018). https://doi.org/10.1016/j.compmedimag.2017.06.003
4. Yamaguchi, Y., et al.: Three modality image registration of brain SPECT/CT and MR images for quantitative analysis of dopamine transporter imaging. In: Gimi, B., Król, A. (eds.) Medical Imaging 2016: Biomedical Applications in Molecular, Structural, and Functional Imaging, San Diego, California, United States, 27 February–3 March 2016, SPIE Proceedings, vol. 9788, p. 97881V. SPIE (2016). https://doi.org/10.1117/12.2217384
5. Jiang, J., et al.: Superpca: a superpixelwise PCA approach for unsupervised feature extraction of hyperspectral imagery. CoRR (2018). http://arxiv.org/abs/1806.09807
6. Reaungamornrat, S., et al.: MIND demons for mr-to-ct deformable image registration in image-guided spine surgery. In: III, R.J.W., Yaniv, Z.R. (eds.) Medical Imaging 2016: Image-Guided Procedures, Robotic Interventions, and Modeling, San Diego, California, United States, 27 February–3 March 2016, SPIE Proceedings, vol. 9786, p. 97860H. SPIE (2016). https://doi.org/10.1117/12.2208621
7. Litjens, G., Debats, O., van de Ven, W., Karssemeijer, N., Huisman, H.: A pattern recognition approach to zonal segmentation of the prostate on MRI. In: Ayache, N., Delingette, H., Golland, P., Mori, K. (eds.) MICCAI 2012. LNCS, vol. 7511, pp. 413–420. Springer, Heidelberg (2012). https://doi.org/10.1007/978-3-642-33418-4_51
8. Harale, A.M., Bairagi, V.K., Boonchieng, E., Bachute, M.R.: Nodules detection in lungs CT images using improved YOLOV5 and classification of types of nodules by CNN-SVM. IEEE Access **12**, 140456–140471 (2024). https://doi.org/10.1109/ACCESS.2024.3466292
9. Jiang, J., Ma, J., Chen, C., Wang, Z., Cai, Z., Wang, L.: Superpca: a superpixelwise PCA approach for unsupervised feature extraction of hyperspectral imagery. IEEE Trans. Geosci. Remote. Sens. **56**(8), 4581–4593 (2018). https://doi.org/10.1109/TGRS.2018.2828029

10. Chu, Q., Zhan, Y., Guo, F., Song, M., Yang, R.: Automatic 3d registration of CT-MR head and neck images with surface matching. IEEE Access **7**, 78274–78280 (2019). https://doi.org/10.1109/ACCESS.2019.2903123
11. Banerjee, J., et al.: Multiple-correlation similarity for.block-matching based fast CT to ultrasound registration in liver interventions. Med. Image Anal. **53**, 132–141 (2019). https://doi.org/10.1016/j.media.2019.02.003
12. Chen, W., et al.: MR-CT image fusion method of intracranial tumors based on res2net. BMC Med. Imag. **24**(1), 169 (2024). https://doi.org/10.1186/s12880-024-01329-x
13. Azimbagirad, M., et al.: Robust semi-automatic segmentation method: an expert assistant tool for muscles in CT and MR data. Comput. methods Biomech. Biomed. Eng. Imaging Vis. **11**(7), 2301403 (2024). https://doi.org/10.1080/21681163.2023.2301403
14. Wei, D., et al.: Synthesis and inpainting-based MR-CT registration for image-guided thermal ablation of liver tumors. CoRR (2019). http://arxiv.org/abs/1907.13020
15. Iantsen, A.: Automated tumor segmentation in multimodal PET/CT/MR imaging. (Segmentation automatique de tumeurs en imagerie multimodale TEP/TDM/IRM). Ph.D. dissertation, University of Western Brittany, Brest, France (2022). https://tel.archives-ouvertes.fr/tel-04573020
16. Mehrabian, H., Richmond, L., Lu, Y., Martel, A.L.: Deformable registration for longitudinal breast MRI screening. J. Digit. Imaging **31**(5), 718–726 (2018). https://doi.org/10.1007/s10278-018-0063-1

Multi-teacher Knowledge Distillation with Triplet Loss for Cross-Modal Object Tracking

Yi Li[1], Lei Liu[2], Mengya Zhang[3], and Chenglong Li[1(✉)]

[1] School of Artificial Intelligence, Anhui University, Hefei 230601, China
lcl1314@foxmail.com
[2] School of Artificial Intelligence, Anhui University of Science and Technology, Hefei 232001, China
[3] School of Computer Science and Technology, Anhui University, Hefei 230601, China

Abstract. Cross-modal sensors have garnered considerable attention for their ability to seamlessly switch between visible (RGB) and near-infrared (NIR) modalities, allowing them to adapt to varying external lighting conditions. This remarkable feature enables these sensors to provide tracking capabilities under all lighting conditions. However, the distinct imaging mechanisms between these modalities result in significant variations in target appearance, posing challenges for existing tracking algorithms. While a few algorithms have addressed this issue, they often rely on complex module designs and numerous parameters. To tackle these challenges, we propose a **M**ulti-teacher **K**nowledged **D**istillation (**MKD**) algorithm. First, we design two teacher networks to learn modality-specific knowledge from data collected from different modalities. Next, we employ a feature-level knowledge distillation mechanism, using cross-modal data to adaptively transfer knowledge from the teacher networks to the student network. To enhance the stability of the distillation process and accelerate convergence, we introduce a triplet loss function to guide learning. Experimental results on the large-scale cross-modal object tracking dataset (CMOTB) demonstrate the superiority of our method across all evaluation metrics, while avoiding the need for complex module designs.

Keywords: Cross-modal Object Tracking · Knowledge Distillation · Triplet Loss · Modal Differences

1 Introduction

Visual object tracking is a fundamental task in computer vision, which involves locating a target object in subsequent video frames based on its initial position and size in the first frame. This task has garnered increasing attention due to its wide range of applications in fields such as video surveillance, intelligent transportation, autonomous driving. However, most existing methods rely on visible image sequences, which can be ineffective in low-light conditions, leading to tracking failures due to the limitations of visible imaging.

Near-infrared (NIR) imaging plays a crucial role in many surveillance cameras [10], enabling automatic switching between visible (RGB) and NIR modalities in response to changes in lighting conditions. Unlike traditional multimodal vision systems, this adaptive imaging system dynamically switches between modalities based on external lighting, ensuring only one modality is active at a time. This eliminates the need for multimodal data alignment, reducing the time and effort required, while also addressing the challenges of building complex multimodal imaging platforms [8]. This advancement significantly aids the development of visual object tracking under varying lighting conditions. However, the data captured by these two modalities exhibit substantial differences in target appearance, as illustrated in Fig. 1, posing considerable challenges for cross-modal object tracking tasks.

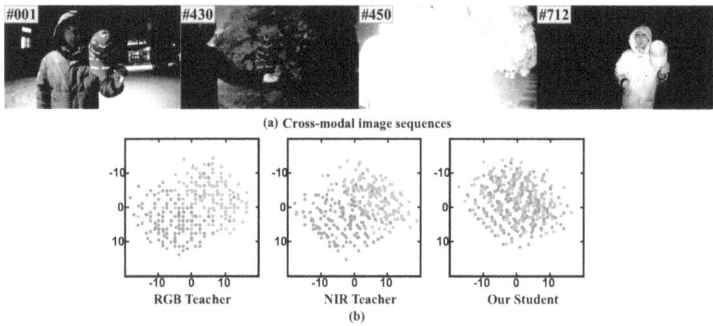

Fig. 1. (a) Illustration of the heterogeneous properties between RGB and NIR modalities. (b) Visualization of cross-modal target features in a sequence.

To address this issue, existing methods [9,11] often use separate branches to learn target-specific representations for different modalities. These methods then employ an adaptive ensemble strategy to manage the modality-agnostic cross-modal tracking process. While these approaches partially mitigate the discrepancies in cross-modal target appearance, they also introduce significant complexity during the training phase and require additional network parameters, which can negatively impact the efficiency of the tracker. In addition, these methods require a multi-stage training process, which prevents end-to-end optimization of the network.

To address the challenge of adapting existing tracking algorithms to variations in cross-modal target appearance, we propose a **M**ulti-teacher **K**nowledged **D**istillation (**MKD**) algorithm. Our approach involves constructing two teacher networks, each designed to learn modality-specific knowledge from data collected from different modalities. This knowledge is then transferred to a unified student network through a feature-level knowledge distillation mechanism, enabling the student network to effectively model cross-modal target appearances. To further improve the stability of the distillation process and accelerate convergence, we incorporate a triplet loss [14] to guide the learning. During the training stage, the dataset is divided into two categories: RGB data and NIR data. The teacher networks are trained using the corresponding data for each category. Subsequently, cross-modal data is employed to train the student network, allowing it to learn cross-modal representations from the teacher networks through feature-level knowledge distillation. This approach effectively addresses the challenge of cross-modal target appearance differences.

To enhance the stability of the student network's learning process and accelerate convergence, we incorporate a triplet loss. This loss function guides the student network to learn image features that closely resemble those of the corresponding teacher networks. During the testing phase, the student network—now capable of effectively modeling cross-modal target appearances—eliminates the need for the teacher networks trained on different modalities. Consequently, we can directly use the student network to achieve robust cross-modal tracking performance. This approach allows for stable modeling of cross-modal target appearance without requiring additional module design. As shown in Fig. 1, there are significant disparities in the distribution of target appearance features across different modalities. However, our method successfully models an effective representation of cross-modal target appearances using a unified student network, effectively mitigating the disparities in cross-modal feature distributions.

We conduct comprehensive experiments on the large-scale CMOTB dataset [9] for cross-modal object tracking. The experimental results show that our method outperforms other algorithms across all evaluation metrics, while also achieving faster tracking speeds.

The main contributions of this paper can be summarized as follows:

- We propose a novel multi-teacher knowledge distillation algorithm that enables the transfer of cross-modal knowledge from multiple teacher networks to a unified student network, allowing the student network to efficiently learn and represent cross-modal target appearances.
- A triplet loss is incorporated to enhance the stability and accelerate the convergence of the distillation learning process.
- Experimental results on the cross-modal object tracking dataset demonstrate that our approach achieves state-of-the-art tracking performance.

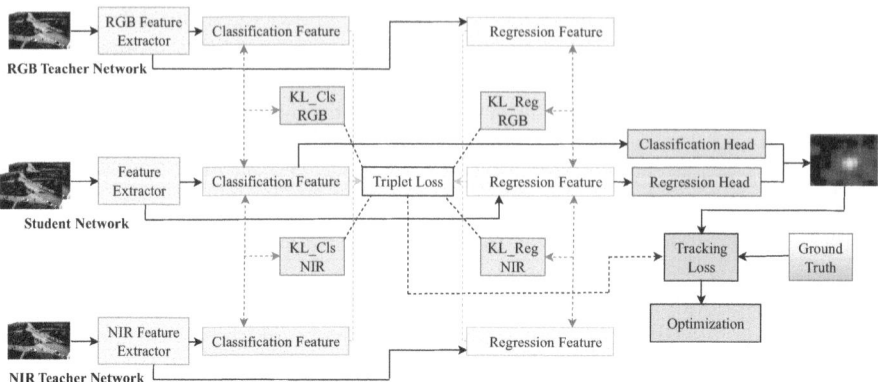

Fig. 2. Overview of our proposed cross-modal object tracking framework.

2 Method

This section begins with an overview of our proposed framework, followed by a brief introduction to the teacher and student trackers used in our approach. Finally, we delve into the details of the proposed knowledge distillation module and the triplet loss.

2.1 Overview

As shown in Fig. 2, the proposed MKD framework consists of two teacher networks, one student network, two knowledge distillation layers, and one triplet loss. Each of the two teacher networks takes an image of its respective corresponding modality as input, and uses modality-specific feature extractors and classification regression heads trained on the corresponding modal data for visual tracking under modality-specific data. Since each teacher network is pre-trained by the corresponding modal data separately, it is able to guarantee an effective target representation in the respective modality.

Similar to the structure of the teacher network, the student network consists of a single-stream network that performs feature extraction and then transmits it to the classification regression heads for visual tracking. However, the student network can handle both visible and near-infrared data as input. Despite this capability, the modal switching introduces significant differences in the target appearance across modalities, making it challenging to maintain a stable target representation during the tracking process.

To address the significant differences in cross-modal features during the feature extraction phase of single-stream student networks, our proposed MKD framework aims to enhance the learning process of the student network from two perspectives: unimodal feature extraction [16] and spatial distance constraint. Firstly, we construct two teacher networks that learn modality-specific knowledge using data collected from different modalities. This allows us to leverage cross-modal data and employ a feature-level knowledge distillation mechanism to adaptively transfer knowledge from the teacher networks to the student network. Furthermore, in order to enhance the stability of distillation learning and accelerate convergence speed, we incorporate a triplet loss to guide the distillation process. This addition helps narrow the representation gap between different modalities, ultimately enabling the student network to achieve competitive tracking results with fewer parameters and higher computational efficiency.

2.2 Teacher and Student Networks

In this section, we will describe the architectures of the employed teacher and student networks, which are both based on the recent deep RGB tracker DiMP [1].

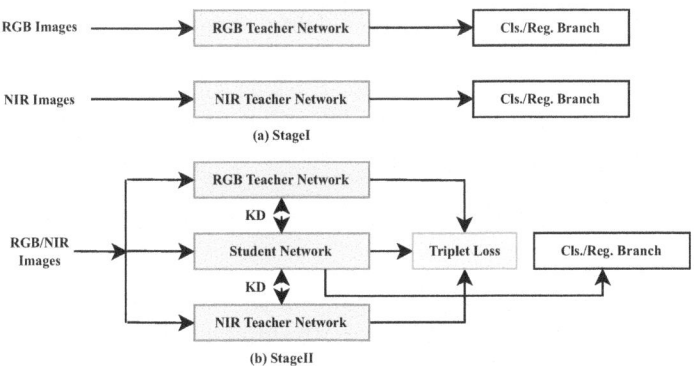

Fig. 3. The visualization of two-stage training method.

Feature Extraction. Both the teacher and student networks utilize ResNet50 [7] as the backbone for extracting RGB or NIR features. During training, knowledge distillation occurs between the corresponding modal teacher network and student network for images of different modalities, based on the modal labels of input images. Similar to the original DiMP tracker, both the teacher network and student network use features from block3 and block4 for regression head, while only features from block4 are used for classification head.

Classification and Regression. The extracted features are then passed to the classification and regression heads, which share the same architecture as those in the original DiMP. Notably, in this stage, both the student and teacher networks employ the original classification and regression heads from DiMP.

2.3 Two-Stage Learning Algorithm

We employ a two-stage learning approach for training, consisting of training the two modality-specific teacher networks and then training the student network, as shown in Fig. 3.

- Stage I: Training two modality-specific teachers. For knowledge distillation, we follow the approach of training the teachers first and then freezing their parameters during student instruction. To obtain well-performing teacher networks for RGB and NIR data respectively, we initially divide the data into two subsets based on modality type. We then use the corresponding subdatasets to train the parameters of the respective teacher networks.
- Stage II: Training the student network. In the training phase of the student network, we utilize a training set comprising cross-modal datasets that include both RGB and NIR data. The parameters of the teacher network remain constant, and based on the input modality state labels, the corresponding teacher network guide the student network. This enables the student network to acquire the knowledge required for tracking different modalities. Even in the absence of modality state information during the testing phase, the student network can effectively track across different modality.

2.4 Loss Function

To ensure stable and reliable cross-modal object tracking, we introduce two additional loss functions to the proposed MKD framework, including knowledge distillation loss and triplet loss.

Knowledge Distillation Loss. In knowledge distillation, the student network can acquire the same or even more knowledge as the teacher network through knowledge transfer. Our proposed method employs a multi-teacher approach by designing teachers for both RGB and NIR modalities. This enables joint instruction for student networks tracking across different modalities. By utilizing feature-level knowledge distillation, we mitigate the issue of cross-modal feature differences. This approach allows student networks to learn cross-modal target representations from a network of teachers with different modalities. The mean square error (MSE) loss [5] is commonly used to measure the difference between the predicted and observed values in regression problems. We consider the features extracted from the student network as predicted values and the features extracted from the corresponding teacher network as true values. Thus, we use MSE to minimize the distance between the two network features in the feature space. The specific formula is as follows:

$$L_{KD} = \frac{1}{n} \sum_{i=1}^{n} \sum_{j=1}^{n} (F_{ij}^T - F_{ij}^S)^2$$

$$s.t. F_{ij}^T \in \{F_{ij}^{T\text{-}N}, F_{ij}^{T\text{-}R}\}$$

(1)

Here, F_{ij}^T represents the feature map of the distillation layer in the teacher network, and F_{ij}^S represents the feature map of the distillation layer in the student network.

Triplet Loss. While knowledge distillation enables the student network to learn cross-modal target representations from different teacher networks, the significant differences between visible and near-infrared modalities can affect the effectiveness of distillation. To enhance the stability of distillation learning and accelerate convergence speed, we incorporate spatial constraints using triplet loss to guide the distillation process.

In triplet loss, we use features extracted from the student network as anchor samples, features extracted from the corresponding teacher network as positive samples, and features extracted from another teacher network as negative samples. This guides the student network to learn features that are more similar to those of its corresponding teacher network, resulting in more robust tracking. Figure 4 illustrates the distance constraint effect. The implementation is shown in the following equation:

$$D_{S\_R} = ||F^{S\_R} - F^{T\_R}||^2 - ||F^{S\_R} - F^{T\_N}||^2$$
$$D_{S\_N} = ||F^{S\_N} - F^{T\_N}||^2 - ||F^{S\_N} - F^{T\_R}||^2 \qquad (2)$$
$$L_{TR} = max(D_{S\_R}|D_{S\_N} + margin, 0)$$

Here, F^{S_R} represents the student network feature map when inputting an RGB image, F^{S_N} represents the student network feature map when inputting an NIR image, F^{T_R} represents the feature map of the RGB teacher, and F^{T_N} represents the feature map of the NIR teacher. The constant margin is set to 0.01. If the input image is RGB, D_{S_R} is used in the computation of L_{TR}; if the input image is NIR, D_{S_N} is used in the computation of L_{TR}.

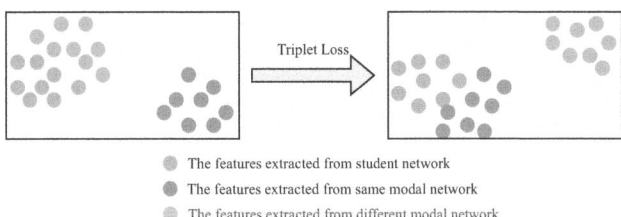

Fig. 4. The triplet loss minimizes the distance between student network and same modal teacher network.

3 Experiments

In this section, we present extensive experimental evaluations on our proposed tracker. We first introduce the experimental setups, then we verify the effectiveness of the main components of the method, and last extensive experiments are conducted to evaluate the proposed tracker against plenty of state-of-the-art trackers on CMOTB benchmark.

3.1 Experimental Setups

Implementation Details. We use the pre-trained model provided with DiMP for parameter initialization. The backbone network uses the ResNet-50 architecture with the learning rate for feature extraction set to 0.00002. In the classification branch, the model predictor learning rate is set to 0.00005, the learning rate of the model optimizer is set to 0.0005 and in the regression branch the network learning rate is set to 0.0002. The training time for 50 epochs on a GeForce RTX 3090 single card was less than 24 h.

Dataset. Our algorithm is evaluated on a large cross-modal object tracking benchmark dataset: CMOTB [9]. CMOTB includes 654 cross-modal object tracking sequences. The total number of video frames reaches over 481K, and the average video length and the maximum length of one sequence are more than 735 and 1.6K frames. This dataset contains most of the real-world challenges in cross-modal object tracking task.

Evaluation Metrics. To evaluate the performance of different trackers, we employ the widely used tracking evaluation metrics [13] including precision rate (PR), normalized precision rate (NPR) and success rate (SR).

3.2 Ablation Study

In order to verify the effectiveness of the main modules of our proposed method, we conduct the following ablation experiments on the cross-modal tracking dataset CMOTB. As can be seen from the results in Table 1, while adding knowledge distillation to the baseline tracker, the tracking performance is improved, which indicates that the baseline method is not well adapted to cross-modal target appearance changes, resulting in poor tracking performance. In contrast, the method in this paper achieves effective modeling of cross-modal features by single-branch feature extraction network through knowledge distillation to learn the representation knowledge of different modalities from the teacher network of different modalities. The tracking performance of the tracker is further improved when the triplet loss is then applied to the student network using knowledge distillation, which guides the student network to learn features that are more similar to those of its corresponding teacher network, resulting in more robust tracking.

3.3 Comparison with State-of-the-Art Trackers

To evaluate the superiority of our proposed method, we compare our method with some existing state-of-the-art trackers, including AiAtrack [6], DiMP [1],

KeepTrack [12], MArMOT-DiMP [9], MArMOT-RT-MDNet [9], MixFormer [4], Ocean [17], SiamBAN [3], TransT [2], and TrDiMP [15], the results are shown in Fig. 5. Our proposed method improves 3.6%, 3.1%, and 2.2% on the three evaluation metrics PR, NPR, and SR, respectively, compared to the benchmark tracker DiMP. Compared to the current optimal cross-modal visual tracker MArMOT, although it is only 1.1%, 0.8%, and 0.6% higher on the three evaluation metrics, our method has a simpler network structure and a smaller number of parameters, which illustrates the effectiveness of our method.

Table 1. The scores of different variants trackers. Where KD denotes modal feature distillation and TR denotes distance constraint.

KD	TR	PR(%)	NPR(%)	SR(%)
×	×	71.4	75.1	62.0
✓	×	73.4	76.8	63.3
✓	✓	**75.0**	**78.2**	**64.2**

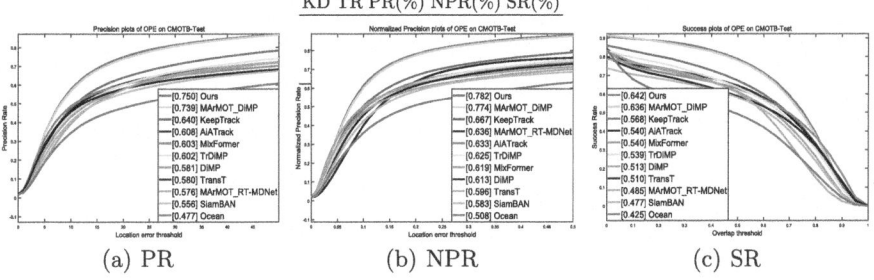

(a) PR (b) NPR (c) SR

Fig. 5. The evaluation results on public CMOTB benchmark, we compare with 10 state-of-the-art trackers. The representative scores of PR/NPR/SR are presented in the legend.

To further illustrate the effectiveness of our method, this subsection shows the visualization results of the tracker associated with the method. As shown in Fig. 6, the output results of our method are shown as red rectangular boxes and the target truth values are shown as green rectangular boxes. The visualization results show that our method is able to achieve more robust tracking compared to other methods when the tracker undergoes mode switching.

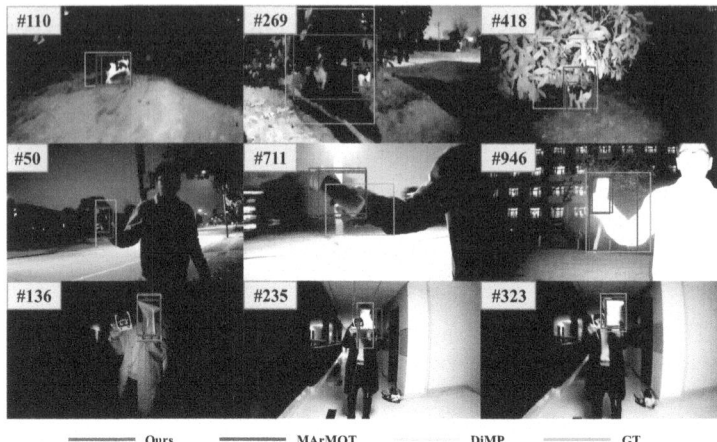

Fig. 6. A comparison of our approach with state-of-the-art trackers in facing the challenge of modality switch, including DiMP and MArMOT. (Color figure online)

4 Conclusion

In this paper, we propose a simple but effective method based on knowledge distillation and triplet loss for cross-modal object tracking. To overcome the significant cross-modal feature differences in the feature extraction phase of single-stream student network, we have adopted a feature-level knowledge distillation strategy. By learning the knowledge from modality-specific teacher networks for specific modalities, we have successfully enabled the single-stream student network to effectively model various cross-modal features without the need for complex module designs and additional network parameters. Additionally, to enhance the stability of the knowledge distillation learning process and accelerate convergence speed, we have also introduced distance constraint techniques based on triplet loss. Extensive experiments on the dataset demonstrate the effectiveness of the proposed method against state-of-the-art trackers. In the future, we will explore more relevant algorithms for more robust tracking.

Acknowledgement. This work was supported in part by the Anhui Provincial Natural Science Foundation under Grant 2408085QF199,Scientific Research Foundation for High-level Talents of Anhui University of Science and Technology under Grant 2024yjrc94.

References

1. Bhat, G., Danelljan, M., Gool, L.V., Timofte, R.: Learning discriminative model prediction for tracking. In: Proceedings of the IEEE/CVF Conference on Computer Vision and Pattern Recognition, pp. 6182–6191 (2019)

2. Chen, X., Yan, B., Zhu, J., Wang, D., Yang, X., Lu, H.: Transformer tracking. In: Proceedings of the IEEE/CVF Conference on Computer Vision and Pattern Recognition, pp. 8126–8135 (2021)
3. Chen, Z., et al.: Siamban: target-aware tracking with siamese box adaptive network. IEEE Trans. Pattern Anal. Mach. Intell. **45**(4), 5158–5173 (2022)
4. Cui, Y., Jiang, C., Wang, L., Wu, G.: Mixformer: end-to-end tracking with iterative mixed attention. In: Proceedings of the IEEE/CVF Conference on Computer Vision and Pattern Recognition, pp. 13608–13618 (2022)
5. Error, M.S.: Mean Squared Error, p. 653. Springer, Boston (2010)
6. Gao, S., Zhou, C., Ma, C., Wang, X., Yuan, J.: Aiatrack: attention in attention for transformer visual tracking. In: Proceedings of the European Conference on Computer Vision, pp. 146–164. Springer, Heidelberg (2022). https://doi.org/10.1007/978-3-031-20047-2_9
7. He, K., Zhang, X., Ren, S., Sun, J.: Deep residual learning for image recognition. In: Proceedings of the IEEE/CVF Conference on Computer Vision and Pattern Recognition, pp. 770–778 (2016)
8. Li, C., et al.: Lasher: a large-scale high-diversity benchmark for rgbt tracking. IEEE Trans. Image Process. **31**, 392–404 (2021)
9. Li, C., Zhu, T., Liu, L., Si, X., Fan, Z., Zhai, S.: Cross-modal object tracking: modality-aware representations and a unified benchmark. In: Proceedings of the AAAI Conference on Artificial Intelligence, vol. 36, pp. 1289–1296 (2022)
10. Li, H., Li, C., Zhu, X., Zheng, A., Luo, B.: Multi-spectral vehicle re-identification: a challenge. In: Proceedings of the AAAI Conference on Artificial Intelligence, vol. 34, pp. 11345–11353 (2020)
11. Liu, L., Zhang, M., Li, C., Li, C., Tang, J.: Cross-modal object tracking via modality-aware fusion network and a large-scale dataset. arXiv preprint arXiv:2312.14446 (2023)
12. Mayer, C., Danelljan, M., Paudel, D.P., Van Gool, L.: Learning target candidate association to keep track of what not to track. In: Proceedings of the IEEE/CVF International Conference on Computer Vision, pp. 13444–13454 (2021)
13. Muller, M., Bibi, A., Giancola, S., Alsubaihi, S., Ghanem, B.: Trackingnet: A large-scale dataset and benchmark for object tracking in the wild. In: Proceedings of the European Conference on Computer Vision, pp. 300–317. Springer, Heidelberg (2018)
14. Schroff, F., Kalenichenko, D., Philbin, J.: Facenet: a unified embedding for face recognition and clustering. In: Proceedings of the IEEE/CVF Conference on Computer Vision and Pattern Recognition, pp. 815–823 (2015)
15. Wang, N., Zhou, W., Wang, J., Li, H.: Transformer meets tracker: exploiting temporal context for robust visual tracking. In: Proceedings of the IEEE/CVF Conference on Computer Vision and Pattern Recognition, pp. 1571–1580 (2021)
16. Zhang, T., Guo, H., Jiao, Q., Zhang, Q., Han, J.: Efficient rgb-t tracking via cross-modality distillation. In: Proceedings of the IEEE/CVF Conference on Computer Vision and Pattern Recognition, pp. 5404–5413 (2023)
17. Zhang, Z., Peng, H., Fu, J., Li, B., Hu, W.: Ocean: object-aware anchor-free tracking. In: Vedaldi, A., Bischof, H., Brox, T., Frahm, J.-M. (eds.) ECCV 2020. LNCS, vol. 12366, pp. 771–787. Springer, Cham (2020). https://doi.org/10.1007/978-3-030-58589-1_46

Enhanced Comprehensive Competition Network for Domain Adaptive Palmprint Recognition

Congcong Jia, Xingbo Dong(✉), Zhe Jin, and Lianqiang Yang

Anhui Provincial Key Laboratory of Secure Artificial Intelligence,
School of Artificial Intelligence, Anhui University, Hefei, China
xingbo.dong@ahu.edu.cn

Abstract. Recent advances in palmprint recognition models using deep learning have shown promise, but performance often degrades significantly when applied to unseen target domains or new datasets. To address this, we propose a palmprint recognition method based on unsupervised domain adaptation that aims to minimize the performance drop across different datasets. Our approach aligns features, followed by a competitive network to enhance feature extraction and model learning, and is further optimized through a tailored loss function. Extensive experiments show that our method maintains strong generalization capabilities on unseen target domains, outperforming existing models in terms of accuracy and robustness across multiple datasets.

Keywords: Palmprint recognition · Domain adaptation · Competitive network

1 Introduction

Palmprint recognition has gained traction as a biometric modality due to its unique features and relatively high accuracy compared to other biometrics [1,2]. With the advent of deep learning techniques, deep learning models have shown impressive performance in palmprint recognition tasks. However, a model trained on a specific palmprint dataset often struggles to perform well on different datasets. This performance drop can primarily be attributed to domain discrepancies, which arise from variations in lighting, sensor quality, and environmental conditions during data acquisition. These inconsistencies create a gap between the source domain (where the model is trained) and the target domain (where the model is tested), leading to reduced accuracy and generalization in practical applications.

The development of domain adaptation has been relatively rapid, with numerous methods emerging in recent years. Its early evolution was influenced by the work of Avrim and Tom [4], who discussed co-training in semi-supervised learning. They introduced the challenge of enhancing algorithm performance by combining small amounts of labeled data with larger volumes of unlabeled data. This foundational work set the stage for subsequent advancements in domain adaptation. Later, David and Blitzer et al. conducted a comprehensive representation analysis of domain adaptation, introducing key concepts and methods to measure distribution differences between domains [3]. They highlighted a "new perspective" for domain adaptation models: minimizing the discrepancy between the source and target domains while maximizing the margins of the training set model [3]. This perspective underpins many contemporary approaches in the field. Recently, [17] proposed a novel method called Joint Pixel and Feature Alignment (JPFA), which employs a two-level alignment on source and target domain datasets to obtain adaptive features, thus enhancing model generalization. Despite these advancements, there remains considerable room for improvement in palmprint recognition methods based on domain adaptation.

To this end, we propose a palmprint recognition method based on unsupervised domain adaptation that aims to minimize the performance drop across different datasets. In our proposed palmprint recognition method, we employ the Comprehensive Competition Network (CCNet) [24], which integrates three learnable Gabor layers and competition blocks to effectively extract competitive features from multi-order textures. Moreover, the network processes images that have undergone a feature alignment module by swapping the low-frequency spectrum and grayscale distributions between the source and target domains. Specifically, after applying the Fourier transform [14], we manipulate the amplitude and phase components, performing an inverse Fourier transform to create new augmented images that share similarities in their feature distributions. Complementing this, we compute histograms and cumulative distribution functions, creating a mapping function that adjusts pixel intensities to enhance visual consistency. Together, these strategies facilitate a robust feature extraction process within CCNet, where competition blocks utilize softmax functions to optimize feature representations, leading to a feature vector that enhances classification performance. The combination of these methodologies ultimately ensures that our model maintains high accuracy and low Equal Error Rate (EER) in palmprint recognition tasks.

2 Related Work

2.1 Palmprint Recognition

Due to the rich texture in palmprints, Li and Zhang et al. used Fourier transform to convert this information into the frequency domain for feature extraction [13]. Han and Lee introduced adaptive Gabor filters for palm vein images, dynamically adjusting parameters to efficiently extract texture information, pioneering a new approach for palm vein recognition [9].

Genovese and Plataniotis et al. combined Gabor filters with Principal Component Analysis (PCA) and Convolutional Neural Networks (CNN) to design PalmNet, a novel network structure. PalmNet demonstrated significant improvements in accuracy and robustness [8]. Shao and Zou et al. proposed a method for cross-dataset palmprint recognition that combines domain adaptation and self-supervised learning, maintaining good performance on unseen datasets [19].

Zhang and Xu et al. introduced a gating mechanism and adaptive feature fusion to dynamically select and integrate important palmprint features, enhancing system robustness [28]. Li and Yuan proposed a method combining palm veins and palmprints, constructing a feature vector space that is invariant to image scaling, rotation, and translation [12]. Wang and Jia addressed the issue of noise and image deformation in extracting directional features from palmprints by proposing a directional response stability measurement method to improve feature stability and recognition accuracy [22].

2.2 Domain Adaptation

Domain adaptation typically involves source and target domains, and the number of these domains has garnered attention from researchers. Dredze and Crammer introduced online multi-domain learning and adaptation methods, applying them in dynamic environments with multiple domains, such as sentiment classification and spam filtering, showing good results. They also explored ways to enhance machine learning models' learning and adaptation capabilities [6].

Influenced by deep learning, Zhang and Yu et al. combined deep learning with domain adaptation, proposing a framework called Deep Transfer Network (DTN). This framework uses deep neural networks to match distributions and performs well in unsupervised domain adaptation tasks, highlighting deep learning's potential in domain adaptation [30]. Pei and Cao et al. solved the distribution differences between multi-source domain adaptations by introducing the multi-adversarial learning framework [15]. Regarding whether the numbers of source and target domains need to be equal, Sun and Shi et al. surveyed multi-source domain adaptation, where there are multiple source domains but only one target domain. By learning features from multiple source domains, the model can better adapt to the target domain, improving generalization ability [21].

Previous methods minimized domain discrepancies but often overlooked intra-class information, leading to poor generalization. To address this, Kang and Jiang et al. proposed the Contrastive Adaptation Network (CAN), which combines contrastive learning with adversarial training to effectively address distribution differences between source and target domain data [10].

3 Methodology

Our method integrates feature alignment, competitive learning with CCNet, and iterative optimization to achieve robust palmprint recognition. First, we align the source and target domain features by swapping low-frequency components in the Fourier domain and adjusting grayscale distributions. These transformed images are then processed by CCNet, which uses learnable Gabor layers and competition blocks to extract competitive multi-order texture features. Finally, the model is optimized using a combination of Cross-Entropy Loss, Contrastive Loss, and

Maximum Mean Discrepancy (MMD) Loss, ensuring effective feature separation, accurate classification, and minimal distribution gap between domains, resulting in strong generalization across datasets. In this section, we outline our approach: feature alignment (Sect. 3.1), model learning with CCNet on transformed images (Sect. 3.2), and iterative optimization using a loss function (Sect. 3.3).

3.1 Feature Alignment

Inspired from [23], we reduce the domain gap between source and target domains by manipulating the frequency components of images. Unlike other domain adaptation methods that often involve complex training processes, frequency domain alignment offers a simple and efficient solution by focusing on the low-frequency components of the image, which typically capture the broad structural information, rather than high-frequency details that tend to vary between domains.

In frequency domain alignment, we first apply the Fourier transform $\mathcal{F}$ to convert a palmprint image from the spatial domain (pixel values) to the frequency domain, where the image is represented by its amplitude and phase components. The amplitude captures the intensity of frequency patterns, while the phase contains spatial information. The key idea behind frequency domain alignment is to swap a portion of the low-frequency components between the source and target domains, making the source image look more like the target image in terms of overall structure.

We select a region of size β (which controls how much of the low-frequency information is swapped) from the phase images of the source and target domains. After performing the swap, we apply the inverse Fourier transform $\mathcal{F}^{-1}$ to reconstruct the aligned image, where the transformed image $x_{D^S \to D^T}$ is obtained as follows:

$$x_{D^S \to D^T} = \mathcal{F}^{-1}(\text{swap}(\mathcal{F}(x_{D^S}), \mathcal{F}(x_{D^T}))). \tag{1}$$

D^S and D^T indelicate source and target domain, respectively. Here, $\beta = 0.1$ is used, which controls how much low-frequency information is swapped. This method is easy to implement, requires no additional training, and works directly on image data to align the source and target domains.

In addition to frequency domain alignment, we apply histogram matching [16,20] to further align the grayscale distributions between the source and target domain images. Histogram matching ensures that the visual appearance, in terms of pixel intensity distribution, of the source domain images closely resembles that of the target domain. This is particularly useful when the two domains have different lighting conditions or contrast levels.

The process of histogram matching involves computing the histograms of both the source and target domain images. A histogram represents the frequency distribution of pixel intensity values in an image, where each bin in the histogram corresponds to a particular intensity level. By calculating the cumulative distribution functions (CDFs) for both the source and target histograms, we can derive a mapping function $T(i)$, which transforms each intensity level in the source domain image to its corresponding value in the target domain.

The transformation is applied pixel-by-pixel, ensuring that the source domain image is modified to match the intensity distribution of the target domain, which can be mathematically expressed as:

$$x_{D^S \to D^T}(h, w) = T(x_{D^S}(h, w)), \tag{2}$$

where $T(i)$ is the mapping function derived from the CDFs. This step effectively aligns the grayscale distribution between the two domains, making the source images more similar to the target in terms of overall appearance and aiding in feature alignment.

3.2 CCNet

The images processed by feature alignment enter the network layers for training. The Comprehensive Competition Network, proposed in [24] integrates three learnable Gabor layers and competition blocks to extract competitive features from multi-order textures. The extracted features are then concatenated and dimensionality reduction is performed through fully connected layers to obtain the feature vector V. The formula for the Gabor Layer is as follows:

$$G(x, y; \sigma, \gamma, \mu, \psi, \theta) = e^{-\frac{\gamma^2 x'^2 + y'^2}{2(2\sqrt{2}\sigma)^2}} \cos(2\pi \mu x' + \psi). \tag{3}$$

The multiple competition blocks mentioned are expressed by the following formula:

$$\begin{aligned} F_c &= \text{Softmax}_c(F_{\text{in}}) \\ F_x &= \text{Softmax}_x(F_{\text{in}}) \\ F_y &= \text{Softmax}_y(F_{\text{in}}). \end{aligned} \tag{4}$$

Subsequently, we perform the integration of the extracted multi-order competition features, which can be represented by the following formula:

$$F_{\text{out}} = w_z \times F_c + w_s \times (F_x + F_y), \tag{5}$$

where F_c, F_x, F_y and $F_{in} \in R^{b \times c \times w \times h}$, F_c, F_x, and F_y represent the results of applying the softmax function along the channel (c), x-axis, and y-axis dimensions of the input feature F_{in}, respectively. F_{out} is the output of the Comp Block, w_z is the weight of the channel competition, and w_s is the weight of the spatial competition. In this work, we set w_z to 0.8 and w_s to 0.1.

3.3 Loss Functions

Training source and target domain images with an integrated adversarial network alone is insufficient, so we introduce three loss functions, L_{ce}, L_{con} and L_{mmd}, to optimize the model further. These loss functions are defined as follows:

Cross Entropy Loss. The cross-entropy loss function can both measure the gap between the predictions and the true labels and optimize the model. It is expressed by the following formula:

$$L_{ce} = -\frac{1}{N}\sum_{i=1}^{N}\sum_{c=1}^{M} y_{i,c}\log(p_{i,c}). \tag{6}$$

Here, N represents the total number of samples, M represents the number of classes, and $y_{i,c}$ is an indicator variable for the true label. For sample i and class c, if sample iii belongs to class c, then $y_{i,c} = 1$; otherwise, $y_{i,c} = 0$. $p_{i,c}$ denotes the predicted probability of sample i belonging to class c by the model.

Contrastive Loss. The introduction of the contrastive loss function aims to bring the representations of similar samples (represented as embedding vectors) closer in the feature space, while pushing dissimilar samples further apart from each other.

$$\mathcal{L}_{con} = -\sum_{i\in I}\frac{1}{|p(i)|}\sum_{p\in P(i)}\log\frac{exp(z_i \cdot z_p/\tau)}{\sum_{a\in A(i)}exp(z_i \cdot z_a/\tau)}, \tag{7}$$

where $\mathcal{I}$ represents the set of samples. For any sample i, $P(i)$ denotes its set of positive samples, which are samples similar to i, while $A(i)$ represents the set of all samples for sample i. The feature representation or embedding vector of sample i is given by z_i, and similarly, z_p denotes the feature representation or embedding vector of sample p where $p \in P(i)$. The temperature parameter τ, The term $exp(z_i \cdot z_p/\tau)$ represents the exponentiated dot product similarity of the feature vectors of samples i and p, adjusted by the temperature parameter τ. Finally, $|P(i)|$ indicates the size of the positive sample set for sample i.

MMD (Maximum Mean Difference) Loss is used to measure the difference in sample distribution between the source and target domains, and its formula is:

$$\mathcal{L}_{mmd} = \frac{1}{n^2}\sum_{i,j}k(\mathbf{x}_i,\mathbf{x}_j) + \frac{1}{m^2}\sum_{i,j}k(\mathbf{y}_i,\mathbf{y}_j) - \frac{2}{nm}\sum_{i,j}k(\mathbf{x}_i,\mathbf{y}_j), \tag{8}$$

where n and m are the number of samples in the source and target domains, respectively. And x_i, x_j and y_i, y_j represent the sample feature vectors in the source domain and the target domain, respectively. The Gaussian kernel function k is used to calculate the similarity between two points in the feature space:

$$k(\mathbf{x}_i,\mathbf{x}_j) = \exp\left(-\frac{\|\mathbf{x}_i - \mathbf{x}_j\|^2}{2\sigma^2}\right), \tag{9}$$

where σ is the bandwidth parameter of the kernel function.

Finally, our proposed loss function is expressed as:

$$\mathcal{L} = \lambda \mathcal{L}_{mmd} + w_{ce} \times L_{ce} + w_{con} \times L_{con}, \tag{10}$$

where $\lambda \in [0,1]$, which can be chosen according to different datasets. In this paper, λ is set to 1, with w_{ce} set to 0.8 and w_{con} set to 0.2.

4 Experiments

4.1 Experiment Settings

We selected 5 datasets with different collection methods for the experiment, they are Tongji [29], PolyU [27], IITD [11], Multi-Spectral (MS) [26] and CasiaM [5], The MS [26] includes four spectral datasets: Red, Green, Blue and NIR, and the CasiaM [5] includes six wavelength datasets. Only three wavelength datasets of 460 nm, 700 nm and 850 nm were selected. Then specifically choosing some of them as source domain datasets and the remaining ones as target domain datasets. The datasets chosen as the source domain are divided into training and testing sets.

4.2 Compare with State-of-the-Art

We conducted a series of experiments to evaluate the performance of our method on several datasets, including MS, Tongji, PolyU, IITD, and CasiaM, focusing on domain adaptation across different domains. The results, summarized in Table 1, Table 2 and Table 3, show that our method consistently outperforms state-of-the-art models, including CCNet, CO3Net, JPFA, and R-ADAH, across most tested domains.

In the MS dataset, our method achieves the lowest error rates across almost all domain pairs. Notably, in the NIR-to-Green and NIR-to-Blue domains, we report error rates of 0.3389% and 0.2867%, respectively, outperforming CO3Net, which records errors as high as 1.5390% and 1.3830%. These results highlight the robustness of our model in aligning features between different spectral bands, a challenging task due to significant differences in the image characteristics between visible and infrared wavelengths.

On the Tongji, PolyU, and IITD datasets, our method also demonstrates strong generalization across different environments. For instance, our method achieves an error rate of 0.772% for the Tongji-to-Green domain, compared to CO3Net's 1.580%. These results suggest that our method is highly effective at adapting to varying imaging setups and environments, making it suitable for real-world applications such as biometric verification and surveillance systems.

The CasiaM dataset results further validate the robustness of our approach. In this dataset, which involves domain adaptation between different wavelength pairs (460 nm, 700 nm, and 850 nm), our method consistently achieves the lowest error rates, such as 1.74% for 700 nm-to-460 nm and 1.44% for 700 nm-to-850 nm, outperforming both CCNet and CO3Net by a large margin. This demonstrates that our method can handle domain shifts between both visible and infrared spectra, which is critical for cross-spectral applications like face recognition in varying lighting conditions. Overall, the experimental results show that our method offers superior performance across a wide range of challenging domain adaptation tasks.

Table 1. EERs of Cross-Dataset setting on MS.

Method	Dataset (D^s/D^t)										
	Red			Green			Blue			NIR		
	Green	Blue	NIR	Red	Blue	NIR	Red	Green	NIR	Red	Green	Blue
CCNet [24]	0.0278%	0.0611%	0.0091%	**0.0026%**	0%	0.2833%	0.0130%	0%	0.3000%	0.0167%	0.5889%	0.6140%
CO3Net [25]	0.1270%	0.1560%	0.0220%	0.0610%	0.0100%	0.6570%	0.1300%	0.0080%	0.8110%	0.0830%	1.5390%	1.3830%
JPFA [17]	0.0340%	0.0730%	0.2500%	–	–	–	0%	0.0550%	1.7700%	–	–	–
Ours	**0.0021%**	**0.0057%**	**0.0056%**	0.0047%	**0%**	**0.1524%**	**0.0094%**	**0%**	**0.1735%**	**0.0081%**	**0.3389%**	**0.2867%**

Table 2. EERs of Cross-Dataset setting on Tongji, PolyU and IITD.

Method	Dataset(D^s/D^t)									
	Tongji						PolyU		IITD	
	Red	Green	Blue	NIR	PolyU	IITD	Tongji	IITD	PolyU	Tongji
CCNet [24]	0.811%	1.402%	1.017%	0.750%	1.570%	6.120%	2.220%	5.730%	2.010%	3.220%
CO3Net [25]	0.878%	1.580%	1.570%	0.783%	1.660%	7.640%	3.580%	8.740%	2.260%	3.720%
R-ADAH [7]	7.000%	6.200%	6.300%	7.700%	–	–	–	–	–	–
Ours	**0.728%**	**0.772%**	**0.616%**	**1.066%**	**1.315%**	**4.094%**	**1.614%**	**4.783%**	**1.360%**	**1.767%**

Table 3. EERs of Cross-Dataset setting on CasiaM.

Method	Dataset(D^s/D^t)					
	460		700		850	
	700	850	460	850	460	700
CCNet [24]	6.70%	3.67%	2.94%	1.39%	4.53%	2.33%
CO3Net [25]	10.72%	4.82%	3.56%	1.50%	5.83%	3.00%
DFMT [18]	10.21% *	–	–	–	–	–
Ours	**2.44%**	**3.39%**	**1.74%**	**1.44%**	**3.50%**	**1.50%**

4.3 Ablation Study

To validate the contribution of each module, we performed ablation experiments using a subset of the data sets. We selected PolyU as the source domain dataset and Tongji as the target domain dataset. The experimental results are shown in Table 4.

Table 4. Ablation Results with Different Components.

Hist. Align	Freq. Align	$\mathcal{L}_{mmd}$	EER	Accuracy
✗	✗	✗	2.22%	98.48%
✓	✗	✗	3.12%	97.88%
✗	✓	✗	3.01%	97.73%
✓	✓	✗	2.81%	97.89%
✗	✓	✓	**1.72%**	98.18%
✓	✓	✓	**1.61%**	**98.52%**

We can observe that, without any components, the model achieved an EER of 2.22% and an accuracy of 98.48%. When histogram alignment and frequency alignment are added individually or jointly, the model shows some overfitting. The result of using any transformation (such as frequency alignment) in combination with $\mathcal{L}_{mmd}$ resulted in an EER of 1.72% and accuracy of 98.18%. When all three components were included, the model achieved the best performance with an EER of 1.61% and an accuracy of 98.52%. These results confirm that the integration of those modules leads to optimal performance in domain adaptation.

5 Conclusion

In this study, we introduced a robust palmprint recognition method that combines feature alignment and competitive learning through CCNet. By addressing domain gaps with Fourier domain manipulation and histogram matching, our approach effectively enhanced model performance. Our experiments confirmed the method's superiority over existing techniques, achieving lower error rates and strong generalization across diverse conditions. Additionally, this work does not consider single source domain generalization, which presents its own set of challenges. Single source domain generalization involves training models on data from only one domain and testing their ability to perform well in various target domains. Addressing this limitation in future research is essential, as improving generalization across multiple domains will enhance the robustness and applicability of palmprint recognition systems in real-world scenarios.

Acknowledgments. This work was supported by the National Natural Science Foundation of China (Nos. 62376003) and Anhui Provincial Natural Science Foundation (No. 2308085MF200).

References

1. Al-Taie, S.A.M., Khaleel, B.I.: Palm print recognition using intelligent techniques: a review. Jurnal Ilmiah Teknik Elektro Komputer dan Informatika **9**(1), 156–164 (2023)
2. Amrouni, N., Benzaoui, A., Zeroual, A.: Palmprint recognition: extensive exploration of databases, methodologies, comparative assessment, and future directions. Appl. Sci. **14**(1), 153 (2023)
3. Ben-David, S., Blitzer, J., Crammer, K., Pereira, F.: Analysis of representations for domain adaptation. Adv. Neural Inf. Process. Syst. **19** (2006)
4. Blum, A., Mitchell, T.: Combining labeled and unlabeled data with co-training. In: Proceedings of the Eleventh Annual Conference on Computational Learning Theory, pp. 92–100 (1998)
5. Chinese Academy of Sciences, Inst. Automation: Biometrics Ideal Test (2005). http://biometrics.idealtest.org/dbDetailForUser.do?id=5. Accessed 28 Jan 2016
6. Dredze, M., Crammer, K.: Online methods for multi-domain learning and adaptation. In: Proceedings of the 2008 Conference on Empirical Methods in Natural Language Processing, pp. 689–697 (2008)
7. Du, X., Zhong, D., Shao, H.: Cross-domain palmprint recognition via regularized adversarial domain adaptive hashing. IEEE Trans. Circuits Syst. Video Technol. **31**(6), 2372–2385 (2020)
8. Genovese, A., Piuri, V., Plataniotis, K.N., Scotti, F.: Palmnet: gabor-pca convolutional networks for touchless palmprint recognition. IEEE Trans. Inf. Forensics Secur. **14**(12), 3160–3174 (2019)
9. Han, W.Y., Lee, J.C.: Palm vein recognition using adaptive gabor filter. Expert Syst. Appl. **39**(18), 13225–13234 (2012)
10. Kang, G., Jiang, L., Yang, Y., Hauptmann, A.G.: Contrastive adaptation network for unsupervised domain adaptation. In: Proceedings of the IEEE/CVF Conference on Computer Vision and Pattern Recognition, pp. 4893–4902 (2019)
11. Kumar, A., Shekhar, S.: Personal identification using multibiometrics rank-level fusion. IEEE Trans. Syst. Man Cybern. Part C (Appl. Rev.) **41**(5), 743–752 (2010)
12. Li, W., Yuan, W.Q.: Multiple palm features extraction method based on vein and palmprint. J. Ambient Intell. Humanized Comput. **15**(2), 1465–1479 (2024)
13. Li, W., Zhang, D., Xu, Z.: Palmprint recognition based on fourier transform. J. Softw. **13**(5), 879–886 (2002)
14. Nussbaumer, H.J., Nussbaumer, H.J.: The Fast Fourier Transform. Springer, Heidelberg (1982)
15. Pei, Z., Cao, Z., Long, M., Wang, J.: Multi-adversarial domain adaptation. In: Proceedings of the AAAI Conference on Artificial Intelligence, vol. 32 (2018)
16. Rolland, J.P., Vo, V., Bloss, B., Abbey, C.K.: Fast algorithms for histogram matching: application to texture synthesis. J. Electron. Imaging **9**(1), 39–45 (2000)
17. Shao, H., Zhong, D.: Towards cross-dataset palmprint recognition via joint pixel and feature alignment. IEEE Trans. Image Process. **30**, 3764–3777 (2021)
18. Shao, H., Zhong, D.: Multi-target cross-dataset palmprint recognition via distilling from multi-teachers. IEEE Trans. Instrum. Meas. (2023)
19. Shao, H., Zou, Y., Liu, C., Guo, Q., Zhong, D.: Learning to generalize unseen dataset for cross-dataset palmprint recognition. IEEE Trans. Inf. Forensics Secur. (2024)
20. Shen, D.: Image registration by local histogram matching. Pattern Recogn. **40**(4), 1161–1172 (2007)

21. Sun, S., Shi, H., Wu, Y.: A survey of multi-source domain adaptation. Inf. Fusion **24**, 84–92 (2015)
22. Wang, H., Jia, W.: Enhance the performance of directional feature-based palmprint recognition by directional response stability measurement. Mach. Intell. Res., 1–18 (2024)
23. Yang, Y., Soatto, S.: FDA: Fourier domain adaptation for semantic segmentation. In: Proceedings of the IEEE/CVF Conference on Computer Vision and Pattern Recognition, pp. 4085–4095 (2020)
24. Yang, Z., Huangfu, H., Leng, L., Zhang, B., Teoh, A.B.J., Zhang, Y.: Comprehensive competition mechanism in palmprint recognition. IEEE Trans. Inf. Forensics Secur. (2023)
25. Yang, Z., et al.: CO3Net: Coordinate-aware contrastive competitive neural network for palmprint recognition. IEEE Trans. Instrum. Meas. (2023)
26. Zhang, D., Guo, Z., Lu, G., Zhang, L., Zuo, W.: An online system of multispectral palmprint verification. IEEE Trans. Instrum. Meas. **59**(2), 480–490 (2009)
27. Zhang, D., Kong, W.K., You, J., Wong, M.: Online palmprint identification. IEEE Trans. Pattern Anal. Mach. Intell. **25**(9), 1041–1050 (2003)
28. Zhang, K., Xu, G., Jin, Y.K., Qi, G., Yang, X., Bai, L.: Palmprint recognition based on gating mechanism and adaptive feature fusion. Front. Neurorobot. **17**, 1203962 (2023)
29. Zhang, L., Li, L., Yang, A., Shen, Y., Yang, M.: Towards contactless palmprint recognition: a novel device, a new benchmark, and a collaborative representation based identification approach. Pattern Recogn. **69**, 199–212 (2017)
30. Zhang, X., Yu, F.X., Chang, S.F., Wang, S.: Deep transfer network: unsupervised domain adaptation. arXiv preprint arXiv:1503.00591 (2015)

MBDR-V2: A Network for MRI Brain Tumor Image Segmentation with Incomplete Modalities

Yanqi Hou[1,2], Longfeng Shen[1,2(✉)], Jiacong Chen[1,2], Liangjin Diao[1,2], Youle Xu[3], and Wei Zhao[4]

[1] School of Computer Science and Technology, Huaibei Normal University, Huaibei, China
longfengshen521@126.com
[2] Institute of Artificial Intelligence, Hefei Comprehensive National Science Center, Hefei, China
[3] China Telecom Corporation Limited, Huaibei Branch, Huaibei, China
[4] Huaibei City People's Hospital, Huaibei, China

Abstract. In clinical practice, modality missingness is a common phenomenon due to various factors. However, most mainstream multimodal brain tumor segmentation methods typically assume that the input multimodal data is complete. When certain modalities are missing, the performance of these methods often degrades significantly, resulting in inaccurate segmentation outcomes. To address the limitations of MBDRes-UNet in handling missing modalities, we propose an improved approach named MBDR-V2, which aims to enhance robustness and segmentation performance in practical applications. The architecture of MBDR-V2 consists of three main components. First, a modality-specific encoder, which incorporates the multi-branch adaptive dilated convolutional residual blocks introduced in MBDRes-UNet, is used to construct four independent encoders. These encoders extract unique modality-specific features while maintaining a lightweight design. Second, the modality-correlated encoder, in which we propose a modality association module. This module integrates the spatial features of different tumor regions with modality-specific features sensitive to these regions through a pre-decoding process. Additionally, at each stage of the modality association module, we introduce segmentation-based regularizers, allowing each encoder to learn fully discriminative features to address the training imbalance caused by missing modalities. Finally, an MBDR decoder is employed to achieve end-to-end segmentation. Extensive experiments conducted on the BraTS 2020 dataset demonstrate that MBDR-V2 outperforms MBDRes-UNet and surpasses state-of-the-art methods for incomplete modality segmentation tasks.

Keywords: Incomplete Multimodal Learning · Brain Tumor Segmentation · Modality Missingness · Multimodal Feature Fusion

1 Introduction

Accurate segmentation of brain tumors is of significant importance in clinical diagnosis, treatment planning, and efficacy assessment. Magnetic resonance imaging (MRI), as a key neuroimaging technique, provides rich multimodal data support. As illustrated in

Fig. 1, common modalities include T1-weighted, T1 contrast-enhanced, T2-weighted, and FLAIR images [1], each revealing different aspects of brain structure and pathological changes. Methods that integrate these multimodal data for segmentation can significantly enhance both the precision and reliability of the segmentation process. However, in practical applications, the acquisition of complete multimodal data often faces challenges.

Modality Missingness refers to the incompleteness of data caused by the absence of certain MRI modalities (such as FLAIR, T1, T1c, and T2) during medical and diagnostic processes. Factors such as equipment limitations, individual patient differences, and inconsistencies in scanning protocols can prevent the consistent acquisition of complete multimodal datasets in clinical settings. This modality missingness problem has a significant impact on the performance of existing segmentation algorithms.

Fig. 1. From left to right: (a) FLAIR, (b) T1, (c) T1c, (d) T2, and (e) Ground Truth. Red represents necrotic and non-enhancing core, yellow represents enhancing core, and green represents peritumoral edema.

To address the issue of modality missingness, researchers have proposed various strategies for incomplete multimodal segmentation. One approach generates specialized data subsets for each missing modality scenario and trains separate models for each, leveraging knowledge transfer from complete to incomplete modalities [2]. However, this approach complicates both the training and inference processes due to the need for handling multiple datasets and models. Another strategy uses generative models, such as Generative Adversarial Networks (GANs), to synthesize missing modalities [3]. Although this can improve segmentation accuracy, training GANs is inherently challenging, especially with 3D images, where mode collapse may occur.

Recent approaches have focused on developing unified models for incomplete multimodal segmentation by projecting available modalities into a common latent space where shared feature representations are learned, followed by projection into the segmentation space. This approach simplifies both the training and inference processes, while enhancing the model's generalization capability compared to handling different modality deficiencies with multiple refinement networks. For example, Wang et al. improve model robustness to missing data through shared and modality-specific feature modeling [4]. mmFormer [5] introduces a Transformer-based approach that captures long-range dependencies using a combination of modality-specific encoders and inter-modality transformers, maintaining segmentation performance despite missing modalities. RFNet [6] employs a region-aware fusion strategy that dynamically adjusts the information flow between modality features, improving the identification of critical regions.

Inspired by these multimodal encoder architectures, we propose an enhancement to MBDRes-UNet[12], termed MBDR-v2, which further improves model performance in incomplete multimodal segmentation tasks and enhances its robustness for clinical use. Our MBDR-v2 retains the encoder-decoder architecture of MBDRes-UNet but differs by utilizing four encoders to form Modality-specific Encoders, extracting distinct features from different modality images. To establish the relationship between image modalities and tumor regions, we introduce a Modality-correlation Module (MCM) in MBDR-v2 to construct a Modality-correlation Encoder. The MCM module captures the complementarity between modalities and optimizes the combination of multimodal features through correlation learning to enhance overall performance. Moreover, we incorporate a segmentation-based regularizer in MBDR-v2, enabling the encoder and decoder to still learn discriminative features even when certain modalities are missing.

We validate the performance of MBDR-v2 on the BraTS 2020 public datasets for multimodal brain tumor segmentation tasks, demonstrating the superiority of our method.

In summary, our contributions are threefold:

1) We propose a network for incomplete multimodal brain tumor segmentation (MBDR-v2).
2) Specifically, we introduce a novel Modality-correlation Module (MCM) that explicitly considers spatial relationships between modalities.
3) To address data imbalance issues, we incorporate a segmentation-based regularizer at each stage of the modality correlation decoder. Each modality encoder generates tumor-discriminative features for segmentation, further enhancing the discriminability of the fused features.

2 Method

In this paper, we introduce MBDR-v2 for incomplete multimodal learning in brain tumor segmentation. We employ an encoder-decoder architecture to build MBDR-v2, with the overall network structure illustrated in Fig. 2. The network primarily consists of Modality-specific Encoder for each modality, a Modality-correlation Encoder, and a decoder derived from MBDRes-UNet.

The four Modality-specific Encoders independently process data from different modalities, capturing the unique features of each. Subsequently, the Modality-correlation Encoder establishes associations between the available information, integrating the cross-modal features. Finally, the Encoder generates a coherent segmentation map, facilitating effective end-to-end processing. Additionally, we incorporate a segmentation-based auxiliary regularizer in the modality correlation decoder. The following sections will provide detailed explanations of each component.

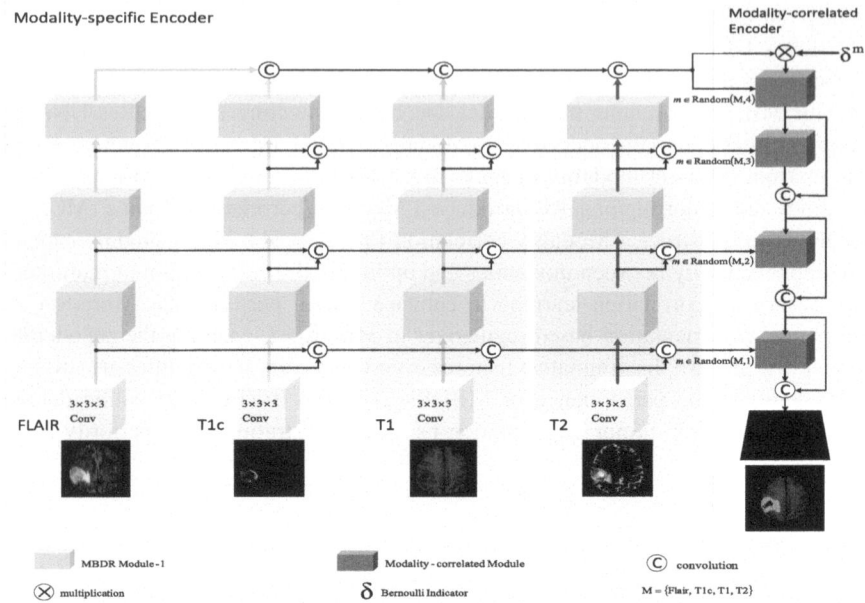

Fig. 2. Overview of the MBDR-v2 network, consisting of four modality-specific encoders, one modality correlation encoder, and an MBDR decoder.

2.1 Modality-Specific Encoder

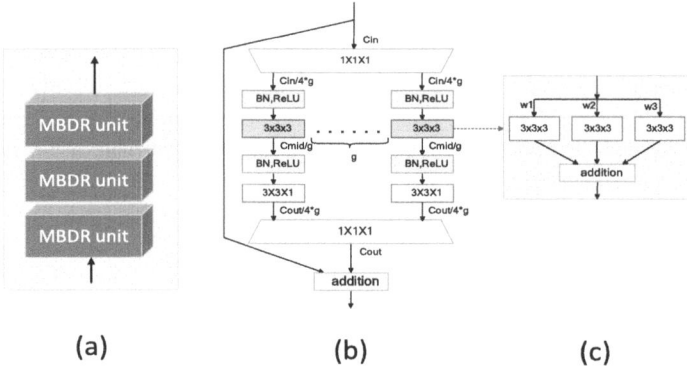

Fig. 3. (a) The MBDR module, (b) a schematic diagram of the MBDR unit structure, and (c) the adaptive dilated convolution layer within the MBDR unit. Here, w1, w2 and w3 represent the weights for each branch of the adaptive dilated convolution layer, and ddd denotes the dilation rate. In this work, we use $g = 8$.

The Modality-specific Encoder is designed to extract features related to each specific modality (T1, T2, T1c, FLAIR) and capture the internal details of each modality. As illustrated in Fig. 2, the network comprises four encoders, with each encoder consisting of a convolutional layer followed by three MBDR modules, each dedicated to extracting features from a specific modality. Specifically, we have constructed a four-level encoder. The first level consists of a convolutional block, which includes a normalization layer, a ReLU activation layer, and a convolutional layer with a kernel size of $3 \times 3 \times 3$. The second and third levels are composed of multi-branch Adaptive Dilated Residual (MBDR) convolution modules. As shown in Fig. 3(a), each MBDR module is composed of a stack of three multi-branch adaptive dilated residual convolution units (MBDR units), which reduce the network's computational cost without compromising accuracy. Between two consecutive blocks, a convolutional layer with a stride of 2 is used to downsample the feature maps. The number of filters at each level of the encoder are 16, 32, 64, and 128, respectively.

2.2 Modality-Correlation Encoder

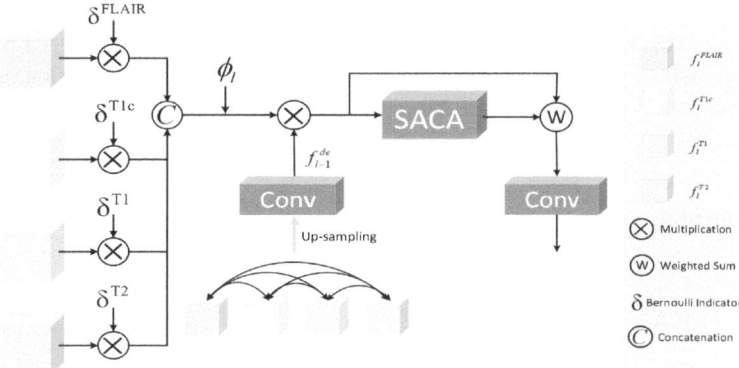

Fig. 4. The schematic diagram of the MCM, ϕ_l represents the position discriminator at layer l, f_{l-1}^{de} features from the preceding layer serve as positional embeddings, while SACA refers to the 3D hybrid attention mechanism.

Modality-correlation Encoder primarily consists of the Modality-correlation Module (MCM). Each MCM not only receives features from the four modality-specific encoders but also incorporates features from the previous layer. This is because the features from the previous layer serve as tumor location embeddings, allowing the module to gain spatial awareness of the overall tumor location. As shown in Fig. 4, before feature fusion, a pre-decoding sequence of the previous layer's features is obtained. This pre-decoding sequence provides preliminary information about the general location of the segmentation targets, specifically the necrotic non-enhancing tumor core (NCR/NET), the edema surrounding the tumor (ED), and the enhancing tumor region (ET). Subsequently, the features from the four modalities are assigned to their respective sensitive spatial regions according to the position discriminator. These features are then integrated through multiplication. Finally, a spatial-channel mixed attention mechanism is introduced as a regularizer, generating attention weights that are applied to the segmentation to obtain modality fusion features with comprehensive location discrimination. The overall representation is as follows:

$$F_l = \text{MCM}\left(\phi_l(\delta^m f_l^m), f_{l-1}^{de}\right) \quad (1)$$

$F_l \in R^{c \times \frac{h \times w \times d}{2^{l-1}}}$, $m \in \text{Randon}(M, 1)$, $M = \{\text{FLAIR}, \text{T1c}, \text{T1}, \text{T2}\}$, l represents the number of layers in the encoder.

2.3 MBDR Decoder

The lightweight decoder from MBDRest-UNet[12] is retained in the decoder to progressively restore the original spatial resolution, thereby generating the segmentation mask or reconstructing the image. The decoder utilizes trilinear interpolation to obtain upsampled features, which are fused with high-resolution features transmitted through skip connections by a fusion module. As shown in Fig. 3(a), these features are passed into a decoding convolution block composed of a multi-branch adaptive residual convolution block (MBDRes block), gradually restoring the feature dimensions. A $1 \times 1 \times 1$ convolution with a stride of 1 is then applied to fuse all channel information, and the final segmentation map is produced through a Softmax function, enabling end-to-end segmentation.

2.4 Segmentation-Based Auxiliary Regularizer

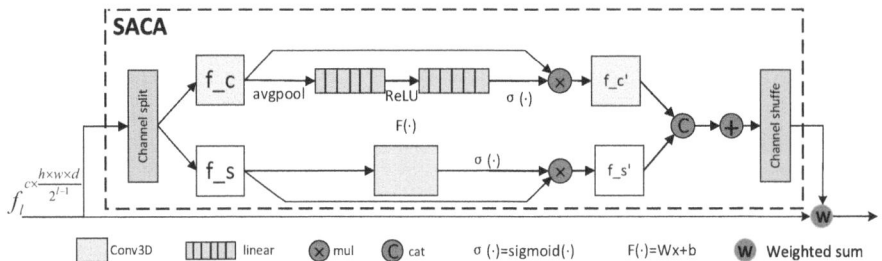

Fig. 5. Illustrates the structure of the hybrid attention module. The channel features are normalized using average pooling, followed by a dot-product operation performed via a 3D convolution layer to achieve spatial attention weighting, resulting in attention-weighted features.

In deep learning models for multimodal data processing, each modality encoder (T1c, T1, T2, FLAIR) is designed to extract useful features from a specific modality to assist with the segmentation task. However, due to the absence of certain modalities, some encoders may fail to effectively identify tumor features, resulting in weaker discriminative patterns and an inability to accurately distinguish tumor regions from the background [7]. To address this, as shown in Fig. 2, we introduce an auxiliary regularizer at each stage of the modality fusion decoder. As depicted in Fig. 5, the module generates feature maps through the pre-decoding operations at each stage and uses the resulting Dice loss as a regularization term, represented as:

$$L = 1 - \text{Dice} = 1 - \frac{1}{C}\sum_{C=1}^{C} \frac{2\sum_i p_i^C g_i^C}{\sum_i p_i^C \sum_i g_i^C} \quad (2)$$

$$L_{total} = \sum_{m \in M}^{l-1} L\left(y_{i,\,m}^{de}, y_{i,\,m}\right) + \sum_{i=1}^{l} L_i^{decoder} \quad (3)$$

where, C represents the number of segmentation classes, g_i^C denotes the classification label values, p_i^C signifies the predicted values, M = {FLAIR, T1c, T1, T2}, $y_{i,m}^{de}$ denotes the segmentation mask of the i-th image in the m modality prediction, and l represents the number of layers in the network.

By enforcing this approach, each modality encoder is compelled to achieve complete discriminative capability for each tumor region location. Consequently, MBDR-V2 can obtain representative fused features, thereby enhancing segmentation performance.

3 Experiments

3.1 Experimental Details

Reassign Number of Columns. We evaluated MBDR-V2 on the BRATS2020 dataset, which includes four different MRI modalities: FLAIR, T1c, T1, and T2. The BRATS2020 dataset contains 369 training subjects, which were randomly split into 258 for training and 111 for testing in a 7:3 ratio. The tumor segmentation labels include background (label 0), necrotic and non-enhancing tumor (label 1), peritumoral edema (label 2), and enhancing tumor (label 4).

Evaluation Metrics. The effectiveness of the model is assessed by calculating both complexity and segmentation accuracy. The segmentation performance of our method is measured using the Dice coefficient, with ET, WT, and TC referring to enhancing tumor (label 1), whole tumor (labels 1, 2, and 4), and tumor core (labels 1 and 4) regions, respectively. Complexity is determined by the number of parameters (Params) and floating-point operations (FLOPs). Params represents the spatial complexity of the network, while FLOPs denotes the temporal complexity. They are defined as follows:

$$\text{Params} = k_h \times k_w \times k_d \times C_{in} \times C_{out} \quad (4)$$

$$\text{Flops} = 2 \times (k_h \times k_w \times k_d \times C_{in}) \times C_{out} \times h \times w \times d \quad (5)$$

In the equation, k_h、k_w and k_d denote the height, width, and depth of the convolutional kernels, respectively. C_{in} and C_{out} represent the number of input and output channels, respectively. Additionally, h, w and d refer to the height, width, and depth of the image, respectively.

Experimental Setup. Experiments were conducted using Python 3.8. The model was trained for 300 epochs with a batch size of 4 on four parallel Tesla T4 GPUs. All experimental networks were implemented using the PyTorch framework. Additionally, the Adam optimizer was employed with a learning rate set to 0.0001.

3.2 Performance Analysis

Comparison with MBDRes-UNet. To evaluate the enhanced architecture's capability in handling incomplete modalities, we conducted a comprehensive comparison of the segmentation performance of MBDRes-UNet[12] and MBDR-V2 on the BRATS2020 dataset. As shown in Table 1, MBDR-V2 demonstrates a substantial improvement across nearly all metrics. The average segmentation results across 15 different modality missing scenarios indicate that MBDR-V2 achieves improvements of 14.6%, 19%, and 17.6% for the WT, TC, and ET classes, respectively.

Table 1. Comparison with MBDRes-UNet on the BRATS2020 dataset.

Modalities				Dice_score(%)					
				WT		TC		ET	
FLAIR	T1c	T1	T2	MBDRes-UNet	MBDR-V2	MBDRes-UNet	MBDR-V2	MBDRes-UNet	MBDR-V2
○	○	○	●	**74.0**	73.9	44.3	**54.9**	28.0	**40.1**
○	○	●	○	48.8	**71.6**	37.5	**79.7**	21.3	**63.3**
○	●	○	○	56.9	**70.8**	55.4	**65.5**	59.8	**67.7**
●	○	○	○	48.2	**88.0**	29.3	**58.6**	14.5	**32.1**
○	○	●	●	76.4	**86.4**	55.6	**76.3**	30.4	**72.1**
○	●	●	○	59.6	**73.0**	58.7	**79.8**	63.9	**69.6**
●	●	○	○	58.3	**88.5**	39.8	**71.8**	18.3	**61.8**
○	●	○	●	77.2	**85.0**	68.8	**74.8**	65.2	**70.0**
●	○	○	●	67.2	**87.2**	47.5	**67.5**	25.9	**45.8**
●	○	●	○	68.6	**88.6**	56.2	**81.8**	61.9	**73.5**
●	●	●	○	72.6	**80.6**	61.7	**81.7**	65.9	**75.4**
●	●	○	●	76.6	**84.6**	66.1	**75.1**	34.2	**39.1**
●	○	●	●	71.1	**89.0**	50.7	**81.7**	28.9	**70.5**
○	●	●	●	80.7	**85.6**	75.6	76.9	71.3	**75.0**
●	●	●	●	**89.9**	88.8	84.4	**84.5**	77.8	76.4
Average				68.4	**83.0**	55.4	**74.4**	44.5	**62.1**
				FLOPs(G)				Params(M)	
MBDRes-UNet				**25.75**				3.85	
MBDR-V2				68.42				**3.42**	

Table 1 Bold indicates the best results. Dice similarity coefficient (DSC) [%] is employed for evaluation with every combination settings of modalities. ● and ○ denote available and missing modalities, respectively.

Comparison with State-of-the-Art Methods. To validate the performance of our proposed model, we compared the segmentation performance of MBDRes-UNet with other advanced networks, including U-HVED [12], RFNet [7], and M^3AE [13] on the BraTS2020 dataset. Due to issues with the encapsulation of certain code models, we only evaluated the computational complexity of RFNet, with comparative results presented in Table 2. The average segmentation results across 15 different modality missing scenarios indicate that our method achieves the highest ET metric at 62.1%. While the overall averages for the WT and TC metrics are slightly lower than those of RFNet [7] and [13], the network parameters and computational complexity of MBDR-V2 are approximately half that of RFNet. This efficiency enables MBDR-V2 to demonstrate superior performance in clinical applications while being more resource-efficient.

Table 2. Comparison with state-of-the-art methods.

Modalities				Dice_score(%)											
				WT				TC				ET			
FLAIR	T1c	T1	T2	U-HVED	RFNet	M^3AE	Ours	U-HVED	RFNet	M^3AE	Ours	U-HVED	RFNet	M^3AE	Ours
○	○	○	●	77.5	**85.1**	84.3	73.9	45.0	66.9	**69.4**	54.9	18.7	43.0	**47.6**	40.1
○	○	●	○	57.9	**73.6**	73.1	71.6	62.7	80.3	**82.9**	79.7	55.5	67.7	**73.7**	63.3
○	●	○	○	56.9	**74.8**	74.3	70.8	40.7	65.2	**66.1**	65.5	8.9	32.3	37.1	**67.7**
●	○	○	○	77.0	85.8	**88.2**	88.0	41.3	62.6	**66.4**	58.6	17.8	35.5	**35.6**	32.1
○	○	●	●	81.7	85.6	85.8	**86.4**	74.4	82.4	**84.2**	76.3	63.3	70.6	**75.3**	72.1
○	●	●	○	65.8	**77.5**	76.7	73.0	68.9	81.3	**83.4**	79.8	59.5	68.5	**74.7**	69.6
●	●	○	○	83.7	**89.0**	88.5	88.5	51.7	**72.2**	70.8	71.8	16.3	38.5	41.2	**61.8**
○	●	○	●	80.5	85.4	**86.2**	85.0	52.9	71.1	71.8	**74.8**	19.3	42.9	48.7	**70.0**
●	○	○	●	85.2	89.3	**89.4**	87.2	51.4	**71.8**	70.9	67.5	22.1	45.4	45.4	**45.8**
●	○	●	○	84.0	**89.4**	89.2	88.6	71.5	81.6	**84.4**	81.8	61.4	72.5	**75.0**	73.5
●	●	●	○	85.9	**89.9**	88.2	80.6	74.1	82.3	**84.1**	81.7	61.9	71.1	74.0	**75.4**
●	●	○	●	86.5	**90.0**	89.1	84.6	56.1	74.0	72.7	**75.1**	22.6	**46.0**	44.8	39.1
●	○	●	●	87.6	**90.4**	90.2	89.0	75.1	82.6	**84.6**	81.7	62.9	73.1	**73.8**	70.5
○	●	●	●	82.5	**86.1**	85.2	85.6	75.8	82.9	**84.4**	76.9	63.6	70.9	**75.4**	75.0
●	●	●	●	88.0	**90.6**	90.0	88.8	76.2	82.9	**84.5**	84.5	63.0	71.4	75.5	**76.4**
Average				78.7	**85.5**	85.3	83.0	61.2	76.0	**77.4**	74.4	41.1	56.6	59.9	**62.1**

	FLOPs(G)		Params(M)	
RFNet	102.28		8.34	
MBDR-V2	**68.42**		**3.42**	

Table 2 Methods include U-HVED [10], RFNet [6], and M^3AE [11] on the BRATS2020 dataset. Bold indicates the best results. Dice similarity coefficient (DSC) [%] is employed for evaluation with every combination settings of modalities. ● and ○ denote available and missing modalities, respectively.

4 Summary

In this paper, we propose MBDR-V2, a network designed to effectively aggregate various available modalities for incomplete multimodal brain tumor segmentation, building on the MBDRes-UNet architecture. The modality association module (MCM) is designed to address the varying sensitivities of different modalities to brain tumor regions by integrating features from different modalities. MCM enables the extraction of the most spatially representative fused features from the different modalities, facilitating accurate segmentation tasks. Additionally, the introduced regularizer not only improves the feature representation extracted by each modality encoder for every tumor region but also accelerates the training speed compared to RFNet. Extensive experiments demonstrate that our method outperforms state-of-the-art approaches in practical applications.

Acknowledgments. This work was partly supported by the University Synergy Innovation Program of Anhui Province, China (Grant No. GXXT-2022-033), the University Natural Science Research Project of Anhui Province (Grant Nos. 2023AH040056 and 2023AH050333), the Open Laboratory project of Huaibei Normal University (Grant No. 2023SYSKF042), and the 2023 National Innovation and Entrepreneurship Training Program for College Students (Grant No. 202310373017), Anhui Innovation and Entrepreneurship Training Program for College Students (Grant Nos. S202410373003, S202410373046), Huaibei Normal University 2024 Graduate Innovation Fund Project (Grant No. CX2024002).

References

1. Menze, B.H., Jakab, A., et al.: The multimodal brain tumor image segmentation benchmark (BRATS). IEEE Trans. Med. Imaging **34**(10), 1993–2024 (2014)
2. Wang, Y., Zhang, Y., et al.: ACN: adversarial co-training network for brain tumor segmentation with missing modalities. In: International Conference on Medical Image Computing and Computer Assisted Intervention, pp. 410–420. Springer, Heidelberg (2021)
3. Dorent, R., Joutard, S., et al.: Hetero-modal variational encoder-decoder for joint modality completion and segmentation. In: Proceedings of International Conference on MICCAI, pp. 74–82. Springer, Heidelberg (2019)
4. Wang, H., Chen, Y., et al.: Multi-modal learning with missing modality via shared-specific feature modelling. In: Proceedings of the IEEE/CVF Conference on Computer Vision and Pattern Recognition, pp. 15878–15887 (2023)
5. Zhang, Y., He, N., Yang, J., et al.: mmformer: multimodal medical transformer for incomplete multimodal learning of brain tumor segmentation. In: International Conference on Medical Image Computing and Computer-Assisted Intervention, pp. 107–117. Springer, Cham (2022)
6. Ding, Y., Yu, X., Yang, Y.: RFNet: region-aware fusion network for incomplete multi-modal brain tumor segmentation. In: Proceedings of the IEEE/CVF International Conference on Computer Vision, pp. 3975–3984 (2021)
7. Dice, L.R.: Measures of the amount of ecologic association between species. Ecology **26**(3), 297–302 (1945)
8. Kingma, D.P., Ba, J.L.: Adam: a method for stochastic optimization. In: 3rd International Conference on Learning Representations. ICLR 2015-Conference Track Proceedings, vol. 1 (2015)
9. Baid, U., Ghodasara, S., et al.: The rsna-asnr-miccai brats 2021 benchmark on brain tumor segmentation and radiogenomic classification (2021). arXiv preprint arXiv:2107.02314
10. Dorent, R., Joutard, S., et al.: Hetero-modal variational encoder-decoder for joint modality completion and segmentation. In: Medical Image Computing and Computer Assisted Intervention–MICCAI 2019: 22nd International Conference, Shenzhen, China, 13–17 October 2019, Proceedings, Part II, vol. 22, pp. 74–82. Springer, Heidelberg (2019)
11. Liu, H., Wei, D., et al.: M3AE: multimodal representation learning for brain tumor segmentation with missing modalities. In: Proceedings of the AAAI Conference on Artificial Intelligence, vol. 37, no. 2, pp. 1657–1665 (2023)
12. Shen, L., Hou, Y., et al.: HouMBDRes-U-Net: multi-scale lightweight brain tumor segmentation network (2024). https://doi.org/10.48550/arXiv.2411.01896

An Innovative Eco-Friendly Weighing System for Reusable Bags Incorporating K210 and QR Code Technology

Yubin Wei[1], Yufei Li[1(✉)], and Yiwen Zhang[2]

[1] College of Artificial Intelligence, Hefei University of Economics, Hefei 230013, China
yufeili_lyf@163.com
[2] College of Computer Science and Technology, Anhui University, Hefei 230601, China

Abstract. The rapid development of The Internet of Things technologies [6] has revealed opportunities to enhance the efficiency and eco-friendliness of retail operations, particularly in supermarkets. Traditional weighing and checkout systems often involve inefficiencies, resource wastage, and environmental pollution due to plastic bag overuse and manual operations. This paper introduces an Internet of Things-based intelligent weighing and checkout system, utilizing artificial intelligence visual recognition and QR code technologies [15] to promote sustainability and efficiency in supermarket operations.

The proposed system includes reusable QR code-identified bags to reduce plastic waste and artificial intelligence-driven automation for product identification and weighing. With modular design and wireless communication, the system ensures scalability and stability. Each module, including the artificial intelligence visual recognition and the QR code scanning modules, communicates via The Internet of Things protocols, while the server manages the data flow and checkout process.

Cost-effective and highly efficient, this system is designed to be widely applicable across retail environments, reducing operational costs while improving customer experiences. Extensive tests have demonstrated its accuracy and reliability under real-world conditions, showcasing significant improvements over traditional methods. This solution aligns with global efforts to reduce environmental impacts, offering a novel approach to sustainable retail operations.

Keywords: The Internet of Things technology · artificial intelligence visual recognition · QR codes · eco-friendly · weighing system · supermarket checkout system

1 Introduction

The need for more efficient and environmentally sustainable weighing and checkout systems has become increasingly crucial in the contemporary retail environment, especially within supermarkets. Traditional systems have long been plagued by inefficiency, high operational costs, and significant environmental pollution. One of the primary causes of this environmental impact is the overuse of single-use plastic bags, which not only contribute to the ever-growing problem of plastic waste but also pose a threat to marine life and ecosystems. Additionally, manual operations within these traditional systems often lead to slower checkout processes and increased labour requirements.

In response to these challenges, this paper presents an innovative weighing and checkout system that harnesses the power of the Internet of Things (IoT) technology. At the core of this system lies artificial intelligence (AI) visual recognition technology, enabling automated product identification. This not only reduces the reliance on manual labour but also significantly enhances the speed and accuracy of the identification process. Complementing this is a unique QR code-based reusable bag system. By encouraging the reuse of bags through this mechanism, the system effectively curtails plastic consumption, thereby mitigating environmental pollution.

Moreover, the system's modular design ensures its scalability and stability. This modularity allows for easy integration and expansion within various retail environments. The seamless integration of IoT technologies within the system facilitates smooth communication between different modules, enabling efficient data exchange. This, in turn, enhances the overall functionality of the system, making it a comprehensive solution for modern supermarket operations.

The following sections of this paper will thoroughly explore the design, implementation, and testing of this proposed system, with a particular focus on its eco-friendly nature, operational efficiency, and intelligent features. This exploration aims to provide a detailed understanding of how this system can revolutionize supermarket weighing and checkout processes while contributing to environmental sustainability.

2 Project Overview

2.1 Background Analysis

Marine pollution, particularly from plastic waste, presents a critical environmental threat. It is estimated that 60–80% of all marine debris consists of plastic materials, originating from both land and sea sources. Ocean vessels alone contribute an estimated 4 to 6.5 million metric tons of plastic waste annually. This pollution affects 267 marine species globally, including 86% of sea turtles, 36% of seabirds, and 28% of marine mammals, as they often mistake plastic for food or become entangled in it [1]. To mitigate this environmental crisis, reducing plastic consumption and promoting reusable alternatives is of paramount importance.

The Internet of Things is defined as "the network of physical objects—devices, instruments, vehicles, buildings and other items embedded with electronics, circuits, software, sensors, and network connectivity that enables these objects to collect and exchange data [2]." The rapid advancement of IoT technology provides viable solutions across numerous sectors, including retail, offering the potential for more sustainable practices.

The traditional supermarket weighing and checkout systems generate significant waste from single-use plastic bags and label stickers, leading to severe environmental pollution [8]. Therefore, developing an IoT-based eco-friendly weighing and checkout system, incorporating reusable QR code bags holds substantial practical value in reducing plastic waste and promoting environmental sustainability.

2.2 Related Work

In the field of Internet of Things (IoT) applications, there has been some research and practical exploration aiming to improve supermarket weighing and checkout systems. However, these efforts mainly concentrate on enhancing system efficiency and optimizing inventory management, with relatively little consideration of environmental protection issues.

Therefore, based on previous work, this project further introduces artificial intelligence visual recognition technology, combined with QR-code technology and database management system, to enable the reuse of weighing bags, thus achieving environmental protection and reducing the generation of plastic waste.

2.3 Application Prospects

Nowadays, the role and value of IoT technology in ecological protection are attracting more and more attention [4]. Considering society's current emphasis on environmental protection and efficiency, coupled with the rapid development of IoT technology, this work has broad application prospects. Firstly, in supermarkets and retail locations, this system can be widely used in daily operations, improving weighing and checkout efficiency, reducing operational costs, and minimizing environmental pollution. Secondly, as IoT technology continues to proliferate and its application scenarios expand, this system could be extended to other fields such as logistics and warehousing, realizing broader application value. Moreover, through continuous optimization and system upgrades, this work is expected to achieve more innovations and breakthroughs in the future, contributing to the development of the IoT application field.

3 Design and Implementation

3.1 System Plan

This system primarily consists of an intelligent scale (AI visual recognition module (K210), wireless communication module (ESP8266), scanner module, HX711AD high-precision pressure sensor module, and Android upper computer, a checkout system, and a server (MQTT server + database management system) (see Fig. 1).

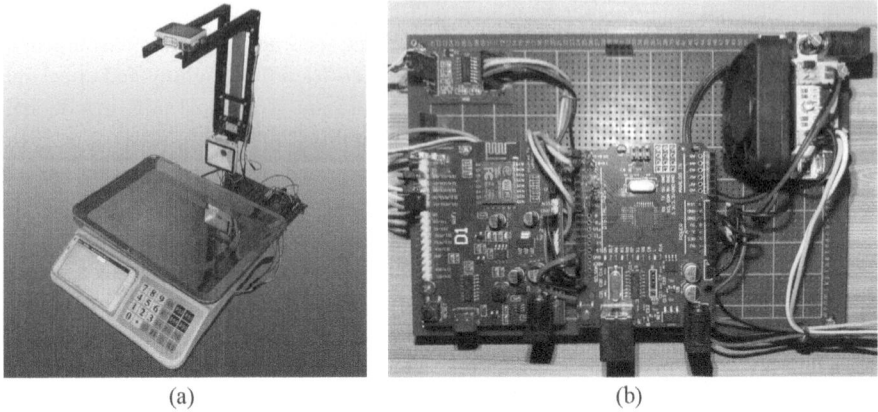

Fig. 1. (a) Smart scale overall, (b) Smart scale circuit board.

3.2 K210 Module

The K210 module is responsible for object detection and recognition. Once the camera captures an image of the fruit, the K210 neural network processor processes the image in real-time, producing detection results. Figure 2 displays the object detection outputs captured by the camera for three types of fruit: green apple, red apple, and mango.

Figure 2 shows the detection of a fruit object, with the bounding box highlighting the identified fruit. The K210 module has successfully recognized the fruit type based on the image features. The final result of the detection confirms the classification and provides the associated details such as the fruit type and detection accuracy.

These images highlight the K210 module's ability to recognize different types of fruits with high accuracy, confirming the classification and providing associated details such as fruit type and detection accuracy. The real-time object detection capability of the K210 module ensures the system's overall functionality and efficiency in retail environments (see Fig. 2).

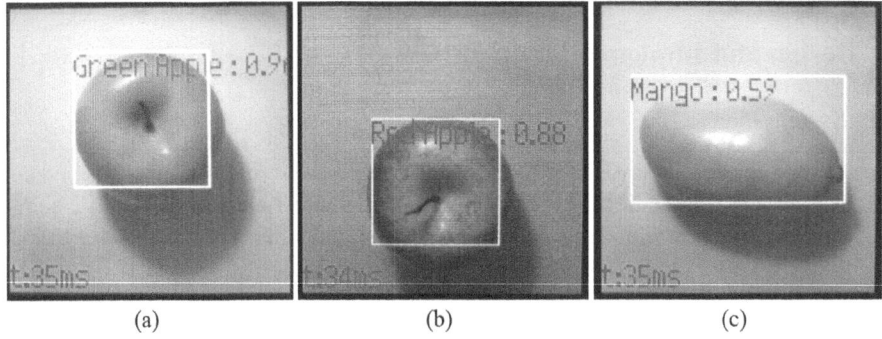

Fig. 2. (a) Recognition result for green apple, (b) Recognition result for red apple, (c) Recognition result for mango.

3.3 QR Code-Identified Bag

The QR code-identified bag is a crucial element of the eco-friendly supermarket weighing and checkout system. These bags are intended for repeated use by customers. When customers finish their shopping and pay for the loose-weighed items using these bags for the first time, they can take the bags home. On their subsequent visits to the supermarket to purchase loose-weighed goods, they can bring the same bags. During the weighing process, the unique QR code on the bag is scanned. When the bag is used next time to hold commodities and be weighed, the latest commodity records will be uploaded. At the cashier, the latest commodity data will be used for settlement. (see Fig. 3).

Fig. 3. QR code-identified bag.

3.4 Android-Based Weighing Interface

The Android-based upper computer interface for the weighing system offers comprehensive functions. Figure 4 shows the weighing interface with elements like the bag's QR code ID, product weight, unit price, and total amount. Users can select product

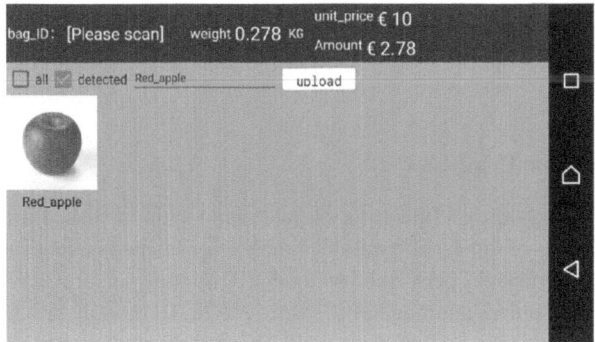

Fig. 4. Android-based Weighing Interface.

categories or let the system auto-identify. Manual input is available if needed. After product selection, click "upload" to finalize weighing. The system can upload weighing information to the server for real-time processing and storage. This interface integrates product selection, accurate weighing, and efficient data transmission, suitable for retail and supermarket environments (see Fig. 4).

3.5 Point-of-Sale (POS) Software Interface

Figure 5 shows the interface of the point-of-sale (POS) software, which is used for processing transactions at the checkout. The interface includes a scanning function to read the QR code on the reusable bag, retrieving the associated product information from the database.

In the interface, the scanned items are listed along with their weight, unit price, and total cost. The cashier can review all items before proceeding with payment. There is also an option to modify or remove items if necessary. After verifying the transaction details, the cashier can finalize the payment, and the system will automatically update the database for future reuse of the bag's QR code.

This user-friendly interface ensures a smooth and efficient checkout process, minimizing errors and enhancing the overall customer experience (see Fig. 5).

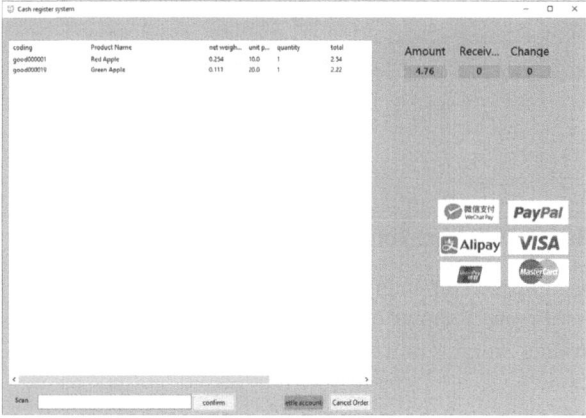

Fig. 5. Point-of-Sale (POS) Software Interface.

3.6 Implementation Principles

Figure 6 shows the system architecture, created using Wondershare EdrawMax, which illustrates the process of image recognition and data transmission using various modules.

The image recognition is performed by the K210 module's neural network processor. When the camera captures an image of the fruit, the K210 module processes the image in real-time and outputs the recognition result. This information is then processed by the Arduino Uno and transmitted to the ESP8266 module.

The QR code scanning module captures the bag's QR code, obtaining the bag's unique ID. The ESP8266 module then uploads this data to the MQTT server via Wi-Fi. The ESP8266 establishes a connection with the MQTT server to enable data publishing and subscribing. Both the K210 module and the HX711AD high-precision pressure sensor send the recognition and weight data through the ESP8266 to the MQTT server.

This workflow highlights the integration of image recognition and weight measurement with efficient data transmission, ensuring accurate product identification and environmental sustainability through bag reuse (see Fig. 6).

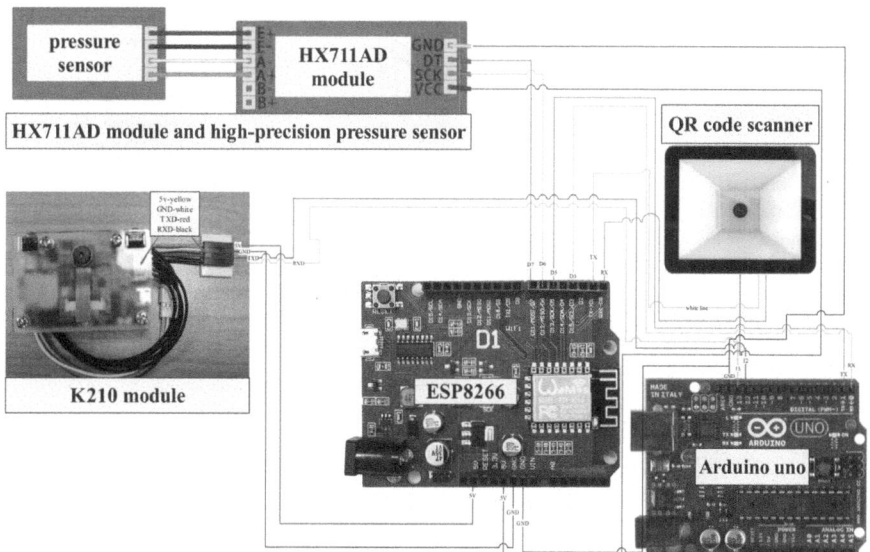

Fig. 6. Smart scale circuit diagram.

3.7 Software Workflow

Figure 7 illustrates the workflow of the system, created using Wondershare EdrawMax. The green lines represent the customer process, while the blue lines depict the system operations.

In the customer process (green lines), customers select fruits, weigh them, and proceed to the checkout counter. It should be noted that the QR code-identified bags are meant to be reused. When customers first use the bags for weighing and paying for their purchases, they can take the bags home. During subsequent shopping trips for loose-weighed items, they bring the same bags back. On the system side (blue lines), the electronic scale first captures the fruit's name, type, and weight. The QR code of the reusable bag is then scanned by the scale operator, after which the system packages this information into JSON format and sends it to the MQTT server. The server operating system is Linux, Python, and MySQL environment are installed and the Paho MQTT library is used. The server parses the JSON data and stores it in the database.

At checkout, the QR code on the bag is scanned, and the point-of-sale software retrieves the corresponding product details from the database using the QR code identifier. This allows for the completion of the checkout process. If the customer reuses the same bag during their next visit, the new product information will overwrite the old data, enabling the reuse of the bag and supporting environmental sustainability.

The diagram highlights the seamless integration of customer interaction and system automation, contributing to the efficient and eco-friendly operation of the weighing and checkout process (see Fig. 7).

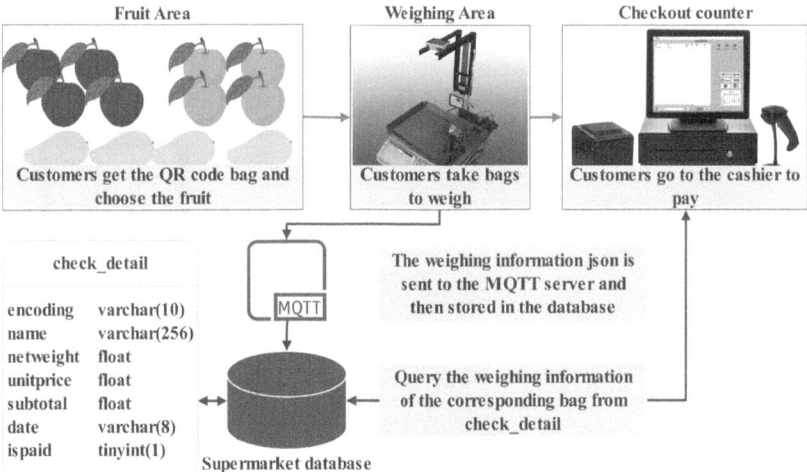

Fig. 7. Software Flowchart.

3.8 Features

This system achieves the following main functions: automatic recognition and display of fruit types, real-time collection and display of fruit weight, scanning of the bag's QR code, and uploading of the weighing details to the server. After user confirmation, the system automatically sends the weighing information to the MQTT server in JSON format. The checkout system reads the checkout information from the database in real time and completes the settlement process.

4 Results and Discussion

The IoT-based supermarket weighing and checkout system performed well in real-world tests, achieving 96% accuracy in AI visual recognition and maintaining a weighing error margin of less than 0.05 kg. Data transmission via MQTT was fast, with real-time updates ensuring smooth operation.

The use of QR code-identified reusable bags reduced plastic waste by 30% during the pilot phase, supporting sustainability efforts. Additionally, the system's cost-effective components and modular design allow for easy expansion in different retail environments.

A key challenge lies in raising environmental awareness among customers [7] and encouraging their participation in reusing the bags. The success of this system's environmental impact depends largely on whether customers embrace the habit of reusing bags. Therefore, effectively communicating the importance of sustainability and guiding customers to adopt eco-friendly practices is crucial to the long-term success of the system.

To increase customer participation in reusing the QR code bags, several strategies could be implemented. For instance, supermarkets could offer incentives such as discounts or loyalty points for customers who bring their reusable bags. Additionally, educational campaigns could be launched to raise environmental awareness and highlight the benefits of reusing bags.

Future improvements could focus on enhancing the recognition algorithm, as well as introducing features like contactless payment and real-time analytics.

5 Conclusion

This work combines AI visual recognition, QR codes, database management and weighing checkout systems, creating an eco-friendly, efficient and intelligent system. It reduces plastic bag consumption via QR-code-identified bag reuse and improves supermarket operation convenience and efficiency. Test results prove its performance and potential in retail.

For large-scale use, long-term operational metrics are crucial. We'll monitor system reliability long-term, establish a maintenance schedule and develop diagnostic tools. Cost-effectiveness will be analyzed deeply. Future work includes optimizing the recognition model, exploring more application scenarios and evaluating these metrics to ensure the system's sustainable implementation in retail and contribute to IoT technology progress.

References

1. Durak, S.G.: Investigation and evaluation of the effect to environmental pollution of plastic shopping bags. Türk Bilimsel Derlemeler Dergisi **2**, 20–24 (2016)
2. Gokhale, P., Bhat, O., Bhat, S.: Introduction to IOT. Int. Adv. Res. J. Sci. Eng. Technol. **5**(1), 41–44 (2018)
3. Zhu, W.Y., Wong, W.K., Morsalin, S., Wang, S.H., Sheu, M.H.: Software and hardware integration system design with fruit identification for smart electronic scale applications. In: 2021 IEEE International Conference on Consumer Electronics-Taiwan (ICCE-TW), pp. 1–2. IEEE (2021)
4. Almalki, F.A., et al.: Green IoT for eco-friendly and sustainable smart cities: future directions and opportunities. Mob. Netw. Appl. **28**(1), 178–202 (2023)
5. Atmoko, R.A., Riantini, R., Hasin, M.K.: IoT real time data acquisition using MQTT protocol. J. Phys. Conf. Ser. **853**(1), 012003 (2017)

6. Luo, Z., Wang, H.: Research on intelligent supermarket architecture based on the internet of things technology. In: 2012 8th International Conference on Natural Computation, pp. 1219–1223. IEEE (2012)
7. Saha, A., Kuruppuge, R.H.: Determinants of consumer awareness of green products: a study of customers of super markets. Mediterr. J. Soc. Sci. **7**(6), 349–355 (2016)
8. Genon, J., Mabunay, J., Opsima, J., Zamora, R., Repaso, J., Sasan, J.M.: Exploring the alternative solutions and strategies of toledo city government for the damaging impact of single-use plastic bag in the environment. ScienceRise **1**(78), 3–11 (2022)
9. Hossain, M.S., Chisty, N.M.A., Hargrove, D.L., Amin, R.: Role of Internet of Things (IoT) in retail business and enabling smart retailing experiences. Asian Bus. Rev. **11**(2), 75–80 (2021)
10. Jayaram, A.: Smart retail 4.0 IoT consumer retailer model for retail intelligence and strategic marketing of in-store products. In: Proceedings of the 17th International Business Horizon-INBUSH ERA-2017, Noida, India 9 (2017)
11. Shekhawat, S.: Smart retail: how AI and IoT are revolutionising the retail industry. J. AI, Robot. Workplace Autom. **2**(2), 145–152 (2023)
12. Vadakkepatt, G.G., et al.: Sustainable retailing. J. Retail. **97**(1), 62–80 (2021)
13. Lai, K.H., Cheng, T.C.E., Tang, A.K.: Green retailing: factors for success. California Manage. Rev. **52**(2), 6–31 (2010)
14. Albăstroiu, I., Felea, M.: Enhancing the shopping experience through QR codes: the perspective of the Romanian users. Amfiteatru Econ. J. **17**(39), 553–566 (2015)
15. Soon, T.J.: QR code. Synthesis J **2008**, 59–78 (2008)
16. Tiwari, S.: An introduction to QR code technology. In: 2016 International Conference on Information Technology (ICIT), pp. 39–44. IEEE (2016)

Focal Consistency Network for Developmental Stage Classification of Embryos with Time-Lapse Embryo Video Datasets

Yiming Li[1,3], Hua Wang[1,3], Jingfei Hu[2(✉)], and Jicong Zhang[1,3(✉)]

[1] School of Biological Science and Medical Engineering, Beihang University, Beijing, China
jicongzhang@buaa.edu.cn
[2] School of Medical Informatics Engineering, Anhui University of Chinese Medicine, Hefei, Anhui, China
[3] Hefei Innovation Research Institute, Beihang University, Hefei, China

Abstract. In the field of assisted reproduction, time-lapse technology can collect embryo images across multiple focal planes, which helps embryologists stage embryos and dynamically evaluate their quality, thereby improving the success rate of transplantation. Clinical practitioners rely on integrated information from various focal planes, as each plane encompasses information from all cells, considering the influence of depth of field. However, existing methods predominantly focus on single-plane image acquisition, either neglecting comprehensive information or failing to exploit internal correlations. To address this issue, we propose a method named Focal Consistency Network (FC-Net) for processing time-lapse embryo video datasets and classifying embryo developmental stages. The FC-Net comprises a classification head and a multi-focal consistency head. While the classification head learns the categories of images from different focal planes at the same time, the multi-focal consistency head ensures consistency between the predictions of other focal planes and the main focal plane, facilitating the model's learning of more stable feature information. The method demonstrates significantly superior performance on publicly available time-lapse embryo video datasets compared to other models, achieving a success rate increase of 3% points. Furthermore, visual analysis of the results confirms the alignment of the predicted embryo developmental stage results with the actual scenario, further validating the effectiveness and superiority of the proposed method.

Keywords: Time-lapse · Multi-focal consistency · Classification

1 Introduction

Infertility is one of the most pressing health issues worldwide [1]. It is characterized by the inability to achieve conception through regular unprotected sexual intercourse [2]. In-vitro fertilization (IVF) stands as a prevalent approach for

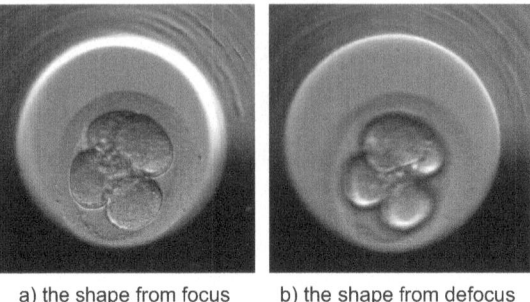

a) the shape from focus   b) the shape from defocus

Fig. 1. Different focal image.

addressing infertility [3]. During IVF, embryos undergo cultivation in a controlled laboratory setting in vitro until they reach the blastocyst stage. Subsequently, a human blastocyst exhibiting high implantation potential is selected for transfer through visual assessment of its dynamic features.

In clinical practice, observers tend to habitually observe the developmental stage and status of embryos on the optimal focal plane. However, embryos are living cells, and in three-dimensional space, many details cannot be fully captured and comprehended on a single focal plane. For example, as shown in Fig. 1(a), which is the optimal focal plane, it's uncertain whether the embryo has completed the transition from three cells to four cells, as the structure at the top left could be a fragment or a complete cell. However, when adjusting to the image at a certain distance away from the optimal focal plane shown in Fig. 1(b), it becomes evident that the structure at the top left is indeed a complete cell, and its morphological status can be observed clearly. Nonetheless, the obscured region at the top right highlights the necessity for multiple focal planes for correction. Consequently, we realize that each focal plane contains all necessary information, albeit with varying emphasis. Therefore, leveraging the information from different focal planes to mutually supervise and optimize each other enables the network model to better understand and learn the features of embryo images at different developmental stages, facilitating inference of developmental timing. Additionally, the final predictions can integrate outputs from multiple focal planes for a comprehensive assessment.

For embryo image staging, traditional approaches typically only select the optimal focal plane image as input for training and detection in neural networks [4,5]. In clinical practice, data from different focal planes play an important role in the evaluation and classification of embryonic development stages, which helps to systematically and comprehensively evaluate the stages of embryonic development.

Therefore, in this work, we propose a novel Focus Consistency Network (FC-Net) for embryo image staging based on collected multi-focal image sequences. The FC-Net consists of two modules, including a classification head and a multi-focal consistency head. The classification head ensures accurate classification of a specific focal image at a particular moment, while the multi-focal consistency head ensures consistency of predicted results across multiple focal images at a specific moment. By treating images from different focal planes at a specific moment as perturbations and complements to the optimal focal plane, the model learns specific information from different focal planes and removes image disturbances (potentially caused by lighting effects, motion artifacts, etc.), thus enhancing the model's performance and obtaining a more stable feature space, thereby improving the model's robustness. The optimal focal plane is determined by the machine through self-assessment using traditional algorithms. Therefore, the multi-focus consistency network proposed in this paper can to some extent compensate for errors introduced during the machine capture process. As shown in Fig. 1, the focal plane identified by the machine as optimal is not necessarily the clearest.

In summary, the main contributions of this paper are as follows:

- We first utilize multi-focus information and propose a novel Focal Consistency Network to identify the developmental stages of embryo images.
- The FC-Net comprises a classification head and a focal-consistency head, where the classification head learns the developmental stage of embryo images, and the focal-consistency head ensures consistency of predicted results across multiple focal images at specific moments.
- The first decomposition of time-lapse data to obtain a dataset for embryo developmental stage identification, which containing comprising 704 embryos, totaling 63832×3 images with developmental stage labels.

2 Method

Figure 2 illustrates in detail the specific structure of our proposed Focal Consistency Network for cross-sectional detection tasks at different developmental stages of embryos.

2.1 Classification Head (CH)

In this process, we adopt an iterative input strategy, where the reconstructed multi-focal image sequences are sequentially fed into a standardized backbone network for processing. In this study, we chose the well-known VGG network as our backbone structure. Each focal plane image, after processing through the backbone, generates corresponding prediction results, which are then compared and analyzed against the labels.

The classification head component operates based on the images obtained from the embryo's optimal focal plane X_0 at a specific moment and images X_i (where $i = 1, 2$) obtained from different out-of-focus focal planes, following a sequence from the center to the sides. After learning, the network generates seven prediction outputs Y_i (where i also ranges from 1 to 2) for each type of focal plane image. To guide the network learning process, we compute the cross-entropy loss function (CE loss) between each output Y_i and its corresponding ground truth Y. This measurement assesses and optimizes the disparity between the model's predicted results and the actual annotations, thereby driving the training and optimization of the entire network.

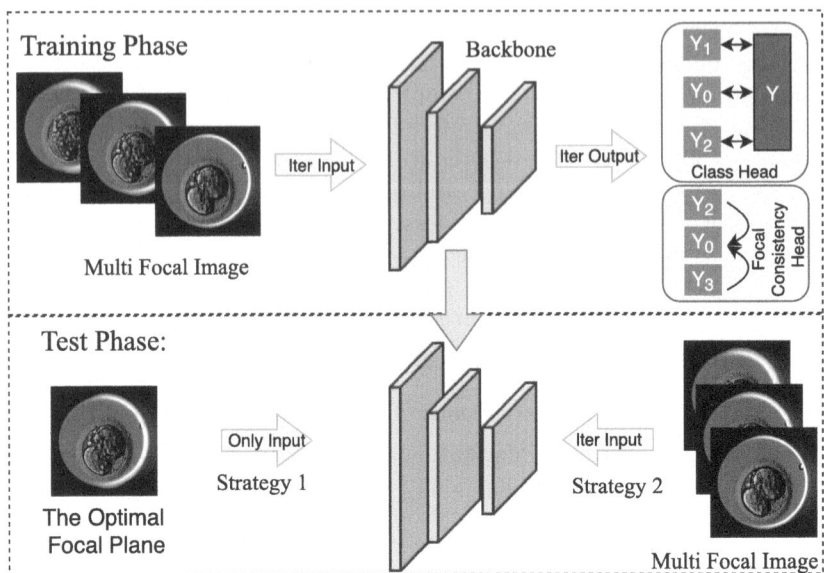

Fig. 2. The details of the proposed FC-Net. During the training process, multi-focus images are iteratively fed into the backbone. Loss is obtained through the classification head and multi-focus consistency head, after which the entire network undergoes training. Trained networks can be utilized in two inference modes. The first mode involves testing only the optimal focal plane image, while the second mode mirrors the training process by iteratively inputting data into the pre-trained model. Multiple outputs are then weighted and combined to produce the final result.

2.2 Focal Consistency Head (FCH)

It is worth noting that in the design of the FC-Net network, besides considering the absolute error between the prediction outputs Y_i corresponding to each focal plane image X_i and the ground truth Y, special attention is paid to the relative consistency of prediction results across different defocus depths. To address this relative loss, the network architecture adopts a contrastive learning approach to ensure that features extracted from different defocused layers are closely related and mutually referenced.

In practical terms, images X_i from non-optimal focal planes are paired with the image X_0 from the optimal focal plane, leveraging the contrastive learning mechanism to encourage X_i to inherit key feature information from X_0. This helps reduce the inconsistency in feature distribution across different focal planes. Additionally, for consecutive defocused images X_{i+1}, their prediction results Y_{i+1} are aligned and calibrated with adjacent prediction results Y_i from the optimal focal plane using mean square error (MSE) loss. This aims to enable the network to finely learn and integrate information from each focal plane to accurately assess the embryo's status.

With this design, FC-Net not only comprehensively captures and understands the subtle structural changes of the embryo across multiple focal planes but also significantly improves the accuracy of identifying critical cross-sectional slices of embryo development. This innovative training strategy also enhances the overall generalization capability and stability of the network, ensuring precise and consistent recognition performance even in the face of various complex focal plane conditions.

2.3 Loss Function

During the overall network updating process, the loss function consists of a combination of weighted cross-entropy loss ($\mathcal{L}_{CH}$) for classification head and weighted mean squared error ($\mathcal{L}_{FCH}$) for focal consistency head, where the weights serve as adjustment parameters for both loss functions.

For an individual sample, the calculation formula for $\mathcal{L}_{CH}$ is given by:

$$\mathcal{L}_{CH} = -\sum_{k=1}^{K} y_k \log(p_k), \quad (1)$$

where y_k represents the ground truth, $y_k = 1$ if the sample belongs to class k, and $y_k = 0$ otherwise. p_k denotes the predicted probability of class k by the model.

On the other hand, MSE is typically utilized in regression problems to gauge the disparity between model predictions and true values. The $\mathcal{L}_{FCH}$ formula is as follows:

$$\mathcal{L}_{FCH} = \frac{1}{n} \sum_{i=1}^{n} (y_i - \hat{y}_i)^2, \quad (2)$$

where y_i represents the predicted distribution for the optimal focal plane, while $\hat{y}_i$ denotes the predicted distribution by the model for out-of-focus images, and n denotes the total number of samples.

Consequently, the loss function for the entire network updating process is the sum of weighted cross-entropy loss and weighted MSE, with the weights serving as adjustment parameters. This is expressed as:

$$\mathcal{L} = \alpha * \mathcal{L}_{CH} + \beta * \mathcal{L}_{FCH}, \quad (3)$$

where $\alpha = 1.0$, $\beta = 0.1$. This formulation allows for flexible adjustment of the relative importance of each loss component during training stage.

3 Experiments

3.1 Dataset and Evaluation Metric

We evaluate the proposed method on the only public embryo dataset [6]. This dataset comprises pre-implantation embryo imaging data captured by the

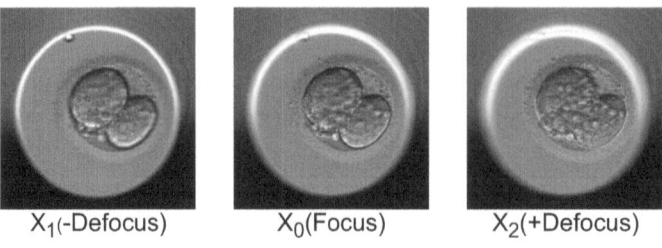

X_1(-Defocus) X_0(Focus) X_2(+Defocus)

Fig. 3. Different Focal Image.

Table 1. The supervised and semi-supervised datasets in this study. The public dataset is labeled and our images are lack of annotations.

Class	0 (1 cell)	1 (2 cell)	2 (3 cell)	3 (4 cell)	4 (5-7 cell)	5 (8 cell)	6 (8+ cell)
Training (493 embryos)	5366	5450	6369	5702	5745	6466	9367
Test (71 embryos)	729	767	1058	814	897	1092	1381
Validation(140 embryos)	1612	1556	1148	1720	1693	1991	2909
All (704 embryos)	7707	7773	8575	8236	8335	9549	13657

embryoscope machine at regular intervals over the first five days of embryonic development. At each time point, multiple focal planes are captured, with the optimal focal plane determined by the machine through self-assessment using traditional algorithms. Therefore, the method proposed in this paper can to some extent compensate for errors introduced during the machine capture process. For this dataset, we uniformly extract data from the optimal focal plane and the two nearest planes at each time point, totaling three images, The images from different focal planes at the same moment are illustrated in Fig. 3. To ensure the accuracy of developmental stage labeling, we extract a certain proportion of images from intermediate time points based on their respective proportions during the developmental period to form the final dataset. Finally, to prevent data leakage, we categorize the dataset primarily based on the embryos and divide it into training, validation, and test sets, as shown in Table 1.

To evaluate the performance of the network, Accuracy is commonly used to test methods, which is defined as:

$$\text{Accuracy} = \frac{TP + TN}{TP + FP + TN + FN}. \quad (4)$$

This indicator is defined based on four parameters: TP (true positive), FP (false positive), TN (true negative), and FN (false negative).

3.2 Implementation Details

We train our model 10 epochs with an initial learning rate of 0.0002, the momentum of 0.9, and the weight decay of 0.0005. We adopt Poly dynamically adjust the learning rate and use the Adam optimizer [7] to update network parameters.

Table 2. Comparison of performances of multi-class classification. Best results are highlighted in bold.

Method	Total (%)	0 (%)	1 (%)	2 (%)	3 (%)	4 (%)	5 (%)	6 (%)
VGG [8]	66.16	91.50	76.01	34.41	49.08	48.38	62.64	96.02
DenseNet [9]	65.90	89.58	74.32	34.22	50.55	44.15	64.84	**97.03**
ResNet [10]	66.51	94.10	**77.84**	36.86	**67.51**	39.91	56.14	93.19
MobilNet [11]	57.43	89.58	71.45	24.10	49.69	31.44	47.62	87.40
FC-Net (one)	69.93	92.73	76.92	**43.76**	59.66	**53.07**	66.58	93.70
FC-Net (multi)	**70.05**	**92.73**	77.05	43.67	60.39	52.29	**67.13**	93.92

The whole framework is implemented on PyTorch (version ≥ 1.10). All models were trained and tested on a 32 GB NVIDIA Tesla V100 GPU. We scale the input size of images to 256×256 before being fed into the models during training and testing. We augment each sample with random scaling, followed by random rotating, random cropping, and random horizontal flipping in the training stage.

3.3 Quantitative Results

In this work, we first compare our proposed method with classical models such as VGG [8], DenseNet [9], and ResNet [10]. Additionally, due to the clever utilization of multi-focus information in our method, we also test whether the multi-focus information is adequately utilized. As illustrated in Fig. 2 during the testing phase, we first input only a single optimal focal plane image to obtain the prediction result, referred to as FCNet (one). Furthermore, we test the iterative input of multiple focal plane images to obtain results for different focal planes, followed by weighted summation to achieve consistency results, referred to as FCNet (multi). From the Table 2, it can be observed that classical methods start to weaken at 3+ cells, primarily due to the increased difficulty in recognizing more cells, given their transparency and overlapping nature. Moreover, the high results for 8+ cells in classical methods can be attributed to the higher proportion of 8+ cell samples in the training data, indicating a weaker model learning capability. Consequently, to minimize loss, the model tends to favor categories with more samples. Furthermore, it is evident that the results of our FCNet (one) and FCNet (multi) are very similar, further indicating that our model has learned a stable feature space and effectively mined the inherent features of the images.

The proposed method leverages information from different focal planes to some extent, thereby promoting consistency among results obtained from various focal planes. Hence, we compared the results of the validation set with those of the test set. From the Table 3, it can be observed that the pure VGG method achieves an overall accuracy of 74.34% on the validation set, whereas the test set results are only 66.16%, showing a difference of 8.18%. Moreover, except for the 8+ cell category, where the decrease is more pronounced, this somewhat

Table 3. Comparison of performances of multi-class classification both validation and test phase.

Method	Total (%)	0 (%)	1 (%)	2 (%)	3 (%)	4 (%)	5 (%)	6 (%)
VGG-Validation	74.34	92.68	81.61	47.65	64.71	54.76	67.20	92.78
VGG-Test	66.16	91.50	76.01	34.41	49.08	48.38	62.64	96.02
FC-Net (multi)-Validation	76.50	92.31	84.37	58.19	70.00	52.75	63.69	92.85
FC-Net (multi)-Test	70.05	92.73	77.05	43.67	60.39	52.29	67.13	93.92

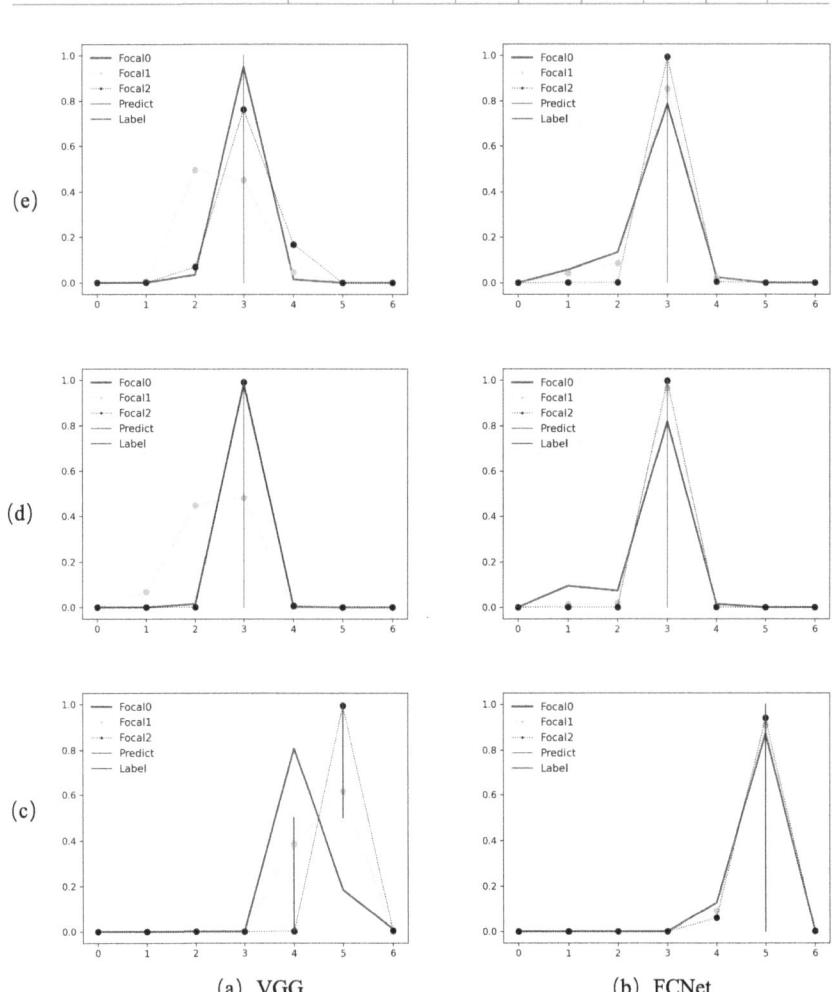

Fig. 4. The trained VGG and FCNet models' prediction probability distributions across different focal planes are displayed. The horizontal axis represents the categories, while the vertical axis indicates the corresponding probabilities.

Table 4. Comparison of performances of multi-class classification. Best results are highlighted in bold.

Method	Total (%)	0 (%)	1 (%)	2 (%)	3 (%)	4 (%)	5 (%)	6 (%)
VGG	66.16	91.50	76.01	34.41	49.08	48.38	62.64	96.02
VGG+CH	68.76	92.73	72.36	38.56	56.21	**56.63**	63.83	**96.38**
VGG+FCH	68.55	92.18	74.84	35.35	**68.51**	43.03	67.03	95.80
FC-Net (multi)(CH+FCH)	**70.05**	**92.73**	**77.05**	**43.67**	60.39	52.29	**67.13**	93.92

indicates the model's poor learning capability, tending towards predicting the category with a larger sample size. However, after incorporating the multi-focus consistency strategy proposed in this paper, it can be observed that the overall decline in the model is smaller, with decreases in each category being more balanced and gradual. Additionally, the categories of single-cell and 5+ cells show some improvement (recognizing more cells is more difficult due to their transparency and overlapping), compared to the pure VGG method, demonstrating superior performance. This once again underscores the effectiveness and robustness of our approach.

3.4 Qualitative Results

Multi Focal Compare. To further validate the advantages of our proposed consistency strategy, we conducted tests using multi-focus images on both the trained VGG and our proposed FCNet models to observe if the results exhibit consistency. Firstly, it is notable that compared to the VGG model in Fig. 4, our model demonstrates more consistent prediction results across multi-focus images, with significantly overlapping probability peaks. This confirms that FCNet learns features that are more stable, exhibiting stronger resistance to interference, and also confirms its ability to more effectively exploit the intrinsic information within images. From the figures, it can be observed that in Fig. 4(d) and (d), although the VGG model correctly classifies, there is an error labeled as "plane data," and the probability distribution across focal planes predicts the occurrence of the second and third categories, indicating a potential confusion-induced misclassification, consistent with the issue presented in our Fig. 1. Meanwhile, in Fig. 4(c), it can be observed that VGG misclassifies at the optimal focal plane but correctly classifies in the out-of-focus image, indicating that the machine's captured focal plane may not necessarily be of the highest quality, in line with our hypothesis. In Fig. 3, it can be observed that the clearest image is not X_0, but rather X_2.

3.5 Ablation Studies

To validate the effectiveness of the proposed method, which cleverly leverages multi-focus information, the FC-Net comprises two modules to achieve the aforementioned objectives: the classification head and the multi-focus consistency head. The classification head ensures accurate classification of a specific focal image at a particular moment, while the multi-focus consistency head ensures consistency in the prediction results of multiple focal images at the same moment.

Hence, we conducted experiments comparing with VGG, VGG with only the classification head utilizing multi-focus images (VGG+CH), VGG with only the focal consistency head utilizing multi-focus images (VGG+FCH), and FC-Net (include both the classification head and focal consistency head on top of VGG). We adopted testing scheme two, namely FCNet (multi), to evaluate the results. From the Table 4, it can be observed that the performance of the pure VGG model is not satisfactory. However, upon incorporating CH and FCH, the overall accuracy performance improves significantly by up to 8% points.

4 Conclusions

In this paper, we propose a Focal Consistency Net for classification of human blastocysts at different stages. This architecture is based on our proposed classification head and focal consistency head, leveraging the multi-focus nature of the images to compensate for errors inherent in the machine capture process. Moreover, the perturbations between different focal planes encourage the network to learn a more stable feature space. Ablation studies demonstrate the superiority of our proposed module. Experimental results validate that our method achieves the best classification performance compared to existing state-of-the-art methods while maintaining competitiveness.

However, due to the limited availability of publicly accessible datasets, there is an urgent need to address the issue of validation on multi-source datasets. Additionally, in the future, maximizing the utilization of unlabeled data to further improve model performance based on existing methods will be necessary.

References

1. Zegers-Hochschild, F., et al.: The international committee for monitoring assisted reproductive technology (icmart) and the world health organization (who) revised glossary on art terminology, 2009. Hum. Reprod. **24**(11), 2683–2687 (2009)
2. Mélodie, V.B., Christine, W.: Fertility and infertility: definition and epidemiology. Clin. Biochem. 2–10 (2018)
3. Zhao, J., et al.: Effects of abnormal zona pellucida on fertilization and pregnancy in ivf/icsi-et. J. Reprod. Contracept. (2015)
4. Khan, A., Gould, S., Salzmann, M.: Deep convolutional neural networks for human embryonic cell counting. In: European Conference on Computer Vision (2016)
5. Rad, R.M., Saeedi, P., Au, J., Havelock, J.: Cell-net: embryonic cell counting and centroid localization via residual incremental atrous pyramid and progressive upsampling convolution. IEEE Access (2019)
6. Gomez, T., et al.: Towards deep learning-powered IVF: a large public benchmark for morphokinetic parameter prediction (2022)
7. Kingma, D., Ba, J.: Adam: a method for stochastic optimization. Comput. Sci. (2014)

8. Simonyan, K., Zisserman, A.: Very deep convolutional networks for large-scale image recognition. arXiv preprint arXiv:1409.1556 (2014)
9. Huang, G., Liu, Z., Van Der Maaten, L., Weinberger, K.Q.: Densely connected convolutional networks. In: Proceedings of the IEEE Conference on Computer Vision and Pattern Recognition, pp. 4700–4708 (2017)
10. He, K., Zhang, X., Ren, S., Sun, J.: Deep residual learning for image recognition. In: Proceedings of the IEEE Conference on Computer Vision and Pattern Recognition, pp. 770–778 (2016)
11. Sandler, M., Howard, A., Zhu, M., Zhmoginov, A., Chen, L.C.: Mobilenetv2: inverted residuals and linear bottlenecks. In: Proceedings of the IEEE Conference on Computer Vision and Pattern Recognition, pp. 4510–4520 (2018)

Chest X-ray Image Rib Segmentation via Disentanglement Enhancement Network

Lili Huang[1], Shiqi Li[1], Lingma Sun[2(✉)], and Chuanfu Li[3]

[1] The School of Computer Science and Technology, Anhui University, Hefei, China
[2] The School of Artificial Intelligence and Big Data, Hefei University, Hefei, China
sunlm@hfuu.edu.cn
[3] The Department of Chinese Medicine, Anhui University of Chinese Medicine, Hefei, China

Abstract. Rib segmentation is one of the most challenging tasks in the field of medical image segmentation, and also an important tool to assist doctors in diagnosis. However, the challenge of this task is that ribs overlap each other and have low contrast and blurred edges. To address these issues, we propose a novel Disentanglement Enhancement Network called DENet, which disentangles and enhances appearance representations of ribs via the segmentation difficulties and location priors respectively, for robust rib segmentation. In particular, we design a difficulty-guided representation disentanglement module to focus on the most challenging ribs by using the separate decoders for these challenging ribs. To leverage the relations among ribs, we design a location-aware mutual enhancement module, which enables the information exchange and enhancement among different ribs according to the location priors. The experimental results show that mDice of our method is improved by 2% compared to previous methods. Extensive experiments on the dataset demonstrate the effectiveness of our DENet against state-of-the-art medical image segmentation methods.

Keywords: Chest X-ray Image · Deep Learning · Disentanglement Method · Rib Segmentation

1 Introduction

Due to its simplicity and low level of radiation dose [13], X-ray is still the main modality for radiological exam currently, despite the computed tomography has better results than X-ray. And the number of chest X-ray images accounted for more than one-third of the total number of images in the hospital radiology department [3]. Consequently, chest X-ray is an important screening tool for several lung disease diagnosis, such as pulmonary effusion, pulmonary nodule, tuberculosis, and bone fractures [5]. There are heart, ribs, clavicle, spine and other organ in chest X-rays. Among these, ribs could affect the results of disease diagnosis and surgery planning [17], since the ribs always coincide with other soft tissues in chest. Therefore, it is crucial to segment rib accurately in chest X-ray for subsequent clinical diagnosis.

In recent years, with the great development of deep neural network [6] and attention mechanism [14] in computer vision, most existing methods [2] often exploit the advanced network framework and overlook the domain knowledge. Despite the effort of recent research, there are still some challenges remain. Firstly, the floating ribs are neglected. The VinDr-RibCXR [8] dataset contains 245 CXR images for rib segmentation. For each chest X-ray image, there are corresponding 20 labeled images, and each labeled image contains a different rib, which does not contain annotated images of floating ribs. Whereas, the primary role of floating ribs is to protect the heart and lungs from injury, and it is vital to segment and determine if there is a fracture in floating ribs. Moreover, floating ribs are more blurred and have lower contrast than others ribs in chest X-rays, which makes the segmentation task extremely difficult. Secondly, the ribs often overlap each other, however, this method ignores the spatial relationship of neighbouring ribs, resulting an ambiguous boundaries and confused segmentation.

To address the aforementioned issues, we propose a novel DENet (shown in Fig. 1), which is based on difficulty guidance and information interaction between ribs. Specifically, We draw on the concept of divide and conquer, and divide the ribs into easy ones and difficult ones to segment and decode them separately, which takes floating ribs as difficult ones since they are much harder than others. Accordingly, a difficulty-guided representation disentanglement module is devised to focus on the challenging ribs by using the separate decoders. In addition, motivated by the clinical knowledge of radiologists, which utilize the prior location knowledge to locate the adjacent ribs, especial for the floating ones, the proposed network exploits the location prior correlation and further advocates a location-aware mutual enhancement module, which applies contiguous ribs to enable the information exchange and enhancement among different ribs. We summarize the main contributions of this paper below:

- We propose a difficulty- representation disentanglement module to segment each rib. The advantage of splitting the most challenging ribs separately is that the decoder only serves itself and does not need to consider the other ribs.
- We propose a location-aware mutual enhancement module that can utilize the information interaction between ribs to enhance rib features.
- The experimental results show that our method is superior to state-of-the-art methods.

2 Related Work

2.1 Rib Segmentation

Rib segmentation is the process of separating the 24 ribs present in a chest X-ray image. The ribs and lung fields overlap with each other, resulting in low-contrast images. Moreover, the ribs also overlap with each other, further complicating the task of rib segmentation [9].

The task of rib segmentation has been extensively studied by researchers. For instance, Loog et al. [7] design a supervised iterative approach based on statistical classification. The approach involves obtaining a rough rib segmentation result, followed by continuous iteration to update the result. However, the method blurs the edges of the ribs, which presents a challenge for accurate segmentation. Wang et al. [16] propose a multi-task dense connection network that relies on multi-scale feature fusion. However, the model may easily overfit due to the large number of Unet parameters involved. Wang et al. [15] present a rib segmentation model that is based on multi-scale networks and unpaired sample enhancement. However, the method still struggles to effectively handle overlapping areas. Candemir et al. [1] propose a rib detection method that utilizes rib-bone atlases to calculate the posterior rib boundaries. However, this method is limited to segmenting only the posterior ribs rather than the entire rib cage. In our novel DENet, we inherit the backbone of deep networks as a encoder to extract rich features, since these deep learning networks have good performance on other tasks. Nevertheless, to address the aforementioned challenges, we design our novel DENet using a disentanglement approach.

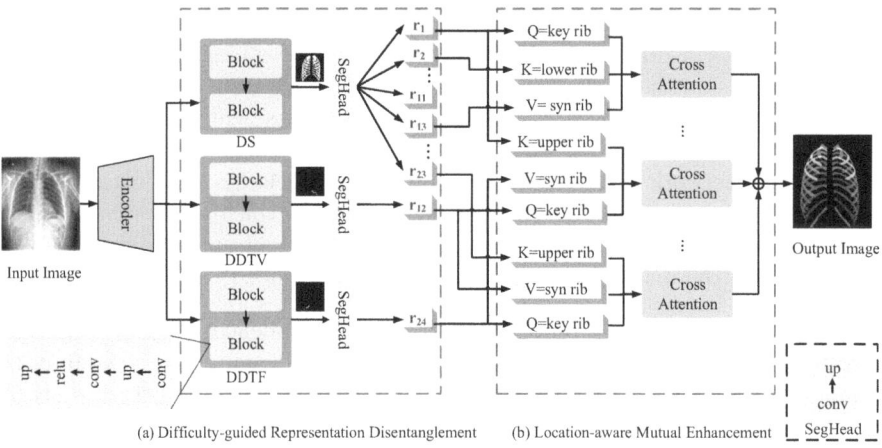

Fig. 1. The architecture of our network. Our network consists of two parts: (a) Difficulty-guided Representation Disentanglement and (b) Location-aware Mutual Enhancement. Decoder_Simple (DS), Decoder_Difficult12 (DDTV), and Decoder_Difficult24 (DDTF) respectively represent three decoders. r_k (k = 1, 2,...,24) represents the feature representation of the k-th rib. Difficulty-guided Representation Disentanglement (DGRD, Sect. 3.1) allows the most challenging ribs to be decoded with a separate decoder. Location-aware Mutual Enhancement (LAME, Sect. 3.2) can realize the information interaction between different ribs.

3 Method

The proposed model architecture is shown in Fig. 1, which is composed of two consecutive parts: a difficulty-guided representation disentanglement module and a location-aware mutual enhancement module. Given a chest X-ray image x∈ $R^{H \times W \times C}$, our goal is to predict 24 images of ribs, each containing a different rib. In the following, the details of proposed module are presented.

3.1 Difficulty-Guided Representation Disentanglement

In chest X-rays, the bottom pair of ribs (i.e., the 12th and 24th ribs) is particularly challenging due to blurred boundaries and lower contrast compared to other ribs. The network can not learn the specific features about these difficult ribs, if a unified decoder is applied for segmentation. Thus, motivated by the spirit of representation disentanglement, a difficulty-guided representation disentanglement module is designed to decode different rib features (shown in Fig. 1(a)). Based on the prior that floating ribs are more difficulty to segment, the difficulty-guided representation disentanglement module (DGRD) consists of three decoders (named Decoder Simple (DS), Decoder Difficult12 (DDTV), and Decoder Difficult24 (DDTF) respectively depending on the information decoded). We use DDTV decoder for the 12th rib, DDTF for the 24th rib, and DS for the remaining 22 simple ribs considering the number of parameters. And each decoder is followed by a segmentation head. The input of the DGRD is the features extracted from an encoder [2].

The three independent decoders in DGRD have similar structure without sharing weights. This simple yet effective design can learn differential features via the representation disentanglement. Each decoder is a multi-cascaded upsampler that decodes hidden features through multiple cascaded upsampling operations, changing the resolution of hidden features from $\frac{H}{P} \times \frac{H}{P}$ to H × W. A decoder block consists of an upsampling operation, two convolution layers, a RELU layer, and another upsampling operation. The decoder can be expressed as follows:

$$f_k = UP(RELU(\sigma(\sigma(UP(f))))), \tag{1}$$

where UP(·) is upsampling operation, RELU(·) is a RELU layer, σ is a 3×3 convolution layer, and f is the feature of the last layer of the encoder.

After each decoder, we segment each rib using a segmentation heads to obtain feature representation of ribs r_k (k = 1, 2,...,24). Our three segmentation heads are similar in structure but do not share weights. The segmentation head consists of a convolution layer and an upsampling operation. The difference between our three segmentation heads is the output channel of the convolution layer. The output channel number of the first segmentation head is 22 (the 22 simple ribs), the output channel number of the second segmentation head is 1 (the 12th rib) and the output channel number of the third segmentation head is 1 (the 24th rib).

3.2 Location-Aware Mutual Enhancement

Each rib has individual variability, the shape, width and length of ribs are variable from the top to the bottom. Nevertheless, the consecutive ribs are more similar than distant ones. In chest X-ray, the floating ribs can also be located assisted by the adjacent ribs for a radiologist. Inspired by these, a location-aware mutual enhancement module is proposed to boost the representation of the ribs caused by the overlap. As mentioned before, the consecutive ribs have similar shape, this attribute can be directly used to strengthen the rib representation. Therefore, we utilize the cross attention mechanism to model the relationship between the ribs. K, Q, and V come from the three ribs, Q from the rib to be segmented, V from the symmetric rib, and K from the nearest rib above it (for the top pair of ribs, the K from the nearest rib below them). We take each rib in turn as a Q and augment it. Considering that the 12th and 24th ribs have poor feature representations and are not suitable to be used as guidance information, we choose the rib above the ribs to be segmented as feature guidance information in our LAME module (see Fig. 2).

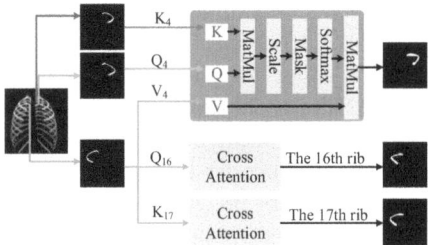

Fig. 2. The 16th rib takes on different roles in different cross attention, acting as V in the 4th cross attention, Q in the 16th cross attention, and K in the 17th cross attention. The subscript of Q, K, and V represent the sequence number of the cross-attention block. For the 4th cross attention block, K from the 3rd rib, Q from the 4th rib, and V from the 16th rib.

Our location-aware mutual enhancement consists of 24 cross-attention blocks, and each cross-attention block can be written as follows:

$$r'_k = \text{soft max}(\frac{QK^T}{\sqrt{d_k}})V, \tag{2}$$

$$r = r'_1 \oplus r'_2 \oplus ... \oplus r'_{24}, \tag{3}$$

where r'_k is the feature representation of the k-th (k = 1, 2,..., 24) rib after being enhanced. K, Q, V are the rib features that come from the output of DGRD, d_k is the length of K and r is the feature representation of all 24 ribs.

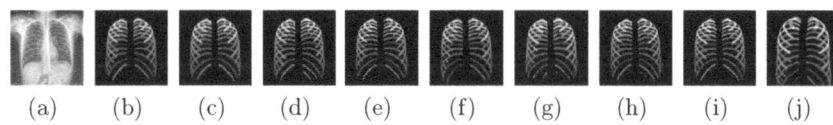

(a)   (b)   (c)   (d)   (e)   (f)   (g)   (h)   (i)   (j)

Fig. 3. The visual rib segmentation results of our method and state-of-the-art methods in AHU-RibCXR Dataset. (a) Original Image. (b) Ground Truth. (c) Ours. (d) AttenUNet. (e) TransUNet. (f) M^2SNet. (g) UNet. (h) UNet++. (i) DA-TrUNet. (j) Missformer.

Table 1. Dataset division of VinDr-RibCXR and AHU-RibCXR Dataset.

	Total	Training Set	Verification Set	Testing Set
VinDr-RibCXR	245	147 (60%)	49 (20%)	49 (20%)
AHU-RibCXR	1254	752 (60%)	251 (20%)	251 (20%)

4 Experiment

In this section, we present two datasets as well as experimental details.

4.1 Dataset

To verify the effectiveness of DENet, we train and test on publicly available datasets VinDr-RibCXR [8] and private dataset AHU-RibCXR, respectively. And the dataset division of VinDr-RibCXR and AHU-RibCXR Dataset indicated in Table 1.

VinDr-RibCXR Dataset. The VinDr-RibCXR dataset contains 245 CXR images for rib segmentation. Since the resolution of the VinDr-RibCXR dataset is not uniform, we resize the original and labeled images to the agreed resolution 448 × 448. Since VinDr-RibCXR does not contain annotated images of floating ribs, we are unable to verify the validity of the DGRD module on this dataset, we only verify the effectiveness of the LAME module.

AHU-RibCXR Dataset. The AHU-RibCXR Dataset contains 1254 CXR images for rib segmentation. For each chest X-ray image, there are corresponding 24 labeled images, and each labeled image contains a different rib. We resize the original and labeled images to the agreed resolution 448 × 448.

4.2 Implementation Details

To verify the effectiveness of our proposed network, we set up experiments on the workstation with 24 GB Quadro RTX 6000 GPU with mini-batch size 4. And our code is written in Python 3.9.

4.3 Evaluation Metric

The Dice similarity coefficient is used to evaluate the overlap between the prediction result and the label, and then the experimental results are evaluated.

We calculate the mean Dice Score of 24 ribs as mDice and the mean Dice Score of the two most challenging ribs (i.e. the 12th rib and the 24th rib) as cDice. We show the Dice Score of the 12th rib as 12thDice and the Dice Score of the 24th rib as 24thDice.

Table 2. Comparison of segmentation results with state-of-the-art methods in AHU-RibCXR Dataset and VinDr-RibCXR Dataset.

Method	AHU-RibCXR				VinDr-mDice
	mDice	12thDice	24thDice	cDice	
UNet [11]	0.786	0.592	0.604	0.598	0.545
UNet++ [19]	0.827	0.603	0.649	0.626	0.572
TransUNet [2]	0.845	0.660	0.671	0.665	0.618
M^{2}SNet [18]	0.821	0.619	0.642	0.630	0.544
Atten_UNet [10]	0.826	0.599	0.646	0.622	0.615
DA-TransUNet [12]	0.854	0.656	0.682	0.669	0.538
Missformer [4]	0.819	0.597	0.640	0.618	0.621
Ours	**0.865**	**0.700**	**0.716**	**0.708**	**0.629**

4.4 Comparison with State-of-the-Art Methods

To affirm the effectiveness of our proposed network architecture, we compare it with conventional medical segmentation algorithms including UNet [11], UNet++ [19] and also with the state-of-the-art methods including TransUNet [2], M^{2}SNet [18], Atten_UNet [10], DA-TransUNet (DA-TrUNet) [12] and Missformer [4]. To ensure a fair comparison, we make sure that the data partitioning and parameter configurations for the compared methods are in alignment with our own experiment.

The experimental results are illustrated in Table 2. Our proposed method proves to be more effective than several state-of-the-art methods, delivering the best segmentation performance for AHU-RibCXR Dataset (shown in Fig. 3). The mean Dice Score for all 24 ribs is 86.5%, while the mean Dice Score for the two most challenging ribs hits 70.8%. Our method achieve the best performance. Compared with the above method, our method is designed for different segmentation difficulties of the ribs, and has stronger specificity. Moreover, our method reasonably utilizes the interrelation between the ribs that other methods neglect. Therefore, our propose method achieved better performance.

The experimental results in VinDr-RibCXR dataset are illustrated in Table 2, while respective visual segmentation results can be observed in Fig. 4. Our

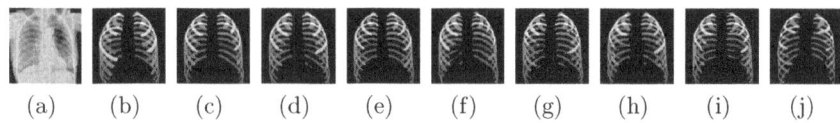

Fig. 4. The visual rib segmentation results of our method and state-of-the-art methods in VinDr-RibCXR dataset. (a) Original Image. (b) Ground Truth. (c) Ours. (d) AttenUNet. (e) TransUNet. (f) M^{2}SNet. (g) UNet. (h) UNet++. (i) DA-TrUNet. (j) Missformer.

proposed method proves to be more effective than several state-of-the-art methods, delivering the best segmentation performance for AHU-RibCXR Dataset. The mean Dice Score of DENet for all 20 ribs is 62.9%, 1.5% higher than the baseline model. Our method achieves the best performance as it enhances the features of each rib compared to other methods.

Table 3. Ablation experiments of DGRD and LAME in AHU-RibCXR Dataset.

	DGRD	LAME	mDice	cDice
Base			0.845	0.665
Base+DGRD	✓		0.855	0.682
Base+LAME		✓	0.861	0.694
Ours	✓	✓	0.865	0.708

4.5 Ablation Study

To verify the effectiveness of DGRD and LAME, an ablation experiment is conducted, and the outcomes are detailed in Table 3. We use TransUNet [2] as our baseline and verify the validity of our method.

Effectiveness of DGRD Module. In this experiment, we empirically evaluate the contribution of DGRG module. Here we train the our model with and without DGRD module, and report the results in the fist two rows of Table 3. The results show that mDice of our model with the DGRD module increased by 1% than the one without it and cDice increased by 1.7%. In this module, we decouple the 24 ribs according to the segmentation difficulty, divide them into easy and difficult parts (i.e. the 12th and 24th ribs), and decode them with three different decoders. Therefore, by decoding difficult ribs with separate decoders, mDice and cDice achieves better results.

Effectiveness of LAME Module. In LAME module, we enhance each rib separately by cross attention. In AHU-RibCXR Dataset, we train the our model with and without LAME module, and show the results in the first and third rows of Table 3. The results show that mDice of our model with the LAME module increased by 1.6% than the one without it and cDice increased by 2.9%. In VinDr-RibCXR dataset, we also train the our model with and without LAME module, and show the results in the first and third rows of Table 3. The method with the addition of the LAME module is 1.1% higher than the mDice of the baseline model [2]. The cross-attention mechanism can make good use of the relatively fixed positional information between ribs as well as the feature information of similar ribs. In this module, to improve segmentation performance, we effectively utilize this information and therefore mDice and cDice achieve better results.

5 Conclusion

This paper introduces our novel disentanglement enhancement network designed for chest X-ray images, in addition to a finely annotated rib segmentation dataset. Our approach is unique in that it utilizes a difficulty-guided representation disentanglement module and a location-aware mutual enhancement module to achieve a finer rib segmentation. With the DGRD module, the tough-to-segment ribs are decoded by a separate decoder, while the LAME module facilitates the exchange of information between all ribs, ultimately contributing to a superior segmentation outcome. Our experimental results demonstrate that our method surpasses other state-of-the-art methods in rib segmentation. However, our approach still has limitations, as we aim to further develop one model to segment multiple organs simultaneously. Moving forward, we aspire to extend our work to other organs present in chest X-ray images, such as the clavicle, lung, and scapula, among others.

Acknowledgments. This work was supported in part by the National Natural Science Foundation of China (No.62106006), in part by the Natural Science Foundation of Anhui Higher Education Institution of China under Grant KJ2020A0040, and in part by Program for Scientific Research Innovation Team in Colleges and Universities of Anhui Province (2022AH010095).

References

1. Candemir, S., et al.: Atlas-based rib-bone detection in chest x-rays. Comput. Med. Imaging Graph. **51**, 32–39 (2016)
2. Chen, J., et al.: Transunet: transformers make strong encoders for medical image segmentation. arXiv preprint arXiv:2102.04306 (2021)
3. Hassantabar, S., Ahmadi, M., Sharifi, A.: Diagnosis and detection of infected tissue of covid-19 patients based on lung x-ray image using convolutional neural network approaches. Chaos, Solitons & Fractals **140**, 110170 (2020)

4. Huang, X., Deng, Z., Li, D., Yuan, X.: Missformer: an effective medical image segmentation transformer. arXiv preprint arXiv:2109.07162 (2021)
5. Jacques, T., Cardot, N., Ventre, J., Demondion, X., Cotten, A.: Commercially-available AI algorithm improves radiologists' sensitivity for wrist and hand fracture detection on x-ray, compared to a CT-based ground truth. Eur. Radiol. **34**(5), 2885–2894 (2024)
6. LeCun, Y., Bengio, Y., Hinton, G.: Deep learning. Nature **521**(7553), 436–444 (2015)
7. Loog, M., Ginneken, B.: Segmentation of the posterior ribs in chest radiographs using iterated contextual pixel classification. IEEE Trans. Med. Imaging **25**(5), 602–611 (2006)
8. Nguyen, H.C., Le, T.T., Pham, H.H., Nguyen, H.Q.: Vindr-ribcxr: a benchmark dataset for automatic segmentation and labeling of individual ribs on chest x-rays. arXiv preprint arXiv:2107.01327 (2021)
9. Ogul, B.B., Sümer, E., Ogul, H.: Unsupervised rib delineation in chest radiographs by an integrative approach. In: International Conference on Computer Vision Theory and Applications, vol. 2, pp. 260–265. SCITEPRESS (2015)
10. Oktay, O., et al.: Attention u-net: learning where to look for the pancreas. arXiv preprint arXiv:1804.03999 (2018)
11. Ronneberger, O., Fischer, P., Brox, T.: U-net: convolutional networks for biomedical image segmentation. In: Medical Image Computing and Computer-Assisted Intervention–MICCAI 2015: 18th International Conference, Munich, Germany, October 5-9, 2015, Proceedings, Part III 18, pp. 234–241. Springer (2015)
12. Sun, G., et al.: Da-transunet: integrating spatial and channel dual attention with transformer u-net for medical image segmentation. Front. Bioeng. Biotechnol. **12**, 1398237 (2024)
13. Tabik, S., et al.: COVIDGR dataset and COVID-SDNet methodology for predicting COVID-19 based on chest x-ray images. IEEE J. Biomed. Health Inform. **24**(12), 3595–3605 (2020)
14. Vaswani, A.: Attention is all you need. In: Advances in Neural Information Processing Systems (2017)
15. Wang, H., Zhang, D., Ding, S., Gao, Z., Feng, J., Wan, S.: Rib segmentation algorithm for x-ray image based on unpaired sample augmentation and multi-scale network. Neural Comput. Appl. **35**(16), 11583–11597 (2023)
16. Wang, W., et al.: MDU-Net: a convolutional network for clavicle and rib segmentation from a chest radiograph. J. Healthc. Eng. **2020**(1), 2785464 (2020)
17. Zhang, Y., Miao, S., Mansi, T., Liao, R.: Task driven generative modeling for unsupervised domain adaptation: application to x-ray image segmentation. In: International Conference on Medical Image Computing and Computer-assisted Intervention, pp. 599–607. Springer (2018)
18. Zhao, X., et al.: M^2snet: multi-scale in multi-scale subtraction network for medical image segmentation. arXiv preprint arXiv:2303.10894 (2023)
19. Zhou, Z., Rahman Siddiquee, M.M., Tajbakhsh, N., Liang, J.: Unet++: a nested u-net architecture for medical image segmentation. In: Deep Learning in Medical Image Analysis and Multimodal Learning for Clinical Decision Support: 4th International Workshop, DLMIA 2018, and 8th International Workshop, ML-CDS 2018, Held in Conjunction with MICCAI 2018, Granada, Spain, September 20, 2018, Proceedings 4, pp. 3–11. Springer (2018)

Instance-Level 3D Model Reassembling from CLuttered Fragments

Longteng Jiang[1], Yijian Liu[1], Feixiang Lu[2(✉)], Chenming Wu[2], and Xin Jin[1]

[1] Beijing Electronic Science and Technology Institute, Beijing, China
[2] Baidu Rescarch, Beijing, China
lufeixiang@baidu.com

Abstract. Automatically parsing and reassembling fragmented 3D models is a critical challenge in the field of 3D modeling. However, existing methods often rely heavily on manual feature engineering, limiting their flexibility and performance. In this work, we present a novel and compact approach called ReassemblingNet that directly parses and reassembles instance-level 3D models from fragmented pieces. ReassemblingNet leverages a deep neural network architecture that efficiently analyzes cluttered 3D fragments and predicts the necessary transformation matrices to enable seamless reassembly. To support the training of this model, we curated a large-scale dataset named PvBreaks, which contains 2,800 fragmented pieces obtained by dissecting 20 pot and vase models from the Stanford Shape Benchmark. Through extensive experimentation, we demonstrate that our ReassemblingNet approach can effectively capture the intricate features present in both virtual and real-world 3D fragments, leading to successful reassembly.

Keywords: 3D fragments · Automatic reassembly · Instance-level modeling

1 Introduction

Reassembling fragmented 3D objects, such as statues and murals from archaeological sites, poses significant challenges in graphics and vision due to their delicate nature and deterioration from long-term burial and environmental factors. The restoration of these cultural treasures depends heavily on leveraging their geometric properties, as their color and texture may have degraded over time.

In this work, we initially curated a set of 20 pot and vase models from the Stanford Shape Benchmark dataset [2], resulting in a dataset called PvBreaks comprising 2800 3D fragments. 20 pot and vase models and cutting results are shown in Fig. 2. The labelling process utilized Ncuts cut data. To automate the reassembly process, we devised a pipeline dubbed ReassemblingNet. ReassemblingNet employs a dynamic fragment graph that iteratively determines the

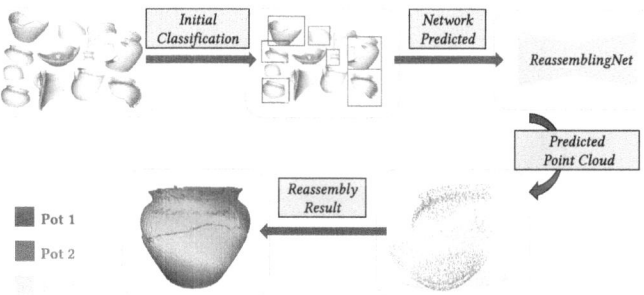

Fig. 1. Overview of Our Approach. The process can be summarized as follows: (1) Cloud point fragments are classified within the pipeline. (2) Our proposed ResassemblingNet predicts the transformation matrix needed for fragment assembly. (3) Post-processing steps are applied to achieve precise fragment assembly in the real world.

poses and relationships among fragments, enabling it to predict the rigid rotation and translation for each input point cloud fragment. PointNet++ [12] is employed to classify point cloud fragments and select potential matches belonging to the same model class. These selected fragments are subsequently fed into ReassemblingNet to predict their rotation and translation parameters. Our complete approach pipeline is shown in Fig. 1. In addition, we conduct comprehensive training and evaluation experiments of our proposed method on the recently introduced Breaking Bad dataset [13], which also comprises a large-scale collection of broken objects. The experimental results demonstrate that our method significantly outperforms the baseline approach (DGL proposed by [17]) by a substantial margin. Furthermore, our method also exhibits excellent performance when applied to real-world fragment data. The contributions of this paper can be summarized as follows:

– We propose a novel method for parsing and reassembling instance-level 3D models from 3D fragments. We achieve a 94.25% accuracy in semantic parsing, 43% accuracy in instance parsing, an RMSE value of 70.23 for rotation prediction, and an RMSE value of 0.47 for translation prediction.
– Our proposed ReassemblingNet can directly predict the rotation quaternions and translation vectors required for reassembling broken fragments.
– To enhance the training process of ReassemblingNet, we curate a synthesized dataset consisting of 2,800 pot fragments.

2 Related Work

2.1 Point Cloud Feature Extraction

PointNet [11] stands as a pioneering work in point cloud neural networks. It introduces a network architecture capable of effectively handling unstructured sets of 3D points, enabling efficient and robust feature learning from point clouds.

Fig. 2. Constructed PvBreaks Dataset: Fragments Extracted via Model Cutting. Derived from the Princeton Benchmark dataset, our dataset showcases visually distinct fragments, each represented by a unique color. The diversity within the dataset is highlighted through these fragment representations. (Color figure online)

However, PointNet primarily focuses on global features and may not fully capture local features. To address this limitation, Qi et al. proposed PointNet++ [12], which leverages query ball grouping and hierarchical PointNet to capture local structures. However, few works achieve automatic prediction and reassembly through deep extraction of features from point cloud fragments like ours.

2.2 3D Fragments Reassembly

Research indicates that the technique of reassembling fragmented antiques with unknown or irregular shapes involves three fundamental steps: accurately and comprehensively representing fracture faces, matching corresponding fracture faces, and determining global poses using observations such as loop closures and overlap constraints. Zheng et al. [18] proposed a method that establishes a local coordinate system based on the normal and tangent directions of sampled contour points. In contrast, Aiger et al. [1] utilized 4-point congruent sets as local descriptors and established matching correspondences through invariant intersection ratios of diagonals under affine transformation. Several approaches have been proposed to enhance the performance of these methods, including Super4PCS [10], GD-4PCS [3], and V4PCS [16]. At present, more work is done using geometric features to achieve reassembling, but we are committed to using deep feature extraction for automated reassembly.

2.3 Shape Reassembly

Several approaches have been developed for shape assembly leveraging network-based methods. Noteworthy studies in this field are discussed in [4,15,17]. However, most of these methods are primarily designed for models with simple surfaces, assuming a certain level of geometric regularity.

Alternatively, efforts such as 3D-ORGAN [5], MendNet [6], DeepMend [8], and DeepJoin [7] utilize deep learning techniques to tackle challenges in shape repair. The shape reassembly task is based on the unique semantic features of the parts, while in the 3D fragment reassembly task, the fragments do not have unique semantic features, or the semantic features of the fragments are not as clear as the parts.

3 Dataset Construction

The Princeton Benchmark dataset [2] consists of a segmented compilation comprising 380 mesh models classified into 19 distinct object categories. Our specific focus lies within a particular category that includes 20 pot and vase models of various types, intended for cutting purposes. To simulate the fragmentation effect on these models, we utilize the normalized cut (Ncuts) [14] to partition the meshes into multiple irregular pieces, which are then subjected to a reassembly process.

Our approach uses a point cloud model, which is converted into a weighted undirected graph represented as $\mathcal{G} = (\mathcal{V}, \mathcal{E})$. Here, $\mathcal{V}$ represents the set of points in the model, while $\mathcal{E}$ represents the set of edges. These edges capture the adjacent relationships between points, and their similarity values are utilized as edge weights. Within this framework, we establish a distinct set of edges denoted as $\mathcal{C}$, achieved by partitioning the point set $\mathcal{V}$ into two mutually exclusive sets: $\mathcal{S}$ and $\mathcal{T}$. These sets satisfy the conditions $\mathcal{S} \cup \mathcal{T} = \mathcal{V}$ and $\mathcal{S} \cap \mathcal{T} = \emptyset$. Notably, all endpoints of edges within $\mathcal{C}$ reside in either $\mathcal{S}$ or $\mathcal{T}$. This collection of edges $(\mathcal{S}, \mathcal{T})$ is referred to as a "cut" or cut set of the graph $\mathcal{G}$. Our primary objective is to identify the cut set with the minimum sum of edge weights, known as the minimum "cut" or Ncuts of the model. Furthermore, we extend the concept of Ncuts to encompass scenarios involving k segments, where $k > 2$.

Suppose A_k is the set of k point cloud fragments and V is the set of the whole model, the k-way Ncut criterion is defined as follows:

$$\text{Ncuts}_k = \sum_{i=1}^{k} \frac{\text{cut}(A_k, V - A_k)}{\text{assoc}(A_k, V)}, \qquad (1)$$

and,

$$\text{cut}(S, T) = \sum_{u \in S, v \in T} w(u, v), \qquad (2)$$

where $w(u,v)$ is the weight of the edge, which in this case is the Euclidean distance between two points.

We define $assoc(S, V)$ as:

$$\text{assoc}(S, V) = \sum_{u \in S, v \in V} w(u, v). \qquad (3)$$

The PvBreaks dataset was generated by subjecting a set of 20 models to the Ncuts algorithm, resulting in the creation of 14 separate fragments from each model. This cutting process was repeated ten times, resulting in a total of 2,800 fragments. To introduce randomness and diversity, the fragment located at the center underwent random rotations along the X, Y, and Z axes. The rotation quaternion was calculated based on the inverse of the random rotation quantity.

4 Parsing and Resembling

4.1 Pre-classification

We initiate the classification process by categorizing the models in the dataset into four distinct classes. In the first step, the fragment set, represented as $\mathcal{F} = \{f_1, f_2, \cdots, f_N\}$, consisting of N 3D fragments, undergoes an initial classification using PointNet++. This process selects a subset of fragments denoted as $\mathcal{F}_i = \{f_{i1}, f_{i2}, \cdots, f_{im}\}$, which belong to the i-th class. It is important to note that $m < n$ and $\bigcup_i \mathcal{F}_i = \mathcal{F}$. Progressing to the second step, the chosen fragment subset $\mathcal{F}_i = \{f_{i1}, f_{i2}, \cdots, f_{im}\}$ is then fed into the ReassemblingNet. This network is responsible for predicting the rotation quaternion denoted as R and the translation vector denoted as t.

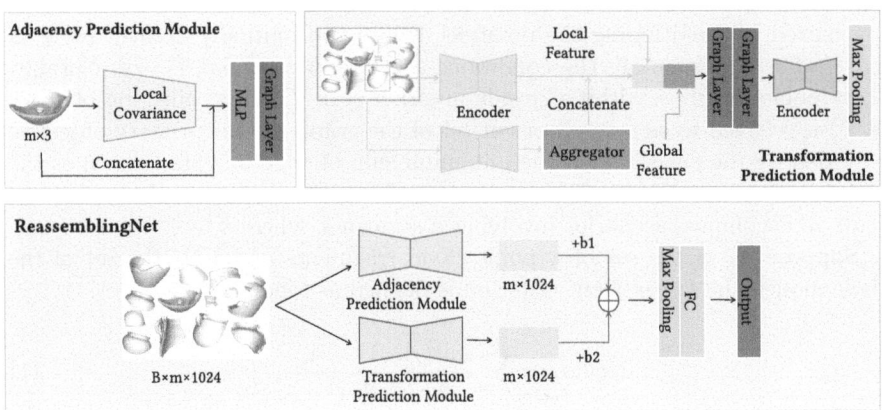

Fig. 3. The ReassemblingNet pipeline consists of two main components: the adjacency prediction module and the transformation prediction module. The adjacency prediction module captures adjacency features, while the transformation prediction module combines these features with maximum pooling and a fully connected layer to generate anticipated rotation and translation.

4.2 Network Design

ReassemblingNet, a crucial component of our framework for assembling point cloud fragments as depicted in Fig. 3, consists of two main modules: the adjacency prediction module and the transformation prediction module.

To capture the structural information in disordered point cloud data, we utilize the local covariance matrix for each point. In the input point aggregation $\zeta = \{S_1, S_2, \ldots, S_N\}$, each point cloud fragment S_i ($i = 1, 2, \ldots, N$) contains a set of points denoted as $S_i = \{p_1, p_2, \ldots, p_m\}$. Each point p_j ($j = 1, 2, \ldots, m$) has 3D feature coordinates represented as $p_j = \{x_{j1}, y_{j1}, z_{j1}\}$. The local covariance matrix is calculated for each point p_j ($j = 1, 2, \ldots, m$) in the point cloud fragment, resulting in a vector of size 1×9. The local covariance is determined by considering the three-dimensional positions of the single-hop neighboring points (including p_j) in the K-NN (K-Nearest Neighbors) graph. We apply a three-layer perceptron that simultaneously processes each feature line. If the adjacency matrix M for the K-NN graph exists, and X serves as the input and output of the layer, the following equations hold:

$$(M_{\max}(X))_{pq} = \text{ReLU}(\max_{k \in N(p)} x_{kq}),$$
$$Y = A_{\max}(X)K, \tag{4}$$

where K represents the feature mapping matrix, and $\max_{k \in N(p)} x_{kq}$ is a signature that aggregates the local neighborhood topology information. By concatenating the graph-based max-pooling layers, the network spreads the topology information over wider areas. During the training of the transformation prediction module, we incorporate the remaining fragments from other models into the overall feature input $T_{k \times m \times 3}$ (where k represents the number of fragments belonging to the same model). Similar to the adjacency prediction module, the local covariance matrix is calculated as the first step.

For a given model divided into k pieces, denoted as $T = \{t_1, t_2, \ldots, t_k\}$, each fragment represents a node in the graph. We define a single layer of graph propagation as follows:

$$p_i^{(l+1)} = F_{rel}\{g_{enc}(p_i^{(l)}), F_{agg}[g_{enc}(p_j^{(l)})]\}, \tag{5}$$

where $p_i^{(l+1)}$ is the output of the $(l+1)$-th layer node, the function aggregates all node features, and the relationship operator represents the relationship between all nodes and node aggregation features. The specific implementation process is as follows:

$$g_{enc} = \{\text{Conv1D} \to \text{BatchNorm1D} \to \text{ReLU}\}, \tag{6}$$

and F_{agg} is the maximum pooling operation along a particular axis, and F_{rel} represents the concatenation of aggregation features with node features. We merge local and overall features by utilizing consecutive layers to extract topological information. The resulting 1×1024 vector is then obtained through the encoder and maximum pooling. Finally, a fully connected layer is applied to the 1024-dimensional feature vector, resulting in a 1×7 vector that includes the rotation quaternion and translation vector.

4.3 Training Loss

We train the model using three loss functions: the pose regression loss $\mathcal{L}_{\text{pose}}$, the chamfer distance loss $\mathcal{L}_{\text{chamfer}}$, and the point-to-point MSE loss $\mathcal{L}_{\text{point}}$. Let q_i represent the ground-truth $SE(3)$ pose, denoted as $\{(R_i^*, T_i^*)\}$. The overall loss function is defined as:

$$\mathcal{L} = \mathcal{L}_{\text{pose}} + \lambda_{\text{chamfer}}\mathcal{L}_{\text{chamfer}} + \lambda_{\text{point}}\mathcal{L}_{\text{point}}, \qquad (7)$$

where

$$\mathcal{L}_{\text{pose}} = \sum_{i=1}^{N} \|T_i - T_i^*\|_2^2 + \lambda_{rot} \left\| R_i^\top R_i^* - I \right\|_2^2,$$

$$\mathcal{L}_{\text{chamfer}} = \sum_{i=1}^{N} CD(R_i P_i, R_i^* P_i) + \lambda_{\text{shape}} CD(S, S^*), \qquad (8)$$

$$\mathcal{L}_{\text{point}} = \sum_{i=1}^{N} \sum_{j} \left\| R_i P_i^j - R_i^* P_i^j \right\|_2^2$$

Let P_i^j represent the j-th point in the point cloud of the i-th fracture piece. The term $CD(R_i P_i, R_i^* P_i)$ calculates the chamfer distance between the point cloud transformed by the predicted rotation and the one transformed by the true rotation. The hyperparameter λ_{shape} controls the contribution of shape similarity in the loss. The point-to-point MSE loss $\mathcal{L}_{point}$ measures the minimum l_2 distance between the point clouds transformed by the predicted and ground-truth rotations. The hyperparameters $\lambda_{chamfer}$, λ_{point}, λ_{rot}, and λ_{shape} in the equations balance the contributions of each loss component in the overall function.

5 Experimental Results

We assess the effectiveness of our model by measuring the root mean square error ($RMSE$) between the predicted and ground-truth rotation R and translation T. The rotation is represented using Euler angles, which serve as an evaluation criterion. Additionally, we employ the shape chamfer distance (CD) and part accuracy (PA) metrics, as outlined in the evaluation protocol presented in [9], to evaluate the performance of our model.

To address the issue of unbalanced sample sizes, we employed PointNet++ [12] for the classification task. The classification accuracy results can be found in Table 1. We compared ReassemblingNet with the state-of-the-art shape assembly method, specifically DGL [17]. Table 2 shows the experimental results. Figure 4 shows the processed results of the complete pipeline generation. We also provide additional qualitative results in Fig. 5.

In order to assess the effectiveness and significance of each module, we conducted ablation experiments, the results of which are outlined in Table 3. These experiments validate the initial classification and the effectiveness of each module within the pipeline.

Table 1. Evaluation of Initial Classification Accuracy: Assessing PointNet++'s precision in identifying the instance category to which the fragments belong. The evaluation involved testing the network's performance with varying numbers of fragments as input.

Fragments (# pieces)	Classification Accuracy (%)
420	43.76
1,400	84.28
2,800	94.25

Table 2. Evaluation of Fragment Reassembly: A comparison of reassembly results between DGL and ReassemblingNet on the selected Breaking Bad subset. The reported results are averaged across all tested models.

Method	RMSE(R) ↓	RMSE(T) ↓	CD ↓	PA ↑
DGL	100.44	**0.33**	0.030	0.30
Ours	**70.23**	0.47	**0.023**	**0.43**

Table 3. Results of the Ablation Study on PvBreaks dataset, evaluating the reassembly performance of different modules. 'Init. Cls.' refers to the initial classification module, 'Trans. Pred.' represents the transformation prediction module, and 'Adj. Pred.' represents the adjacency prediction module.

Init. Cls.	Trans. Pred.	Adj. Pred.	RMSE(R) ↓	RMSE(T) ↓	CD ↓	PA ↑
✓	✓		90.76	0.57	0.0329	0.277
	✓	✓	146.70	1.03	0.0612	0.088
✓	✓	✓	**70.23**	**0.47**	**0.0233**	**0.430**

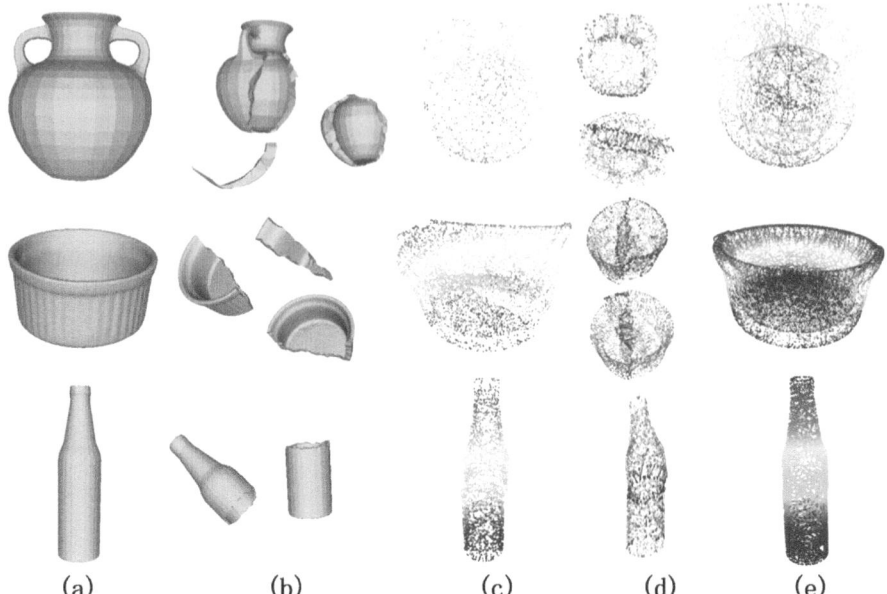

Fig. 4. The complete pipeline generates processed results presented in five columns. These columns showcase: (a) the original models, (b) the fragmented model pieces, (c) the reassembly results obtained from ReassemblingNet predictions, (d) the processed results after boundary extraction and registration, and (e) the final reassembly results following optimization.

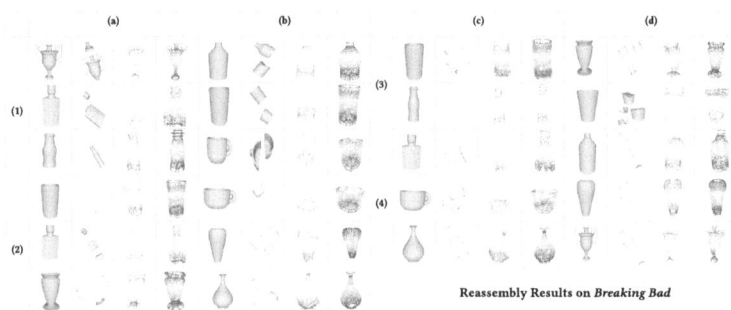

Fig. 5. Qualitative results on the Breaking Bad dataset are shown with five columns. Columns (a) to (d) display the recombination results of different model groups, with each row in (1), (2), (3), and (4) representing different model types. Each set of images sequentially shows the original model, model fragments, reassembly results from ReassemblingNet predictions, and the final processed results after boundary extraction and registration.

6 Conclusion

In conclusion, this paper introduces ReassemblingNet, a novel and efficient approach for automatically parsing and reassembling cluttered fragments in 3D modeling. By leveraging a deep network architecture, ReassemblingNet directly parses fragments and predicts transformation matrices, enabling efficient and seamless reassembly of instance-level 3D models. The extensive experiments conducted on the PvBreaks dataset demonstrate the effectiveness of the approach in capturing intricate features in both virtual and real 3D fragments. With its impressive performance, ReassemblingNet accomplishes the reassembly task without the need for additional fine-tuning or data mixing, showcasing its practicality and potential impact in various domains requiring 3D model reconstruction.

Future work includes extending the ReassemblingNet framework to address tasks involving rich texture information and handling missing fragments. Furthermore, the ReassemblingNet framework demonstrates broader generalization capabilities in complex and challenging scenarios, which gradually reduces the need for post-processing and achieving complete automatic reassembly.

Acknowledgments. We thank the ACs and reviewers. This work is partially supported by the Natural Science Foundation of China (62072014), the Fundamental Research Funds for the Central Universities (3282024049).

Disclosure of Interests. The authors have no competing interests to declare that are relevant to the content of this article.

References

1. Aiger, D., Mitra, N.J., Cohen-Or, D.: 4-points congruent sets for robust pairwise surface registration. In: ACM SIGGRAPH 2008 papers, pp. 1–10 (2008)
2. Chen, X., Golovinskiy, A., Funkhouser, T.: A benchmark for 3D mesh segmentation. ACM Trans. Grap. (TOG) **28**(3), 1–12 (2009)
3. Ge, X.: Non-rigid registration of 3D point clouds under isometric deformation. ISPRS J. Photogramm. Remote. Sens. **121**, 192–202 (2016)
4. Harish, A.N., Nagar, R., Raman, S.: RGL-NET: a recurrent graph learning framework for progressive part assembly. In: 2022 IEEE/CVF Winter Conference on Applications of Computer Vision (WACV), pp. 647–656. IEEE (2022)
5. Hermoza, R., Sipiran, I.: 3D reconstruction of incomplete archaeological objects using a generative adversarial network. In: Proceedings of Computer Graphics International 2018, pp. 5–11 (2018)
6. Lamb, N., Banerjee, S., Banerjee, N.: MendNet: restoration of fractured shapes using learned occupancy functions. In: Computer Graphics Forum, vol. 41, pp. 65–78. Wiley Online Library (2022)
7. Lamb, N., Banerjee, S., Banerjee, N.K.: DeepJoin: learning a joint occupancy, signed distance, and normal field function for shape repair. ACM Trans. Graph. (TOG) **41**(6), 1–10 (2022)

8. Lamb, N., Banerjee, S., Banerjee, N.K.: Deepmend: learning occupancy functions to represent shape for repair. In: Computer Vision–ECCV 2022: 17th European Conference, Tel Aviv, Israel, October 23–27, 2022, Proceedings, Part III, pp. 433–450. Springer (2022)
9. Li, Y., Mo, K., Shao, L., Sung, M., Guibas, L.: Learning 3D part assembly from a single image. In: Vedaldi, A., Bischof, H., Brox, T., Frahm, J.-M. (eds.) ECCV 2020 Part VI. LNCS, vol. 12351, pp. 664–682. Springer, Cham (2020). https://doi.org/10.1007/978-3-030-58539-6_40
10. Mellado, N., Aiger, D., Mitra, N.J.: Super 4PCS fast global pointcloud registration via smart indexing. In: Computer Graphics Forum, vol. 33, pp. 205–215. Wiley Online Library (2014)
11. Qi, C.R., Su, H., Mo, K., Guibas, L.J.: Pointnet: deep learning on point sets for 3D classification and segmentation. In: Proceedings of the IEEE Conference on Computer Vision and Pattern Recognition, pp. 652–660 (2017)
12. Qi, C.R., Yi, L., Su, H., Guibas, L.J.: Pointnet++: deep hierarchical feature learning on point sets in a metric space. In: Advances in Neural Information Processing Systems, vol. 30 (2017)
13. Sellán, S., Chen, Y.C., Wu, Z., Garg, A., Jacobson, A.: Breaking bad: a dataset for geometric fracture and reassembly. In: Thirty-sixth Conference on Neural Information Processing Systems Datasets and Benchmarks Track (2022)
14. Shi, J., Malik, J.: Normalized cuts and image segmentation. IEEE Trans. Pattern Anal. Mach. Intell. **22**(8), 888–905 (2000)
15. Willis, K.D., et al.: Joinable: learning bottom-up assembly of parametric cad joints. In: Proceedings of the IEEE/CVF Conference on Computer Vision and Pattern Recognition, pp. 15849–15860 (2022)
16. Xu, Y., Boerner, R., Yao, W., Hoegner, L., Stilla, U.: Pairwise coarse registration of point clouds in urban scenes using voxel-based 4-planes congruent sets. ISPRS J. Photogramm. Remote. Sens. **151**, 106–123 (2019)
17. Zhan, G., et al.: Generative 3D part assembly via dynamic graph learning. Adv. Neural. Inf. Process. Syst. **33**, 6315–6326 (2020)
18. Zheng, S., Huang, R., Li, J., Wang, Z.: Reassembling 3D thin fragments of unknown geometry in cultural heritage. ISPRS Ann. Photogrammetry, Remote Sens. Spatial Inf. Sci. **2**, 393–399 (2014)

Brain-Inspired Action Generation with Spiking Transformer Diffusion Policy Model

Qianhao Wang[1,2], Yinqian Sun[1], Enmeng Lu[1], Qian Zhang[1,2,4(✉)], and Yi Zeng[1,2,3,4(✉)]

[1] Brain-inspired Cognitive Intelligence Lab, Institute of Automation, Chinese Academy of Sciences, Beijing, China
{q.zhang,yi.zeng}@ia.ac.cn
[2] School of Artificial Intelligence, University of Chinese Academy of Sciences, Beijing, China
[3] Key Laboratory of Brain Cognition and Brain-inspired Intelligence Technology, CAS, Shanghai, China
[4] Center for Long-term Artificial Intelligence, Beijing, China

Abstract. Spiking Neural Networks (SNNs) has the ability to extract spatio-temporal features due to their spiking sequence. While previous research has primarily foucus on the classification of image and reinforcement learning. In our paper, we put forward novel diffusion policy model based on Spiking Transformer Neural Networks and Denoising Diffusion Probabilistic Model (DDPM): Spiking Transformer Modulate Diffusion Policy Model (STMDP), a new brain-inspired model for generating robot action trajectories. In order to improve the performance of this model, we develop a novel decoder module: Spiking Modulate Decoder (SMD), which replaces the traditional Decoder module within the Transformer architecture. Additionally, we explored the substitution of DDPM with Denoising Diffusion Implicit Models (DDIM) in our framework. We conducted experiments across four robotic manipulation tasks and performed ablation studies on the modulate block. Our model consistently outperforms existing Transformer-based diffusion policy method. Especially in Can task, we achieved an improvement of 8%. The proposed STMDP method integrates SNNs, dffusion model and Transformer architecture, which offers new perspectives and promising directions for exploration in brain-inspired robotics.

Keywords: Brain-indpired robotics · Spiking Neural Networks · Transformer · Diffusion Policy

1 Introduction

Spiking Neural Networks transmit information through spike sequences, which encode both temporal and spatial information. This grants SNNs spatio-temporal feature extraction capabilities [1]. The most common models include the Integrate-and-Fire (IF) neuron, the Leaky Integrate-and-Fire (LIF) neuron [2], and the Hodgkin-Huxley (HH) neuron [3]. Since SNNs are modeled in a more biologically plausible manner, their training processes differ slightly from those of artificial neural networks (ANNs).

Behavior Cloning (BC), a branch of imitation learning, directly learns a policy by mapping states to actions based on expert demonstration trajectories [4]. But, it suffers from compounding errors, resulting in poor generalization and suboptimal performance in practical scenarios. Action Chuncking with Transformers (ACT) [5] employs a Variational Autoencoder (VAE) for imitation learning, generating action trajectories, while diffusion policy [6] uses diffusion models for the same purpose. Due to the current transformer-based spiking neural network models performing poorly in generating action trajectories. We propose the Spiking Transformer Modulate Diffusion Policy Model and the Spiking Modulate Decoder to enhance the accuracy of robotic manipulation tasks.

In this work, we build upon the diffusion policy method [6] and integrate it with SNNs to tackle the aforementioned challenges. Diffusion policies offer strong generative capabilities, enabling them to learn diverse patterns from data, while SNNs' spatio-temporal feature extraction ensures more coherent action trajectories. We propose a novel diffusion policy model based on Spiking Transformer Neural Networks, along with a novel decoder module.

Our key contributions are as follows:

- We present the Spiking Transformer Modulate Diffusion Policy Model, and its core module of this model: the Spiking Modulate Decoder module.
- Additionally, We implemented diffusion policies utilizing a spiking Transformer architecture, and develop the Spiking Diffusion Transformer (SDIT), which is based on Diffusion with Transformer (DIT) framework.
- We conducted experiments on four robotic manipulation tasks, demonstrating that our proposed STMDP model significantly enhanced performance. Specifically, in the Can task, our model outperformed the current best Transformer-based Diffusion Policy model by 8%.

2 Related Work

SNNs are composed of various spiking neuron models, each differing in their complexity of neuron modeling. SNNs have been widely applied in image processing [7,8] and reinforcement learning [9–11]. Common training methods include Spike-Timing-Dependent Plasticity (STDP) [12,13], ANN-to-SNN conversion [14,15], and direct training methods [16,17]. In our study, we utilized the LIF neuron model and the direct training method. Cao [18] introduced the SDDPM, which improved image generation with low energy consumption. Hou [19] proposed the SDP, enhancing robotic manipulation performance. While their work is based on the Unet architecture, our approach uses a transformer architecture. Furthermore, we introduce a more general module: SMD.

BC is attractive due to its simplicity, easy of training, and interpretability. However, it suffers from compounding errors (distributional drift and cumulative errors) and limited behavior patterns [20]. To address these limitations, numerous methods have been proposed, such as data augmentation to enhance data diversity and mitigate compounding errors [21], and the use of mixture Gaussian models [22,23] or implicit models [24] to capture more diverse behavior patterns.

The Denoising Diffusion Probabilistic Model (DDPM) [25] improves generative performance by predicting noise instead of the noisy image itself, reducing the complexity of prediction. DDIM [26] constructs a posterior distribution such that its marginal distribution is Gaussian, eliminating the need for the forward process to follow a Markov chain. Recent work has explored replacing the typical Unet-based diffusion model with transformer-based diffusion models, yielding better generative results [27]. Our work differs in that we modify the decoding module rather than the encoding module, allowing for more flexible handling of conditional inputs, and our model is primarily designed for action trajectory generation instead of image generation.

3 Method

3.1 Preliminaries

Spiking Neural Networks: SNNs imitate the signal transmission process in the brain and transmit information in sparse spike sequences. We use the LIF neuron model for modeling, which is suitable for deep learning tasks. The dynamic process can be described as Eq. (1). In practical applications, we discretize it to facilitate forward process and training. It is described as Eq. (2) and Eq. (3).

$$\tau \frac{dV(t)}{dt} = -(V(t) - V_{reset}) + I(t) \quad (1)$$

$$u^i_{t+1,n+1} = k u^i_{t,n+1}(1 - o^i_{t,n+1}) + \sum_{j=1}^{l(n)} w^{ij}_n o^j_{t,n+1} \quad (2)$$

$$o^i_{t+1,n+1} = f(u^i_{t+1,n+1} - V_{th}) \quad (3)$$

where $I(t)$ represents the incoming synaptic current, $V(t)$ denotes the membrane potential, τ is the time constant, V_{reset} is the resetting potential. When the membrane potential exceeds the threshold V_{th}, the neuron fires a spike, and the membrane potential is reset to the resetting potential.

The training method of our model uses direct training, the derivation process of direct training is described in Eq. (4) and Eq. (5).

$$\frac{\partial L}{\partial o^i_{t,n}} = \sum_{j=1}^{l(n+1)} \frac{\partial L}{\partial o^i_{t,n+1}} \frac{\partial o^i_{t,n+1}}{\partial o^i_{t,n}} + \frac{\partial L}{\partial o^i_{t+1,n}} \frac{\partial o^i_{t+1,n}}{\partial o^i_{t,n}} \quad (4)$$

$$\frac{\partial L}{\partial u_{t,n}^i} = \frac{\partial L}{\partial o_{t,n}^i}\frac{\partial o_{t,n}^i}{\partial u_{t,n}^i} + \frac{\partial L}{\partial o_{t+1,n}^i}\frac{\partial o_{t+1,n}^i}{\partial o_{t,n}^i} \tag{5}$$

where n represents the number of layers in the network, t denotes the timestep, i refers to the $neuron_i$ in a given layer, k is a constant. $f(x)$ is a step function, V_{th} represents the threshold value, L denotes the loss function, and $l(n)$ indicates the number of neurons. $o_{t,n}^i$ represents the output(0/1) of the i_{th} neuron in the n_{th} layer at the t_{th} timestep. $u_{t,n}^i$ represents the membrane potential of the neuron and w_n^{ij} denotes the weight.

Because of the $f(x)$ is undifferentiable, we use a surrogate function $g(x)$ to solve this problem during the back-propagation process. The surrogate function is described as Eq. (6).

$$g(x) = \begin{cases} 0, & x \leq -\frac{1}{\epsilon}, \\ -\frac{1}{2}\epsilon^2|x|, & x<|\frac{1}{\epsilon}|, \\ 1, & x \geq \frac{1}{\epsilon} \end{cases} \tag{6}$$

where $g(x)$ represents the surrogate function, ϵ is a constant, we set $\epsilon = 2$.

DDPM and DDIM: We constructed STMDP spiking neural based DDPM and also explored replacing DDPM with DDIM. Before introducing the architectures, we briefly review some diffusion model concepts involved in the research. The diffusion model includes a forward process $q(x_t \mid x_0) = \mathcal{N}(x_t; \sqrt{\overline{\alpha}_t}x_0, (1-\overline{\alpha}_t)I)$, which adds noise to the data; and a reverse process $q_\theta(x_{t-1} \mid x_t) = \mathcal{N}(\mu_\theta(x_t), \Sigma_\theta(x_t))$, which recovers data from random noise. Diffusion models are trained to learn q_θ, using the variational lower bound [28] for training, simplified to minimize the KL divergence between q_θ and $q(x_{t-1} \mid x_t, x_0)$. This paper employs a fixed variance $\epsilon_\theta(x_t) = \beta I$. Finally, the predicted mean is converted to predicted noise through the reparameterization process $L_{t-1}(\theta) = \|\epsilon_\theta(x_t) - \epsilon\|^2$. The DDPM forward process is a Markov process, whereas DDIM eliminats the dependency on the forward Markov process and can generate high-quality samples in fewer sampling timesteps.

3.2 Spiking Transformer Modulate Diffusion Policy

The STMDP model differs from the DIT architecture, which is primarily based on the Transformer encoder structure [29]. In contrast, our proposed model preserves the encoder-decoder architecture of the Transformer. Additionally, STMDP is built upon a spiking neural network. The encoding strategy involves direct encoding, where the input is replicated across eight time steps, and decoding is performed by averaging the inputs over these steps to generate the output. The overall architecture of the model is illustrated in Fig. 1.

The diffusion policy process consists of two phases: a training phase and a generation phase. During the training phase, a noise prediction model is trained, and in the generation phase, the trained model is employed to progressively reduce noise and generate robot-executable action trajectories guided by visual

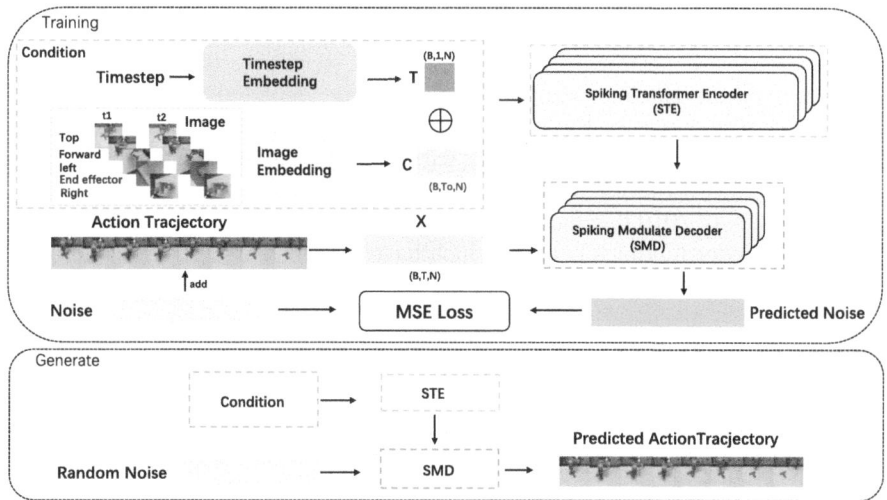

Fig. 1. The architect of spiking transformer modulate diffusion policy module.

input. For the training phase, a fixed time step is embedded to high dimension vector as T. Image frames are extracted from demonstration videos, and multi-angle images are encoded as high dimension vector C. The timestep embedding and image embedding are then combined and input into the Spiking Transformer Encoder (STE) module, followed by processing in the SMD module for guidance. After adding noise to the action trajectory, it is encoded into the X vector and decoded by the SMD module together with the output of STE. The decoder module outputs a prediction of the noise in the action trajectories, which is updated using the Mean Square Error (MSE) with target noise Y. During the generation phase, images and inference time steps serve as conditions; however, the input to the SMD module is random noise, which is iteratively denoised to produce robot-executable action trajectories. This can be represented:

$$M = SSA(T, C) + FFN(SSA(T, C)) \tag{7}$$

$$L = \|Y - SMDM(M, X)\|^2 \tag{8}$$

where SSA stands for Spiking Self Attention module and FFN stands for Feed Forward neural Network module and M stands for the output of the STE module. Y is noise and $SMDM$ stands for SMD module.

3.3 Base Block

In the Spiking Encoder Module, the core component is the spiking multi-head self-attention block, which combines a LIF layer with a multi-head self-attention mechanism. The remaining blocks retain the fundamental structure of the standard encoder block. In the SMD module, the base block includes a spiking multi-head self-attention block, a spiking modulated cross-attention block, and a modulated feed-forward network block. The detailed structures of the basic blocks within the Spiking Encoder and SMD modules are depicted in Figs. 2, respectively. A key difference in the SMD module is that modulation is applied to the cross-attention module, contrasting with the DIT structure, which primarily modulates the self-attention mechanism in the encoder. This design is inspired by the architecture of BERT, which is based on encoder modules, and GPT, which is built on decoder modules-both of which have demonstrated strong generative capabilities.

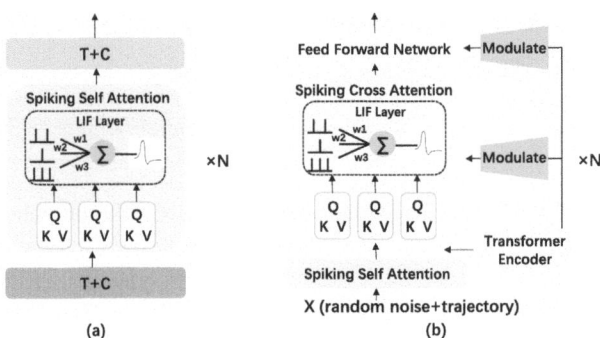

Fig. 2. The architect of base block. (a) represents the architecture of Spiking Encoder; (b) represents the architecture of SMD.

4 Experiment

4.1 Datasets and Experimental Settings

The proposed STMDP model is evaluated across four robotic manipulation tasks: PushT, Can, Square and ToolHang (Square and Can is mh version, ToolHang is ph version). These tasks cover a diverse range of actions, including pushing and grasping, as shown in Fig. 3, offering a thorough assessment of the model's performance. PushT: Move the T-shaped block to the specified position on the desktop; Can: Grasping a can of soda and placing it into a slot; Square: Lifting an object with a square hole and fitting it onto a square pillar; ToolHang: Assembling a rack from tools on the table and hanging the remaining tools on the rack.

We present results derived from the average of the last 10 checkpoints (saved every 50 epochs) across 50 initializations of the environment, totaling an average of 500 experiments, while all other settings remain consistent with the previous configurations [6]. When conducting experiments with DDIM model, we adopt the experimental setups: the value of α is set to 0 to make the forward process of its diffusion process deterministic, the inference timestep is set to 50 and is randomly selected (in contrast to 100 in DDPM).

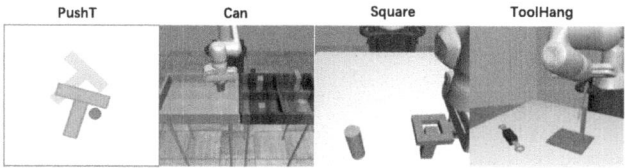

Fig. 3. Tasks used for the experiment.

4.2 Experiment Results

To ensure a comprehensive comparison, the experiments have implemented diffusion policies based on spiking Transformer (SDPT), which utilizing the same Transformer architecture as the diffusion policy based on ANN described (DPT) in [6], and those based on spiking DIT (SDIT) structures as in [27]. The main difference between them is whether there is Spiking Transformer (ST). The Modulate Block (M-Block) plays a critical role in the DIT model. This work also implements a spiking modulate decoder module based on the M-Block and experimentally validates its effectiveness. Additionally, we replace DDPM in our model with DDIM and propose STMDP-I to conduct experiments. The experimental results are shown in Table 1. The STMDP achieved the 75.4% in PushT task, it exceeded the previous best result by nearly 5% (achieved by DPT). As for the STMDP-I, We guess that because the parameters α and inference step are not selected appropriately. While It achieved the best result in the Square task, demonstrating its potential.

Table 1. Execution results of all models on four tasks.

Model	Blocks		Tasks			
	ST	M-Block	PushT	Can	Square	ToolHang
DPT [6]			0.707	0.860	0.720	0.540
DIT [27]		✓	0.657	0.920	0.800	**0.780**
SDPT	✓		0.747	0.220	0.040	0.000
SDIT	✓	✓	0.700	0.920	0.660	0.540
STMDP	✓	✓	**0.754**	**0.940**	0.800	0.540
STMDP-I	✓	✓	0.640	0.740	**0.880**	0.720

Impact of the Modulate Blocks: To validate the effectiveness of our proposed core module (SMD module), we compared the performance of ST, SDIT and SMD on all datasets. SDPT does not have a modulation block, SDIT have a modulation encoder, and our STMDP features a modulation decoder (SMD). The results are shown in Table 2. We achieved the best results in all the other three experiments. Especially in Can and Square tasks, we achieved 94% and

80% respectively. The results show that our SMD module has advantages in action trajectory generation. It can be clearly seen from Fig. 4.

Table 2. The impact of the modulate block.

Model	M-Block	Tasks			
		PushT	Can	Square	ToolHang
SDPT		0.747	0.220	0.040	0.000
SDIT	✓	0.700	0.920	0.660	**0.540**
STMDP	✓	**0.754**	**0.940**	**0.800**	**0.540**

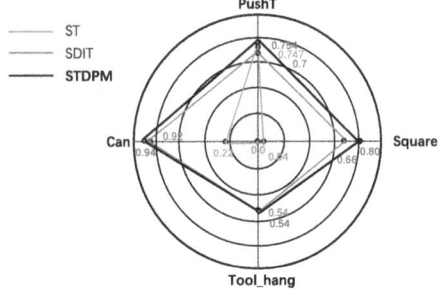

Fig. 4. The results of the ablation study.

5 Conclusion

In this work, we proposed the Spiking Transformer Modulate Diffusion Policy Model, a new diffusion policy architecture based on the Spiking Transformer. This model effectively integrates SNNs with diffusion policy's capability to generate action trajectories. As a pioneering attempt based on the Spiking Transformer architecture in the diffusion policy, it achieves high-quality motion trajectory generation. The Spiking Modulate Decoder module we proposed is modulated based on the Transformer Decoder module, demonstrating the superiority of modulation based on the Decoder. Although the STMDP model has achieved good results, there is still room for further research on the selection of modulation modules and diffusion models. We will also conduct some experiments on the selection of DDIM parameters.

Acknowledgments. This study was supported by the Postdoctoral Fellowship Program of CPSF (Grant No. GZC20232994), the Funding from Institute of Automation,

Chinese Academy of Sciences(Grant No. E411230101), Special Research Assistant Program of the Chinese Academy of Sciences (Grant No. E4S9230501), Frontier Scientific Research Program of Chinese Academy of Sciences (ZDBS-LY-JSC013).

References

1. Qiao, H., Chen, J., Huang, X.: A survey of brain-inspired intelligent robots: integration of vision, decision, motion control, and musculoskeletal systems. IEEE Trans. Cybern. **52**(10), 11267–11280 (2021)
2. Gerstner, W., Kistler, W.M.: Spiking Neuron Models: Single Neurons, Populations. Plasticity. Cambridge University Press, Cambridge (2002)
3. Hodgkin, A.L., Huxley, A.F.: A quantitative description of membrane current and its application to conduction and excitation in nerve. J. Physiol. **117**(4), 500 (1952)
4. Wu, Y.-H., Charoenphakdee, N., Bao, H., Tangkaratt, V., Sugiyama, M.: Imitation learning from imperfect demonstration. In: International Conference on Machine Learning, pp. 6818–6827. PMLR (2019)
5. Zhao, T.Z., Kumar, V., Levine, S., Finn, C.: Learning fine-grained bimanual manipulation with low-cost hardware. arXiv preprint arXiv:2304.13705 (2023)
6. Chi, C., et al.: Diffusion policy: visuomotor policy learning via action diffusion. arXiv preprint arXiv:2303.04137 (2023)
7. Wang, Z., Fang, Y., Cao, J., Zhang, Q., Wang, Z., Xu, R.: Masked spiking transformer. In: Proceedings of the IEEE/CVF International Conference on Computer Vision, pp. 1761–1771 (2023)
8. Zhou, Z., et al.: Spikformer: when spiking neural network meets transformer. arXiv preprint arXiv:2209.15425 (2022)
9. Zhang, D., Zhang, T., Jia, S., Bo, X.: Multi-sacle dynamic coding improved spiking actor network for reinforcement learning. I:n Proceedings of the AAAI Conference on Artificial Intelligence, vol. 36, pp. 59–67 (2022)
10. Chen, D., Peng, P., Huang, T., Tian, Y.: Fully spiking actor network with intralayer connections for reinforcement learning. IEEE Trans. Neural Netw. Learn. Syst. (2024)
11. Oikonomou, K.M., Kansizoglou, I., Gasteratos, A.: A hybrid spiking neural network reinforcement learning agent for energy-efficient object manipulation. Machines **11**(2), 162 (2023)
12. Bi, G., Poo, M.: Synaptic modifications in cultured hippocampal neurons: dependence on spike timing, synaptic strength, and postsynaptic cell type. J. Neurosci. **18**(24), 10464–10472 (1998)
13. Song, S., Miller, K.D., Abbott, L.F.: Competitive Hebbian learning through spike-timing-dependent synaptic plasticity. Nat. Neurosci. **3**(9), 919–926 (2000)
14. Rueckauer, B., Lungu, I.-A., Yuhuang, H., Pfeiffer, M., Liu, S.-C.: Conversion of continuous-valued deep networks to efficient event-driven networks for image classification. Front. Neurosci. **11**, 682 (2017)
15. Sengupta, A., Ye, Y., Wang, R., Liu, C., Roy, K.: Going deeper in spiking neural networks: VGG and residual architectures. Front. Neurosci. **13**, 95 (2019)
16. Lee, C., Sarwar, S.S., Panda, P., Srinivasan, G., Roy, K.: Enabling spike-based backpropagation for training deep neural network architectures. Front. Neurosci. **14**, 497482 (2020)

17. Neftci, E.O., Mostafa, H., Zenke, F.: Surrogate gradient learning in spiking neural networks: bringing the power of gradient-based optimization to spiking neural networks. IEEE Sig. Process. Mag. **36**(6), 51–63 (2019)
18. Cao, J., Wang, Z., Guo, H., Cheng, H., Zhang, Q., Xu, R.: Spiking denoising diffusion probabilistic models. In: Proceedings of the IEEE/CVF Winter Conference on Applications of Computer Vision, pp. 4912–4921 (2024)
19. Hou, Z., Gao, M., Yu, H., Yang, M., Ieong, C-I.: SDP: spiking diffusion policy for robotic manipulation with learnable channel-wise membrane thresholds. arXiv preprint arXiv:2409.11195 (2024)
20. Bansal, M., Krizhevsky, A., Ogale, A.: Chauffeurnet: Learning to drive by imitating the best and synthesizing the worst. arXiv preprint arXiv:1812.03079 (2018)
21. Bojarski, M.: End to end learning for self-driving cars. arXiv preprint arXiv:1604.07316 (2016)
22. Mandlekar, A., et al.: What matters in learning from offline human demonstrations for robot manipulation. arXiv preprint arXiv:2108.03298 (2021)
23. Shafiullah, N.M., Cui, Z., Altanzaya, A.A., Pinto, L.: Behavior transformers: cloning k modes with one stone. In: Advances in Neural Information Processing Systems, vol. 35, pp. 22955–22968 (2022)
24. Wu, J., et al.: Spatial action maps for mobile manipulation. arXiv preprint arXiv:2004.09141 (2020)
25. Ho, J., Jain, A., Abbeel, P.: Denoising diffusion probabilistic models. Adv. Neural. Inf. Process. Syst. **33**, 6840–6851 (2020)
26. Song, J., Meng, C., Ermon, S.: Denoising diffusion implicit models. arXiv preprint arXiv:2010.02502 (2020)
27. Peebles, W., Xie, S.: Scalable diffusion models with transformers. In: Proceedings of the IEEE/CVF International Conference on Computer Vision, pp. 4195–4205 (2023)
28. Kingma, D.P.: Auto-encoding variational bayes. arXiv preprint arXiv:1312.6114 (2013)
29. Vaswani, A., et al.: Attention is all you need. In: Advances in Neural Information Processing Systems (2017)

Single-Stage Dual-Task Joint Learning Framework for Hand Hygiene Assessment

Sizhe Qin[1,2], Zijian Tu[2,3], Deyu Su[2,4], and Zi Wang[3(✉)]

[1] School of Artificial Intelligence, Anhui University, Hefei, China
[2] Institute of Artificial Intelligence, Hefei Comprehensive National Science Center, Hefei, China
[3] School of Biomedical Engineering, Anhui Medical University, Hefei, China
ziwang1121@foxmail.com
[4] Department of Radiology, The First Affiliated Hospital of Anhui Medical University, Hefei, China

Abstract. In recent years, Action Quality Assessment technology has been introduced to evaluate hand hygiene. However, current two-stage methods, which separate segmentation and Action Quality Assessment, suffer from unstable evaluation due to the reliance on the models for each respective task. Additionally, single-stage methods perform poorly in long-duration videos due to extensive background noise interference. Moreover, existing hand hygiene datasets have low annotation density and limited samples. To alleviate these limitations in methodology and data, this paper proposes a Single-stage Dual-task Joint Learning (SDJL) framework and a hand hygiene evaluation dataset (HHA1009). First, this model utilizes the vision transformer with the fixed token as the backbone, significantly improving computational efficiency and prediction accuracy by jointly learning Action Segmentation and quality assessment tasks. Additionally, the HHA1009 dataset captures video sequences in various scenarios and the individual action in each video is scored by multiple raters. The experimental results on the HHA1009 show the superior performance of the proposed SDJL, ablation studies further confirm the effectiveness of each module.

Keywords: Hand Hygiene Assessment · Joint Learning Strategy · Action Segmentation · Action Quality Assessment

1 Introduction

In recent years, Action Quality Assessment (AQA) technology has gradually entered the medical field [5], including hand hygiene, which is one of the most critical measures in infection prevention and control [2,10]. AQA methods show potential in automating the assessment of medical professionals' performance, reducing human error. However, existing studies on hand hygiene assessment have limitations in both methods and data.

(1) Methods. Some two-stage methods [9,13] perform Action Segmentation (AS) to predict the action class, followed by Action Quality Assessment based on the predicted classes. These approaches involve high complexity, and the quality of the evaluation heavily depends on the accuracy of the segmentation, leading to potentially unstable results. On the other hand, the long-duration nature of hand hygiene tasks, involves sequential steps, such as finger friction or palm scrubbing. Thus, the common single-stage models, often perform poorly in long-duration videos due to background noise interference. **(2) Datasets.** The widely used hand hygiene evaluation dataset HHA300 [6], suffers from issues such as a small dataset size. Each video assigned a single overall score, lacking detailed annotations for specific actions. Additionally, the dataset is limited in terms of camera angles and lighting conditions, reducing the generalization ability of models trained on it.

To address the limitations of methods, we propose a Single-stage Dual-task Joint Learning (SDJL) framework and a comprehensive hand hygiene evaluation dataset (HHA1009). Specifically, we use a transformer with the fixed token as the backbone to extract the feature for training. By employing fixed tokens, we reduce background noise, enhancing the model's ability to recognize scores in long-term videos. Furthermore, we combined AS and AQA tasks into the unified framework for joint learning, further improving the computational efficiency and prediction accuracy. **To alleviate the shortcomings of the existing dataset**, we contribute the HHA1009 dataset, which includes 1009 hand hygiene operation videos, significantly increasing the data size and diversity. Each video includes not only an overall score but also fine-grained scores for individual actions, resulting in 28,362 annotations. We also introduced an error annotation mechanism, providing additional textual labels when participants make mistakes. The dataset features over one hundred participants with varying hand hygiene proficiency, covering multiple environments, which are highly consistent with real scenes.

Our experiments on the HHA1009 dataset validate the superior performance of the proposed SDJL, and ablation studies further confirm the effectiveness of each module. In summary, our main contributions are as follows:

- We propose a Single-stage Dual-task Joint Learning (SDJL) framework for AS and AQA tasks, SDJL addresses the limitations of existing two-stage methods, which leads to unstable evaluations due to the interdependence of each task. It also overcomes the background noise challenges faced by single-stage methods.
- We construct a comprehensive hand hygiene dataset HHA1009, which consists of 1009 videos with fine-grained annotations for individual actions. We introduce an error annotation mechanism, greatly enhancing the granularity and diversity of the data. HHA1009 also addresses the limitations of existing datasets in terms of camera angles, scenarios, and participant generalization.
- Experimental results on the HHA1009 dataset demonstrate that our model significantly outperforms existing action evaluation and hand hygiene assess-

ment methods in terms of inference speed and prediction accuracy, particularly in the evaluation tasks involving long-duration sequences.

2 Methodology

We propose a single-stage fixed-token scoring model, referred to as SDJL. This model aims to simultaneously perform Action Segmentation and Action Quality Assessment (AQA), simplifying the task of recognizing and scoring actions through the use of fixed tokens. Unlike traditional two-stage methods, SDJL efficiently handles the joint prediction of action labels and quality scores within a single-stage framework, thus improving the model's performance.

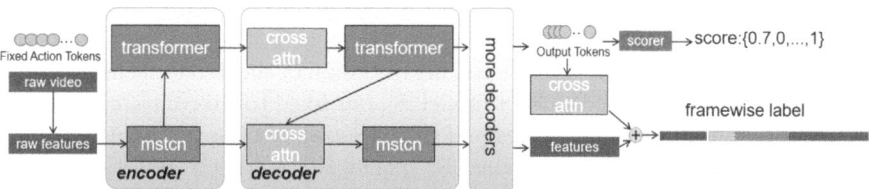

Fig. 1. The pipeline of proposed Single-stage Dual-task Joint Learning (SDJL) framework. Our model uses an end-to-end joint learning architecture, which is faster than typical two-stage methods and more effective for long-sequence videos compared to common single-stage methods.

2.1 Overview

As illustrated in Fig. 1, the proposed Single-Stage Dual-Task Joint Learning (SDJL) framework is an end-to-end system designed to generate action labels for each frame and corresponding action quality scores. The architecture of the SDJL is the encoder-decoder similar to Fact model [8].

The input to the model consists of a pre-extracted feature matrix $X \in \mathbb{R}^{T \times D}$ and a set of learnable fixed tokens $A_0 \in \mathbb{R}^{(n_{class}+1) \times D}$, where T is the number of frames, D is the feature dimension, and n_{class} is the number of foreground action classes, with an additional label representing the background class. The input passes through a cross-attention mechanism that interacts between the token features and the frame features, progressively generating updated token and frame features. The model ultimately outputs a set of token features of length $(n_{class}+1) \times 2048$, denoted as A_f, and frame features of length $T \times 2048$, denoted as F_f. The segmentation results are computed from the average of the relationship matrix generated by the final cross-attention layer and the frame features, where the first $n_{class}+1$ vectors of each feature represent the probability of belonging to the corresponding class.

Formally, the input is represented as:

$$(X, A_0) \rightarrow \text{cross-attention}(X, A_0).$$

The output is:

$$A_f \in \mathbb{R}^{(n_{class}+1) \times 2048}, \quad F_f \in \mathbb{R}^{T \times 2048}.$$

The final segmentation prediction and quality score are derived from the combination of token features and frame features. We use the SCNN's KAS scorer [6] to sum the scores of all tokens except the fixed background class token, generating the Action Quality Assessment result, which ensures consistency in action classification and quality assessment.

2.2 Fixed Token Design

One of the core innovations of SDJL lies in the design of the fixed tokens. Inspired by the token mechanism in the Fact model [8], our fixed token count matches the number of foreground action classes and the background class, without including empty classes. Each token is designed to specifically learn the holistic features of a corresponding action class, effectively capturing action information. Unlike Fact [8], where tokens require complex matching with action classes through the Hungarian algorithm, SDJL directly matches and predicts via fixed token-class correspondence. This design greatly reduces computational complexity while enhancing the model's ability to represent action classes.

2.3 Background Noise Removal

Our SDJL further designs a background noise removal mechanism. In the output stage, the fixed background class token does not participate in scoring, as scoring is limited to the foreground action tokens. When certain foreground action classes cannot be predicted, the token degenerates into a background-like action rather than being removed or set as an empty class. In this case, the model still scores these tokens degenerated into the background, implicitly expressing those poorly performed actions that still have scoring potential. Experimental results show that this approach significantly improves Action Segmentation and enhances the model's robustness in complex scenarios, allowing it to more accurately handle subtle action differences.

2.4 Loss Function

Our loss function consists of four components: Action Segmentation loss, scoring loss, cross-attention loss, and temporal smoothing loss. Unlike Fact [8], our method simplifies the process by fixing tokens to correspond to action classes, eliminating the need to compute token-to-segment matching. The detailed descriptions of each component are provided below.

(1) Action Segmentation Loss. The action segmentation loss comprises frame-level classification cross-entropy loss and temporal smoothing loss. The cross-entropy loss measures the accuracy of the model's frame-level action class predictions:

$$L_{\text{CLS}} = \frac{1}{T} \sum_t -\log(y_{t,c}), \tag{1}$$

where T represents the number of frames in the video, and C is the number of action categories. $y_{t,c}$ denotes the probability that frame t belongs to class c. Additionally, we use cross-entropy loss on the action token's predicted class:

$$L_{\text{act}} = \frac{1}{M} \sum_n -\log(P_b^a(n, a_n)), \tag{2}$$

where $P_b^a(n, a_n)$ is the predicted probability of token n for class a_n.

(2) Temporal Smoothing Loss. The temporal smoothing loss is designed to reduce the over-segmentation problem. This loss is applied to smooth both the predicted action probabilities for each frame and the alignment (attention map) between frames and action tokens. The loss is calculated as follows:

$$L_{\text{ts}} = w \sum_b \left(hp(P_b^f) + hp(\Lambda_b^a) + hp(\Lambda_b^f) \right), \tag{3}$$

where $hp(\cdot)$ is the smoothing function, and w is a weight that balances this loss against the other components in the overall loss function.

(3) Scoring Loss. The scoring loss is designed to evaluate the discrepancy between the predicted action quality scores and the ground truth scores. We use the mean squared error (MSE) as the loss function:

$$L_{\text{MSE}} = \frac{1}{n} \sum_{i=1}^{n} (\hat{y}_i - y_i)^2, \tag{4}$$

where $\hat{y}_i$ and y_i are the predicted and ground truth scores for action i.

(4) Cross-Attention Loss. Cross-attention loss ensures the alignment between action tokens and frame features through their attention weights. This loss helps the model better capture the relationship between action tokens and the corresponding frames:

$$L_{\text{cross}} = \frac{1}{T} \sum_n \sum_{t \in T_n} -\left(\log(\hat{\Lambda}_b^a(t)) + \log(\hat{\Lambda}_b^f(t)) \right), \tag{5}$$

where $\hat{\Lambda}_b^a(t)$ and $\hat{\Lambda}_b^f(t)$ represent the attention weights between action tokens and frames at time t.

Finally, our overall loss function combines several components: the action classification loss L_{CLS}, the action token loss L_{act}, the mean squared error loss L_{MSE}, the cross-attention loss L_{cross}, and the temporal smoothing loss L_{ts}. Thus, the overall loss can be expressed as $L = L_{\text{CLS}} + L_{\text{act}} + L_{\text{MSE}} + L_{\text{cross}} + L_{\text{ts}}$.

3 HHA1009 Hand Hygiene Dataset

3.1 Data Collection and Processing

(1) Data collection. The raw videos in our dataset involved over 100 participants, including both non-professionals and medical professionals. Participants were divided into groups and performed hand hygiene actions under different requirements, scenarios, and angles. Additionally, we incorporated a large amount of freely recorded content, which introduced a variety of non-professional actions, thereby increasing the challenge of distinguishing noisy actions. **(2) Data processing.** After video collection, each video is divided into six action steps (If some motion is not detected, the number may be as low as five, as shown in Fig. 2), and each action will receive several scores from different experts. For incorrectly performed steps, we provide textual explanations for deductions. This fine-grained annotation resulted in 28,362 annotations. Finally, the average of expert ratings on the different actions will be taken as the final score.

		Act. 1	Act. 2	Act. 3	Act. 4	Act. 5	Act. 6
(a) Existing Annotation	final score: **3.5**						
(b) Fine-grained Annotation	final score: **3.1875**						
	Exp. 1	1	0.5	0	1	1	0.5
	Exp. 2	1	0.5	0	0.5	0.5	0
	Exp. 3	0.75	0.25	0	0.25	0.25	0.25
	Exp. 4	1	0.5	0	1	1	1

Fig. 2. Different annotation methods. (a) Exists annotation in HHA300 [6], only includes one score annotation. (b) Fine-grained annotation in HHA1009 includes detailed scores for each action from multiple experts.

3.2 Dataset Description

We constructed the HHA1009 dataset to increase data complexity, scene diversity, and annotation granularity. Our dataset includes various environments such as hospital scenes, dormitory scenes, and restroom scenes, featuring a wide range of lighting conditions, including natural light and indoor lighting from different angles, such as left, right, overhead, close-up, and front-facing views. The dataset contains a total of 1009 videos, with annotations covering 6 action categories (rubbing palm to palm, back of hands, between fingers, backs of fingers, thumbs, and fingertips to palms). The annotations were provided by 22 experts.

Given the distinctive characteristics of our model, we did not conduct comparative experiments on other datasets. Notably, mainstream Action Quality Assessment (AQA) datasets generally lack simultaneous annotations for both segmentation and scoring; their annotations typically consist of one-to-one mappings between videos and scores, with models often designed to minimize score regression errors for pre-classified videos. In contrast, our model prioritizes the integration of global information and the interpretability of long videos, with a critical requirement to jointly leverage scores and segmentation results as constraints. While the model can still perform Action Segmentation (AS) based on frame-level annotations in the absence of scores, it would, in such a case, reduce to a Fact model focused primarily on capturing global features.

Our dataset features two types of segmentation: in the standard split, there are 856 samples in the training set and 153 in the test set. However, since some models cannot handle videos with only negative examples, we removed purely negative examples for comparison purposes. So in this adjusted split, the training set consists of 786 videos and the test set includes 142 videos, all experiments in our paper are conducted under this adjusted split.

4 Experiments

4.1 Experimental Setup

(1) Implementation. Our experiments are conducted on the HHA1009 dataset. The feature dimension for each frame is 2048-dim. We use the Adam optimizer with a learning rate of 0.0001 in training. **(2) Metrics.** We employed commonly used metrics, including frame-wise accuracy (Acc), segmental edit distance (Edit), and segmental F1 scores. For the AQA task, we use Spearman's rank correlation coefficient (ρ) and the Relative L2-distance (R-ℓ2).

4.2 Experimental Results

(1) Results on Action Segmentation Task. As shown in Table 1, our method achieves the best performance in all indicators except Edit. Specifically, compared to the Fact [8], our method improves Acc from 78.8% to **79.9%**, an increase of **1.1%**. On the F1@50 metric, our model achieves a **2.7%** improvement, rising from Fact [8]'s 53.5% to **56.2%**. Moreover, our method outperforms in the F1@10 metrics, our model achieves a **1.7%** improvement over LTContext [1].

Our model achieves such significant improvements mainly due to our joint learning method, where the Action Quality Assessment task enhances the model's understanding of the segmentation task.

(2) Results on Action Quality Assessment Task. As shown in Table 2, our model significantly outperforms all single-stage methods. Specifically, our model achieves **93.4%** on ρ, compared to **87.5%** from the SCNN [6] model, marking an improvement of **5.9%**. Furthermore, the RL2 of our model is reduced to **0.99**, significantly better than SCNN [6]'s **2.24**, showcasing exceptional prediction accuracy. In terms of time occupancy, our model processes each video at **0.182 s**, compared to 0.390 s from SCNN [6], reducing time usage by more than 50%. Our model leverages token representations of class-specific actions, which enables the model to learn latent representations of poorly performed actions, rather than simply relying on the segmentation results.

Table 1. Compared with AS methods on the HHA1009 dataset (in %).

Models	Edit	Acc	F1@50	F1@25	F1@10
UVAST [3]	33.9	72.6	19.5	25.9	30.2
LTContext [1]	**64.2**	73.1	50.1	60.8	65.5
BCN [11]	53.3	75.1	41.0	52.7	57.9
MSTCN [4]	49.9	77.3	43.8	51.0	55.7
DiffAct [7]	63.5	72.2	45.8	57.0	61.7
Fact [8]	58.2	78.8	53.5	62.2	64.6
ASFormer [14]	54.6	76.5	51.5	60.2	62.2
SDJL (Ours)	60.4	**79.9**	**56.2**	**65.9**	**67.2**

Table 2. Compared with AQA methods on the HHA1009 dataset (in %).

Model	Segmentation	ρ ↑	RL2 ↓	Time Occupancy ↓
CoRe [15]	×	57.0	6.96	0.014
CoFInAl [16]	×	52.3	10.2	**0.012**
GDLT [12]	×	55.9	6.6	0.020
SCNN [6]	✓	87.5	2.24	0.390
SDJL (Ours)	✓	**93.4**	**0.99**	0.182

(3) Ablation Study. In this section, we investigated the effects of the fixed token method and background noise removal (BNR). The fixed token method, by introducing inductive bias through fixing the class corresponding to the token, significantly improved the model's action assessment performance, reducing the RL2 from 1.80% to **0.99%**, a decrease of **0.81%**. As for background noise removal, the model was able to retain valuable background information from tokens where no actions were predicted. This not only enhanced segmentation performance from 63.7% to 79.9%, which increased **16.2%** but also improved the model's ability to predict scores for poorly performed actions (Table 3).

Table 3. Ablation Study on HHA1009 dataset (in %).

Model	ρ ↑	RL2 ↓	F1@50 ↑	ACC ↑
SDJL (Ours)	**93.4**	**0.99**	**56.2**	**79.9**
w/o BNR	91.7 (−1.7)	1.29 (+0.20)	49.5 (−6.7)	63.7 (−16.2)
w/o Fixed Token	89.6 (−3.8)	1.80 (+0.81)	53.0 (−3.2)	78.4 (−1.5)

5 Conclusion

This paper proposes a Single-stage Dual-task Joint Learning (SDJL) framework, effectively addressing the limitations of existing two-stage methods by reducing the instability caused by inter-task dependencies and overcoming the background noise challenges faced by single-stage approaches. Additionally, we construct a comprehensive hand hygiene dataset, HHA1009, which consists of 1009 videos with fine-grained annotations for individual actions. By introducing an error annotation mechanism, we significantly enhance the granularity and diversity of the data, while HHA1009 also addresses the limitations of existing datasets in terms of camera angles, scenarios, and participant generalization. Experimental results on the HHA1009 dataset demonstrate that our model significantly outperforms existing hand hygiene assessment methods in terms of prediction accuracy, in both Action Segmentation and Action Quality Assessment, particularly in evaluation tasks involving long-duration sequences.

References

1. Bahrami, E., Francesca, G., Gall, J.: How much temporal long-term context is needed for action segmentation? In: Proceedings of the IEEE/CVF International Conference on Computer Vision, pp. 10351–10361 (2023)
2. Bearman, G., Morgan, D.J., Murthy, R.K., Hota, S.: Infection Prevention: New Perspectives and Controversies. Springer, Cham (2022)
3. Behrmann, N., Golestaneh, S.A., Kolter, Z., Gall, J., Noroozi, M.: Unified fully and timestamp supervised temporal action segmentation via sequence to sequence translation. In: European Conference on Computer Vision, pp. 52–68. Springer (2022)
4. Farha, Y.A., Gall, J.: MS-TCN: multi-stage temporal convolutional network for action segmentation. In: Proceedings of the IEEE/CVF Conference on Computer Vision and Pattern Recognition, pp. 3575–3584 (2019)
5. Gao, Y., et al.: Jhu-isi gesture and skill assessment working set (jigsaws): a surgical activity dataset for human motion modeling. In: MICCAI Workshop: M2cai, vol. 3, p. 3 (2014)
6. Li, C., Zhu, Q., Liu, T., Tang, J., Su, Y.: Hand hygiene assessment via joint step segmentation and key action scorer (2023). https://arxiv.org/abs/2209.12221
7. Liu, D., Li, Q., Dinh, A.D., Jiang, T., Shah, M., Xu, C.: Diffusion action segmentation. In: Proceedings of the IEEE/CVF International Conference on Computer Vision, pp. 10139–10149 (2023)
8. Lu, Z., Elhamifar, E.: Fact: frame-action cross-attention temporal modeling for efficient action segmentation. In: Proceedings of the IEEE/CVF Conference on Computer Vision and Pattern Recognition, pp. 18175–18185 (2024)
9. Parmar, P., Morris, B.: Action quality assessment across multiple actions. In: 2019 IEEE Winter Conference on Applications of Computer Vision, pp. 1468–1476. IEEE (2019)
10. Pittet, D., Boyce, J.M., Allegranzi, B.: Hand Hygiene: A Handbook for Medical Professionals. Wiley Online Library (2017)
11. Wang, Z., Gao, Z., Wang, L., Li, Z., Wu, G.: Boundary-aware cascade networks for temporal action segmentation. In: European Conference on Computer Vision, pp. 34–51. Springer (2020)

12. Xu, A., Zeng, L.A., Zheng, W.S.: Likert scoring with grade decoupling for long-term action assessment. In: Proceedings of the IEEE/CVF Conference on Computer Vision and Pattern Recognition, pp. 3232–3241 (2022)
13. Xu, J., Rao, Y., Yu, X., Chen, G., Zhou, J., Lu, J.: Finediving: a fine-grained dataset for procedure-aware action quality assessment. In: Proceedings of the IEEE/CVF Conference on Computer Vision and Pattern Recognition, pp. 2949–2958 (2022)
14. Yi, F., Wen, H., Jiang, T.: Asformer: transformer for action segmentation. arXiv preprint arXiv:2110.08568 (2021)
15. Zhang, B., Chen, J., Xu, Y., Zhang, H., Yang, X., Geng, X.: Auto-encoding score distribution regression for action quality assessment. Neural Comput. Appl. **36**(2), 929–942 (2024)
16. Zhou, K., Li, J., Cai, R., Wang, L., Zhang, X., Liang, X.: Cofinal: enhancing action quality assessment with coarse-to-fine instruction alignment. arXiv preprint arXiv:2404.13999 (2024)

Enhancing Few-Shot Learning in Spiking Neural Networks Through Hebbian-Augmented Associative Memory

Weiyi Li[1,2], Dongcheng Zhao[1,3], Yiting Dong[1,3,5], Guobin Shen[1,3,5], and Yi Zeng[1,2,3,4,5(✉)]

[1] Brain-inspired Cognitive Intelligence Lab, Institute of Automation, Chinese Academy of Sciences, Beijing, China
yi.zeng@ia.ac.cn
[2] School of Artificial Intelligence, University of Chinese Academy of Sciences, Beijing, China
[3] Center for Long-term Artificial Intelligence, Beijing, China
[4] Key Laboratory of Brain Cognition and Brain-inspired Intelligence Technology, CAS, Shanghai, China
[5] School of Future Technology, University of Chinese Academy of Sciences, Beijing, China

Abstract. This paper presents the Hebbian-Augmented Associative Memory (HAAM) framework, which utilizes synaptic plasticity to improve few-shot learning capabilities in Spiking Neural Networks (SNNs). By addressing the limitations of traditional learning paradigms, the HAAM framework incorporates dynamic synaptic adjustments, facilitating efficient learning from limited labeled data. Experimental evaluations conducted on benchmark datasets, such as Omniglot and MiniImageNet, demonstrate that the HAAM plasticity rule significantly outperforms leading non-pretrained SNN meta-learning methods while remaining competitive with advanced Artificial Neural Network (ANN) techniques. These findings highlight the effectiveness of biologically inspired learning mechanisms in capturing temporal dynamics and enabling rapid adaptability to novel tasks. Moreover, the HAAM framework exhibits robust generalization capabilities, positioning it as a promising solution for real-world applications characterized by data scarcity. This research contributes to the expanding domain of neuro-inspired computation, offering insights for future investigations aimed at optimizing neural architectures to enhance learning efficiency.

Keywords: Hebbian-Augmented Associative Memory · Spiking Neural Networks · Synaptic Plasticity · Few-shot Learning

1 Introduction

Spiking Neural Networks (SNNs), often referred to as the third generation of artificial neural networks, represent a pivotal framework in brain-inspired intelligence research [10]. Characterized by their bio-inspired architecture and functionality, SNNs transmit information through discrete spikes or action potentials, offering substantial advantages in energy efficiency and hardware adaptability [20]. The event-driven nature of SNNs, combined with their low-power computational capabilities, positions them as critical systems for advancing artificial general intelligence (AGI) and enabling efficient real-time processing in edge-computing devices [20]. Despite these advantages, the performance of SNNs, particularly in tasks requiring new knowledge acquisition, often mirrors that of traditional artificial neural networks (ANNs), as they continue to rely heavily on large, labeled datasets for effective training.

In contrast, biological neural networks exhibit remarkable adaptability through synaptic plasticity, a mechanism wherein synapses modify their strength over time in response to experience [2,11]. Synaptic plasticity is fundamental to learning and memory processes, with critical implications for neural activity in regions such as the hippocampus, where it plays a central role in memory encoding and retrieval [7,15,16]. Emerging research demonstrates that certain forms of plasticity can manifest within seconds, facilitating rapid learning or even one-shot memory formation [11,13]. These insights underscore the potential of synaptic plasticity to address few-shot learning challenges within artificial systems.

Motivated by the potential of plasticity rules to enhance memory retention and few-shot learning in artificial systems, we incorporated plasticity mechanisms into several SNN-based variants of Recurrent Neural Networks (RNNs), Long Short-Term Memory networks (LSTMs), and Multi-Layer Perceptrons (MLPs). We evaluated these models using meta-learning experiments on well-established benchmarks, including Omniglot, miniImageNet. In contrast to conventional architectures, these plasticity-enhanced models dynamically update their synaptic weights at each time step, driven by biologically inspired rules. Our findings indicate that integrating plasticity rules not only improves few-shot learning performance but also endows MLPs and RNNs with enhanced temporal memory, achieving comparable accuracy to LSTM models. Moreover, beyond the standard Hebbian learning rule, we introduced a novel plasticity rule—Hebbian-Augmented Associative Memory (HAAM)—which sets a new state of the art (SOTA) for non-pretrained SNN meta-learning models.

In summary, our contributions are as follows:

- We demonstrate the effective integration of plasticity rules into SNN meta-learning models, enhancing temporal processing and memory capabilities.
- We introduce the HAAM rule, which outperforms existing non-pretrained SNN meta-learning models, establishing a new benchmark.
- Our model exhibits strong performance across multiple standard datasets, showcasing the potential of SNNs in few-shot learning tasks.

2 Related Work

SNNs are well-suited for processing temporal information due to their event-driven nature and energy efficiency [1]. However, their reliance on large supervised datasets poses challenges in few-shot learning, similar to traditional ANNs, which struggle to generalize with limited labeled data [29]. To address these challenges, various meta-learning approaches have been developed within the Learning-to-Learn (L2L) framework, aimed at enhancing models' capacity to transfer knowledge across tasks [25].

Meta-learning methods can be categorized into three types: metric-based, memory-based, and optimization-based [26]. Metric-based approaches, such as Siamese Networks [8], Matching Networks [27], and Prototypical Networks [23], learn shared feature representations for few-shot classification via nearest-neighbor comparisons. Memory-based methods, including MetaNet [14] and Memory Networks [22], leverage external memory to facilitate rapid adaptation to new tasks. Optimization-based strategies, like Model-Agnostic Meta-Learning (MAML) [5], Reptile [19], optimize network parameters for efficient fine-tuning.

Recent studies have applied these meta-learning strategies to SNNs, leveraging their energy efficiency for real-world applications. For instance, Rosenfeld et al. [21] introduced OWOML, an online meta-learning model for SNNs, though its adaptability is limited by the absence of local update rules. Jiang et al. [6] proposed Multi-timescale Optimization (MTSO), integrating adaptive gated LSTMs with SNNs to manage diverse neural dynamics. Zhan et al. [30] developed a two-stage metric-based framework for SNNs, enhancing performance through pretraining, albeit requiring large datasets.

In contrast, plasticity-based meta-learning models offer a biologically inspired alternative by enabling task-agnostic learning through dynamic synaptic weight adjustments, reducing the need for explicit supervisory signals [3]. Recent advancements, such as surrogate gradient techniques [17], enable efficient optimization of synaptic plasticity in SNNs. Our work employs an Enhanced Hebbian learning rule within a two-level optimization framework [4] to facilitate rapid weight adjustments, improving performance in few-shot learning tasks.

3 Methods

3.1 Model Architecture

The proposed model comprises three fundamental components: an Encoding Network, a Recurrent Layer, and a Linear Classifier, as illustrated in Fig. 1. The Encoding Network utilizes an SNN-based feature extractor, such as SConv-4 or SResNet-12, to facilitate input processing. The Recurrent Layer serves as the core component, wherein weights are classified into two categories: plastic weights, initialized to zero and updated dynamically according to specified plasticity rules, and static weights, optimized through gradient descent. The Linear Classifier outputs both the predicted label y^t and a regulatory factor η_t, which modulates the plasticity dynamics.

The training mechanism is structured into two interconnected loops: the outer loop optimizes static weights via backpropagation, while the inner loop updates plastic weights in an unsupervised manner, driven by neuronal activities. This architecture fosters a seamless integration of plasticity-driven unsupervised learning with gradient-based supervised learning.

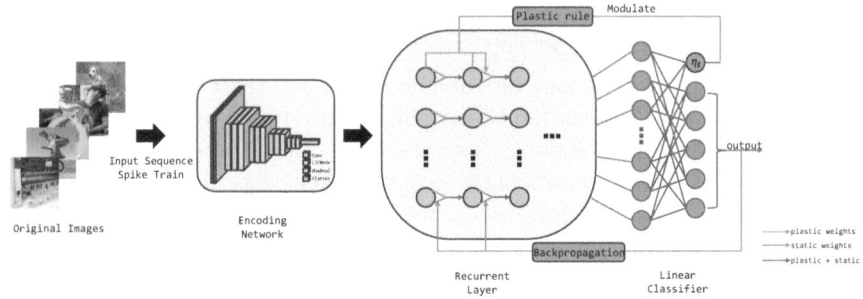

Fig. 1. Architecture of the proposed model, comprising an Encoding Network, a Recurrent Layer, and a Linear Classifier. The plastic weights within the Recurrent Layer are updated in accordance with specified plasticity rules, while static weights are refined through backpropagation based on classifier outputs and ground truth labels.

The forward update equations governing the dynamics of the model are based on the BrainCog framework [28] and the Vanilla LIF model, and are formulated as follows:

$$\begin{aligned} \mathbf{U}^{(t,l)} &= \mathbf{U}^{(t-1,l)} \circ (1 - \mathbf{S}^{(t-1,l)}) + \mathbf{C}^{(t,l)} \\ \mathbf{C}^{(t,l)} &= (\mathbf{W} + \tilde{\mathbf{W}})(\mathbf{S}^{(t,l-1)} - \mathbf{U}^{(t-1,l)}) \\ \mathbf{S}^{(t,l)} &= \mathbf{H}(\mathbf{U}^{(t,l)} - V_{th}) \end{aligned} \quad (1)$$

In this context, $\circ$ denotes element-wise multiplication. $\mathbf{W}$ represents the static weights, while $\tilde{\mathbf{W}}$ refers to the plasticity weights. $\mathbf{U}^{(t,l)}$ indicates the membrane potential at time-step t in layer l, and $\mathbf{C}^{(t,l)}$ represents the input to layer l at time-step t. The output spike vector $\mathbf{S}^{(t,l)}$ is determined using the Heaviside step function $\mathbf{H}(\cdot)$, which generates a spike when the membrane potential surpasses a threshold V_{th}.

3.2 Enhanced Hebbian Plasticity via HAAM Mechanism

Within the Recurrent Layer, the model's plastic weights are updated utilizing a Hebbian-inspired mechanism, which is further enhanced by our proposed HAAM rule. Traditional Hebbian learning captures synaptic updates based solely on instantaneous neuronal activities; however, it neglects the long-term dependencies that are crucial for tasks requiring temporal memory. To address this limitation, we introduce HAAM, which enriches the Hebbian update by incorporating historical synaptic activities, thereby augmenting the model's capacity to learn complex temporal patterns.

The standard Hebbian learning rule updates the plastic weights $\mathbf{W}^{t,l}$ based on the activities of the presynaptic $\mathbf{S}^{t,l-1}$ and postsynaptic $\mathbf{S}^{t,l}$ neurons, as expressed in the following equations:

$$\Delta \mathbf{W}^{t,l}_{hebb} = \mathbf{S}^{t,l-1}(\mathbf{S}^{t,l})^T \tag{2}$$

$$\mathbf{W}^{t,l} = (1 - \eta^t)\mathbf{W}^{(t-1,l)} + \eta^t \alpha^l \circ \Delta \mathbf{W}^{t,l}_{hebb}, \quad \mathbf{W}^{0,l} = 0 \tag{3}$$

Here, α^l is a learnable parameter that allows synapse to have a distinct learning rate. $\Delta \mathbf{W}^{t,l}_{hebb}$ represents the change in synaptic weights of neurons in layer l at time-step t, resulting from the application of the Hebbian learning rule.

In contrast, HAAM extends this framework by aggregating synaptic updates across time steps, thereby enabling the model to more effectively capture temporal dependencies. The HAAM update rule is defined as follows:

$$\Delta \mathbf{W}^{t,l}_{HAAM} = (1 - \lambda) \circ \Delta \mathbf{W}^{t-1,l}_{HAAM} + \lambda \circ \mathbf{S}^{t,l-1}(\mathbf{S}^{t,l})^T \tag{4}$$

$$\mathbf{W}^{t,l} = (1 - \eta^t)\mathbf{W}^{(t-1,l)} + \eta^t \alpha^l \circ \Delta \mathbf{W}^{t,l}_{HAAM} \tag{5}$$

The decay term η^t is introduced to prevent gradient explosion and controls the global plasticity learning rate, computed as follows:

$$\begin{aligned} \eta^t &= \eta_0 \times \text{Sigmoid}(\tilde{\eta}^t) \times \min\left\{1, \frac{\text{max\_norm}}{||\delta^t||_2}\right\}, \text{where} \\ \delta^t &= \text{Concat}(\text{Vec}(\mathbf{S}^{t,l-1} \cdot^\top \mathbf{S}^{t,l}) | l \in P) \end{aligned} \tag{6}$$

Here, η_0 controls the maximum learning rate, $Vec(\cdot)$ denotes vectorization of the matrix, $Concat(\cdot)$ represents the concatenation of vectors. P refers to all plasticity layers in the network. The regularization term $||\delta^t||_2$ is used to scale the internal learning rate η^t to prevent excessively rapid weight updates.

4 Experiments

4.1 Experimental Setup

Datasets. We conducted 5-way 1-shot classification experiments to evaluate the model's rapid learning capabilities on the Omniglot and MiniImageNet datasets.

The **Omniglot** dataset comprises 1,623 handwritten characters from 50 alphabets, each drawn by 20 individuals via Amazon Mechanical Turk [9]. Each image measures 105 × 105 pixels. We allocated 1,200 classes for training, 123 for validation, and 300 for testing.

The **MiniImageNet** dataset [27] contains 100 classes, with 600 images per class, sized 84 × 84 pixels, partitioned into 64 for training, 16 for validation, and 20 for testing.

Implementation Details. Inspired by biological synaptic plasticity [11,12, 18], we investigated how plasticity facilitates rapid learning through one-shot classification. We ensured parameter count parity across models and plasticity rules.

Input images are converted into spike sequences over four time steps. Each trial randomly selects five image classes, represented as image-label pairs (a_t, b_t), where a_t may consist of a single image or five images, and b_t is a one-hot vector of class labels.

During meta-training, b_t corresponds to the label $\tilde{y}_t$; in meta-testing, b_t is set to 0, requiring the model to predict the label. Performance is quantified using the cross-entropy loss function:

$$L = \frac{1}{5} \sum_{t \in \text{test}} \text{CrossEntropy}(y_t, \tilde{y}_t) \tag{7}$$

Accuracy is calculated based on the first image prediction to mitigate bias from prior knowledge. This approach enhances classification efficiency in few-shot scenarios.

Comparison Methods. We employed an LSTM model as the recurrent layer, comparing the HAAM plasticity rule with various SNN and ANN meta-learning methods. ANN methods included ConvNet-4-based approaches: Siamese Networks [8], Matching Networks [27], Prototypical Networks [23], MAML [5], and Relation Networks [24].

For SNN methods, we compared against OWOML-SNN [21], MTSO-SNN [6], CESM, and MESM [30], all sharing the same backbone architecture as our model. OWOML-SNN enables lifelong learning through local updates without backpropagation, while MTSO-SNN integrates an adaptive gated LSTM for efficient knowledge acquisition from limited samples. CESM and MESM employ a two-stage, metric-based pretraining framework for SNNs.

4.2 Results and Analysis

Comparison Results on Different Methods. We evaluated our model on the three previously mentioned datasets. Table 1 and Table 2 present the comparative results on Omniglot and miniImageNet. Our HAAM model consistently outperformed state-of-the-art non-pretrained SNN meta-learning methods. Specifically, in 5-way 1-shot experiments, HAAM achieved improvements of 3.68% on Omniglot and 4.1% on miniImageNet compared to OWOML, along with a 0.78% enhancement over MTSO-SNN on Omniglot.

Notably, on miniImageNet, the HAAM model attained an accuracy of 50.25%, exceeding several advanced ANN meta-learning methods, including the pretrained CESM model. While slightly lower than the pretrained MESM method, HAAM demonstrated a remarkable 14.5% performance gain over the non-pretrained MSME-nopretrain method. This indicates HAAM's robust generalization and efficient learning capabilities, particularly in scenarios with limited resources or where extensive pretraining is impractical.

Table 1. Comparison of different models on Omniglot

Model	Backbone	Fine-tune	Acc
Siamese Net [8]	ANN-ConvNet-4	Y	97.3
Matching Net [27]	ANN-ConvNet-4	N	97.9
Prototypical Networks [23]	ANN-ConvNet-4	N	98.8
MAML [5]	ANN-ConvNet-4	Y	98.7 ± 0.4
Relation Net [24]	ANN-ConvNet-4	N	99.6 ± 0.2
OWOML [21]	SNN-ConvNet-4	N	89.9 ± 0.64
MTSO-SNN [6]	SNN-ConvNet-4	N	92.8 ± 0.68
CESM [30]	SNN-ConvNet-4	Y	90.9 ± 0.38
MESM [30]	SNN-ConvNet-4	Y	94.86 ± 0.32
HAAM(ours)	SNN-ResNet-12	N	92.0 ± 1.91
HAAM(ours)	SNN-ConvNet-4	N	93.58 ± 0.70

Table 2. Comparison of different models on miniImageNet

Model	Backbone	Fine-tune	Acc
Siamese Net [8]	ANN-ConvNet-4	Y	42.89 ± 0.77
Matching Net [27]	ANN-ConvNet-4	N	43.56 ± 0.84
Prototypical Networks [23]	ANN-ConvNet-4	N	49.42 ± 0.78
MAML [5]	ANN-ConvNet-4	Y	48.70 ± 1.84
Relation Net [24]	ANN-ConvNet-4	N	50.44 ± 0.82
OWOML [21]	SNN-ResNet-12	N	46.15 ± 0.34
CESM [30]	SNN-ResNet-12	Y	48.37 ± 0.24
MESM [30]	SNN-ResNet-12	Y	51.54 ± 0.23
MESM-nopretrain [30]	SNN-ResNet-12	N	35.78
HAAM(ours)	SNN-ResNet-12	N	50.25 ± 0.87

Ablation Experiment. To assess the general applicability of the HAAM plasticity rule, we compared the performance of various recurrent layer models—LSTM, RNN, and MLP—on the 5-way 1-shot tasks across the miniImageNet and Ominiglot datasets. The backbone networks employed was ConvNet-4, with results illustrated in Fig. 2. Notably, the HAAM rule demonstrated over a 15% improvement in accuracy across most recurrent layers, supporting our hypothesis that it enhances the temporal capabilities of the network.

Interestingly, RNN and MLP networks, initially exhibiting inferior temporal performance compared to LSTM, achieved accuracy levels comparable to or surpassing those of LSTM when integrated with the HAAM rule. We hypothesize that HAAM provides a stable memory storage mechanism similar to the cell state in LSTM networks, thereby mitigating the original advantage of LSTM.

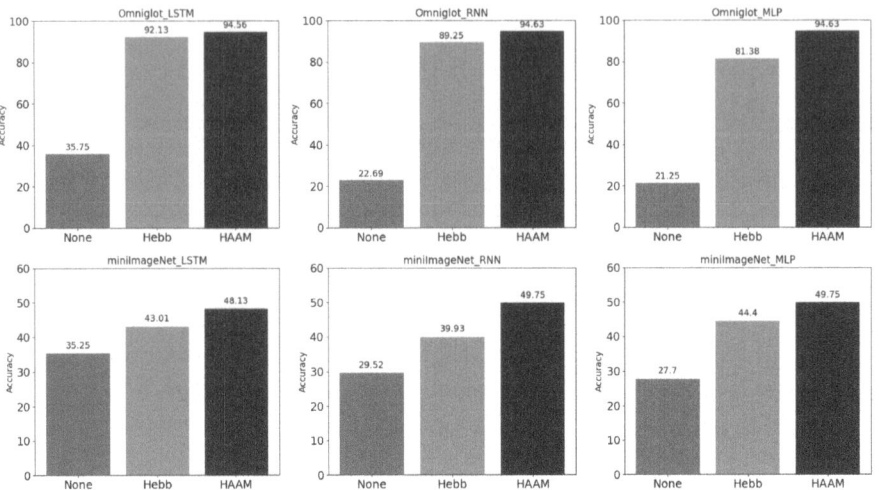

Fig. 2. Performance of various models (test accuracy) on the one-shot image classification task using the Ominiglot and miniImageNet dataset with ConvNet-4.

5 Conclusion

In this study, we presented the HAAM framework, which employs synaptic plasticity to enhance few-shot learning capabilities in SNNs. Our experimental results demonstrate that the HAAM plasticity rule significantly improves classification accuracy across various datasets, surpassing state-of-the-art non-pretrained SNN meta-learning methods while maintaining competitiveness with advanced ANN approaches. By integrating plasticity-driven mechanisms, our model effectively captures temporal dynamics and enhances adaptability to new tasks with minimal labeled data. These findings highlight the promise of biologically inspired learning paradigms in addressing rapid learning challenges in resource-constrained environments. Future research will focus on refining the HAAM framework and exploring its applications in more complex tasks, ultimately aiming to develop more robust neural architectures and advance the field of artificial intelligence.

Acknowledgments. This research was financially supported by a funding from Institute of Automation, Chinese Academy of Sciences (Grant No. E411230101).

References

1. Abbott, L.F., DePasquale, B., Memmesheimer, R.M.: Building functional networks of spiking model neurons. Nat. Neurosci. **19**(3), 350–355 (2016)

2. Abraham, W.C., Jones, O.D., Glanzman, D.L.: Is plasticity of synapses the mechanism of long-term memory storage? NPJ Sci. Learn. **4**(1), 9 (2019)
3. Brown, T.: Language models are few-shot learners. Adv. Neural. Inf. Process. Syst. **33**, 1877–1901 (2020)
4. Duan, Y., Jia, Z., Li, Q., Zhong, Y., Ma, K.: Hebbian and gradient-based plasticity enables robust memory and rapid learning in RNNs. arXiv preprint arXiv:2302.03235 (2023)
5. Finn, C., Abbeel, P., Levine, S.: Model-agnostic meta-learning for fast adaptation of deep networks. In: International Conference on Machine Learning, pp. 1126–1135. PMLR (2017)
6. Jiang, R., Zhang, J., Yan, R., Tang, H.: Few-shot learning in spiking neural networks by multi-timescale optimization. Neural Comput. **33**(9), 2439–2472 (2021)
7. Kim, W.B., Cho, J.H.: Encoding of discriminative fear memory by input-specific LTP in the amygdala. Neuron **95**(5), 1129–1146 (2017)
8. Koch, G., Zemel, R., Salakhutdinov, R., et al.: Siamese neural networks for one-shot image recognition. In: ICML Deep Learning Workshop, vol. 2, pp. 1–30. Lille (2015)
9. Lake, B.M., Salakhutdinov, R., Gross, J., Tenenbaum, J.B.: One shot learning of simple visual concepts. In: Proceedings of the 33rd Annual Meeting of the Cognitive Science Society, pp. 2568–2573 (2011)
10. Maass, W.: Networks of spiking neurons: the third generation of neural network models. Neural Netw. **10**(9), 1659–1671 (1997)
11. Magee, J.C., Grienberger, C.: Synaptic plasticity forms and functions. Annu. Rev. Neurosci. **43**, 95–117 (2020)
12. Martin, S.J., Grimwood, P.D., Morris, R.G.: Synaptic plasticity and memory: an evaluation of the hypothesis. Annu. Rev. Neurosci. **23**(1), 649–711 (2000)
13. Milstein, A.D., et al.: Bidirectional synaptic plasticity rapidly modifies hippocampal representations. Elife **10**, e73046 (2021)
14. Munkhdalai, T., Yu, H.: Meta networks. In: International Conference on Machine Learning, pp. 2554–2563. PMLR (2017)
15. Nabavi, S., Fox, R., Proulx, C.D., Lin, J.Y., Tsien, R.Y., Malinow, R.: Engineering a memory with LTD and LTP. Nature **511**(7509), 348–352 (2014)
16. Nakazawa, K., McHugh, T.J., Wilson, M.A., Tonegawa, S.: NMDA receptors, place cells and hippocampal spatial memory. Nat. Rev. Neurosci. **5**(5), 361–372 (2004)
17. Neftci, E.O., Mostafa, H., Zenke, F.: Surrogate gradient learning in spiking neural networks: bringing the power of gradient-based optimization to spiking neural networks. IEEE Signal Process. Mag. **36**(6), 51–63 (2019)
18. Neves, G., Cooke, S.F., Bliss, T.V.: Synaptic plasticity, memory and the hippocampus: a neural network approach to causality. Nat. Rev. Neurosci. **9**(1), 65–75 (2008)
19. Nichol, A., Schulman, J.: Reptile: a scalable metalearning algorithm. arXiv preprint arXiv:1803.02999, vol. 2, no. 3, p. 4 (2018)
20. Pei, J., et al.: Towards artificial general intelligence with hybrid tianjic chip architecture. Nature **572**(7767), 106–111 (2019)
21. Rosenfeld, B., Rajendran, B., Simeone, O.: Fast on-device adaptation for spiking neural networks via online-within-online meta-learning. In: 2021 IEEE Data Science and Learning Workshop (DSLW), pp. 1–6. IEEE (2021)
22. Santoro, A., Bartunov, S., Botvinick, M., Wierstra, D., Lillicrap, T.: Meta-learning with memory-augmented neural networks. In: International Conference on Machine Learning, pp. 1842–1850. PMLR (2016)
23. Snell, J., Swersky, K., Zemel, R.: Prototypical networks for few-shot learning. In: Advances in Neural Information Processing Systems, vol. 30 (2017)

24. Sung, F., Yang, Y., Zhang, L., Xiang, T., Torr, P.H., Hospedales, T.M.: Learning to compare: Relation network for few-shot learning. In: Proceedings of the IEEE Conference on Computer Vision and Pattern Recognition, pp. 1199–1208 (2018)
25. Thrun, S., Pratt, L.: Learning to learn: introduction and overview. In: Learning to Learn, pp. 3–17. Springer (1998)
26. Vanschoren, J.: Meta-learning: a survey. arXiv preprint arXiv:1810.03548 (2018)
27. Vinyals, O., Blundell, C., Lillicrap, T., Wierstra, D., et al.: Matching networks for one shot learning. In: Advances in Neural Information Processing Systems, vol. 29 (2016)
28. Zeng, Y., et al.: Braincog: a spiking neural network based, brain-inspired cognitive intelligence engine for brain-inspired AI and brain simulation. Patterns **4**(8) (2023)
29. Zhan, Q., Liu, G., Xie, X., Zhang, M., Sun, G.: Bio-inspired active learning method in spiking neural network. Knowl.-Based Syst. **261**, 110193 (2023)
30. Zhan, Q., Wang, B., Jiang, A., Xie, X., Zhang, M., Liu, G.: A two-stage spiking meta-learning method for few-shot classification. Knowl.-Based Syst. **284**, 111220 (2024)

Palmprint Texture Fusion Based on TinyViT for Recognition

Fuchuan Huang, Cunyu Sheng, Jian He, and Wei Jia[✉]

Hefei University of Technology, No. 485, Danxia Road, Hefei, Anhui, China
2024110481@mail.hfut.edu.cn

Abstract. Palmprint recognition, as a biometric technology, is highly valued for its uniqueness and stability. This paper integrates traditional palmprint recognition techniques with deep learning-based methods, proposing a palmprint texture fusion ViT (PTF-ViT), aimed at enhancing the accuracy and robustness of palmprint recognition. We leverage the advantages of the Vision Transformer (ViT) to fuse the texture features of palmprint with original image information through an attention mechanism, thereby enhancing the model's discriminative ability. Experimental results indicate that PTF-ViT performs exceptionally well across various palmprint datasets, exhibiting high Average Recognition Rate (ARR) and low Equal Error Rate (EER), especially demonstrating strong robustness on cross-device datasets. Compared to traditional methods and existing deep learning approaches, PTF-ViT holds significant advantages in the field of palmprint recognition, proving its effectiveness and advancement. This study not only fills the void of ViT in palmprint recognition but also paves a new direction for future research in biometric technology.

Keywords: Palmprint recognition · Transformer · Deep learning

1 Introduction

Biometric technology plays a crucial role in modern society, particularly in identity recognition, authentication, and security applications. It verifies personal identity by analyzing unique physiological or behavioral characteristics. Physiological traits include inherent features such as facial recognition, fingerprints, and palmprints, while behavioral traits encompass dynamic actions like handwriting and gait. Compared to traditional methods like passwords or PIN codes, biometrics offer uniqueness, long-term stability, and stronger resistance to forgery.

Currently, biometric technology research is highly active, with palmprint recognition emerging as a popular field. Compared to facial recognition, palmprint recognition is non-intrusive and offers higher data security. In contrast to fingerprint recognition, palmprints provide a larger and richer feature area, making them harder to forge. Moreover, palmprint recognition is more cost-effective than iris recognition, as low-resolution data can be captured with standard smartphones. These advantages give palmprint recognition broad application prospects. Existing palmprint recognition techniques can be divided into traditional methods and deep learning-based methods. The former relies on handcrafted feature descriptors, while the latter leverages the similarity within dataset samples to train effective neural network models, enabling automatic learning and feature extraction from raw data.

As a key technology in the field of artificial intelligence, deep learning has made significant advancements in various domains, including computer vision, speech recognition, natural language processing, and robotics. [1–4] In 2020, dosovitskiy et al. [5] introduced the Vision Transformer (ViT), bringing a major breakthrough to computer vision. Unlike convolutional neural networks (CNNs), which rely on local receptive fields, ViT excels at capturing global contextual information, enabling more accurate and intelligent predictions.

This paper proposes a palmprint texture fusion method based on TinyViT [6], called Palmprint Texture Fusion ViT (PTF-ViT). It employs an attention-based fusion mechanism to integrate palmprint texture features with original image information, aiming to maximize the extraction of learnable texture representations. This approach effectively enhances the recognition performance of Vision Transformers on palmprint images, addressing the gap in palmprint recognition using Transformers. Comparative experiments were conducted on multiple palmprint datasets against traditional methods and deep learning-based techniques, and the results validate the effectiveness of the proposed method.

2 Related Work

Currently, palmprint recognition primarily includes two types of recognition algorithms: traditional palmprint recognition algorithms and deep learning-based palmprint recognition algorithms.

Traditional palmprint recognition methods extract palm features and encode them into feature vectors, which are then matched or recognized based on the similarity between these vectors. These methods can be further categorized into texture-based and orientation-based approaches. In 2008, Jia et al. [7] introduced a modified finite Radon transform (MFRAT) to effectively detect palm line orientation features, leading to the development of the robust line orientation code (RLOC). In 2020, Lunke et al. [8] combined multiple features such as orientation and texture to propose the learn compact multifeature code (LCMFC). In 2024, Fan et al. [9] introduced the rich orientation code (ROC), which extracts both directional and width information from intersecting palm lines.

In recent years, deep learning-based palmprint recognition algorithms have garnered significant attention. These methods typically utilize convolutional neural networks (CNNs) for feature extraction and recognition. In 2022, Jia et al. [10] designed a new loss function and proposed an efficient and effective palmprint recognition network (EEPNet), employing various strategies to improve recognition performance. In 2023, Dandan et al. [11] proposed an Aligned Multilevel Gabor Convolution Network (AMGNet) by integrating principal line and wrinkle features, while introducing a directional CosAngle loss function.

With the development of Vision Transformer (ViT) and its outstanding performance in computer vision tasks, researchers have begun exploring the application of ViT in biometric recognition. In 2022, Huang et al. [12] proposed the FV Transformer (FVT) for finger vein recognition. In 2024, Grosz et al. [13] integrated global and local features, combining ViT and CNN, to propose a mobile-based, end-to-end palmprint recognition system called Palm-ID.

3 Proposed Method

Based on the excellent performance of TinyViT, this paper introduces a multiscale texture feature extraction module, an attention-selective fusion module, and improves the loss function to enhance model performance. Figure 1 illustrates the network structure of the Palmprint Texture Fusion ViT (PTF-ViT). However, the standard ViT structure fails to fully exploit its potential for image processing, and existing research indicates that single features are insufficient to capture the complete texture information of palmprints. By employing a feature fusion method to blend multiscale texture features with the original image, the impact of differences in image resolution and texture discriminability can be effectively mitigated. To this end, we utilize Local Binary Patterns (LBP) to extract textural features from palmprint images. By simultaneously inputting both the original image and the LBP image into the model, we aim to overcome noise originating from low resolution, such as illumination and blurriness.

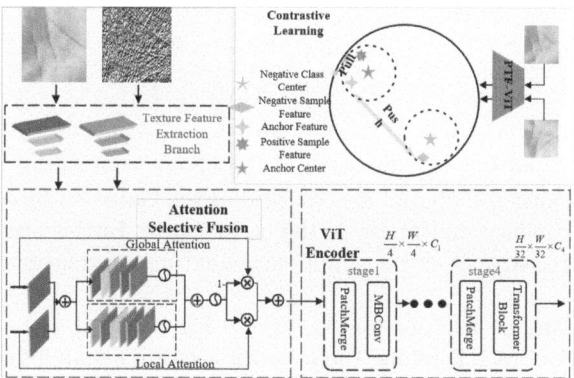

Fig. 1. Structure diagram of PTF-ViT.

3.1 Multiscale Texture Feature Extraction Module

By introducing a learnable Gabor convolution (LGC) [14] layer, we designed a multiscale texture feature extraction (MTFE) module. The structure of the module is illustrated in Fig. 2. The LGC is used to extract texture features, and a coordinate attention (CA) [15] mechanism is employed to emphasize the more discriminative line regions in the palmprint. Finally, a channel-level softmax function is applied to retain the directional sorting features.

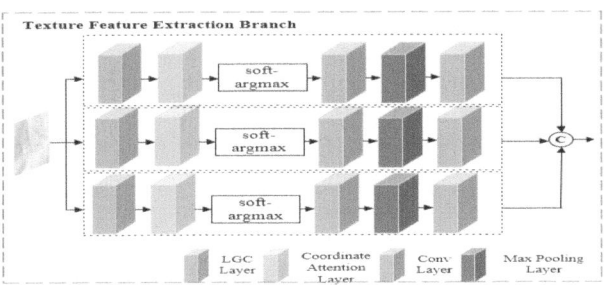

Fig. 2. Multi-scale Texture Feature Extraction Module.

The Gabor convolution kernel in LGC can be expressed by Eq. 1, featuring learnable hyperparameters $\sigma, \gamma, u, \psi, \theta$.

$$G(x, y, \sigma, \gamma, u, \psi, \theta) = -exp\left\{-\frac{\gamma^2 x + y'^2}{2(2\sqrt{2}\sigma)^2}\right\} cos(2\pi u x^2 + \psi) \qquad (1)$$

where $x' = x\cos\theta + y\sin\theta$ and $y' = -x\sin\theta + y\cos\theta$, (x, y) represent the coordinate positions, and θ is the direction of the Gabor convolution kernel. σ and γ are the bandwidth and spatial aspect ratio of the Gaussian filter, while u and ψ denote the frequency and phase of the cosine wave. In this study, the Gabor filter is set to three sizes to extract multi-scale texture information: 7×7, 17×17, and 35×35.

Considering the position of fragile texture features [16], not all textures in the image are effective. Therefore, the network needs to assign different attention levels to different regions; for instance, the core line areas should receive more weight. In this study, the coordinate attention mechanism is used to embed positional information into the channel attention.

3.2 Attention Selection Fusion Module

After obtaining the multi-scale texture features, the Attention Selective Fusion (ASF) [17] module is employed to fuse global and local features, enhancing the discriminative characteristics of the image. The structure of the ASF module is illustrated in Fig. 3. The two feature maps extracted by the Multi-scale Texture

Feature Extraction (MTFE) generate a weight score map through the ASF, which is then used to fuse the features of the original image and the LBP image, resulting in the final feature representation.

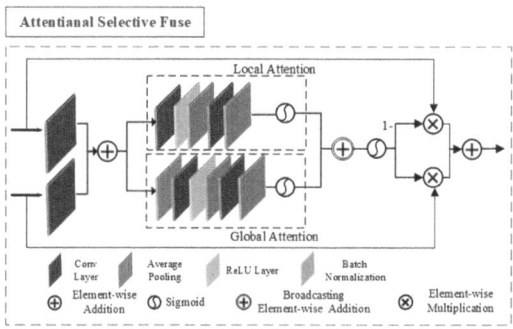

Fig. 3. Attention Selection Fusion Module.

The process of generating the weight score maps W_c and W_l is detailed below. First, the two feature maps X_{lbp} and X_{rgb} are aggregated and processed through the global attention branch and the local attention branch to obtain G and L, respectively. The global attention branch is represented by Eq. 2, while the local attention branch is represented by Eq. 3:

$$G(U) = \sigma(BN(Conv_G^2(ReLU(BN(Conv_G^1(AP(X))))))) \qquad (2)$$

$$L(U) = \sigma(BN(Conv_L^2(ReLU(BN(Conv_L^1(X)))))) \qquad (3)$$

where AP denotes the average pooling operation, BN represents batch normalization, and σ is the commonly used sigmoid activation function.

After obtaining the attention maps from the two branches, $GL(U)$ is derived from the broadcasting addition of $G(U)$ and $L(U)$. Finally, the output feature map U of the ASF module is obtained as shown in Eq. 4.

$$U = X_{lbp} \otimes \sigma(GL(U)) + X_{rgb} \otimes \sigma(1 - GL(U)) \qquad (4)$$

3.3 The Dropout Mechanism of Multi-Head Self-Attention

To enhance the randomness of Multi-Head Self-Attention (MSA) [18] across different blocks and to discard less informative inputs, guiding the model to more accurately search for significant palmprint regions while reducing the model's parameter count, this paper introduces a Multi-Head Self-Attention Dropout mechanism: Multi-Attention Dropping (MAD). This mechanism is added after each stage's MSA layer. As shown in Fig. 4, each self-attention head within the

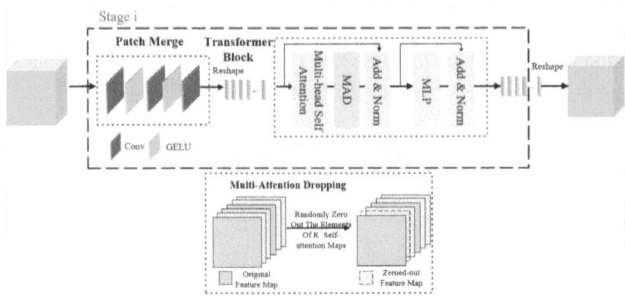

Fig. 4. ViT stage structure diagram and MAD illustration.

MSA layer forms an independent feature space while concurrently focusing on different feature relationships within the input sequence. Notably, the MAD operation is only executed during the training phase and does not occur during inference.

3.4 Loss Function Design

In classification tasks, the cross-entropy loss function is commonly used to minimize the distance between training samples and their corresponding class centers. However, palmprint recognition is more of a one-to-one comparison for identity verification, making the matching distance between different samples equally important for validation. Therefore, we aim for a smaller intra-class distance among samples and a larger inter-class distance. To address this, a contrastive loss function [19] has been proposed.

The final loss function combines the cross-entropy loss $\mathcal{L}_{se}$ and the contrastive loss $\mathcal{L}_{con}$, which can be formulated as Eq. 5. In this study, the weights w_{ce} and w_{con} are set to 0.8 and 0.2, respectively.

$$\mathcal{L} = W_{ce} \times \mathcal{L}_{se} + W_{con} \times \mathcal{L}_{con} \qquad (5)$$

4 Experiment

4.1 Database and Experimental Setting

The datasets used in this experiment include PolyU II [20], PolyU M_B [21], HFUT [22], HFUT-CS [23], TJU [24], and MPD [25]. Detailed information about the datasets is provided in Table 1.

The hyperparameter settings for the experiment are shown in Table 2, and the division of the training and testing sets is based on different collection phases.

Table 1. Detailed Description of the Datasets Used in the Experiments.

Database	Sensing Mode	Palms	Imgs	Phase	Format	Res.	Year
PolyU II	touch	386	7752	2	gray	128 × 128	2003
PolyUM_B	touch	500	6000	2	gray	128 × 128	2010
HFUT	touch	800	16000	2	gray	256 × 256	2012
HFUT-CS	touchless	200	12000	2	gray	256 × 256	2012
TJU	touchless	600	12000	2	gray	128 × 128	2017
MPD	touchless	400	16000	2	RGB	224 × 224	2020

Table 2. Experimental Parameter Settings.

Lr	Weight decay	Momentum	Epochs	Batch size	Image size	Optimizer
0.01	0.05	0.9	300	8	224	AdamW

4.2 Experimental Results

The experiments include traditional encoding methods such as CompCode [20] and OrdinalCode [26], as well as deep learning recognition techniques. The performance evaluation metrics used are the Receiver Operating Characteristic (ROC) curve and the Equal Error Rate (EER), both derived from the Genuine Acceptance Rate (GAR) and the False Acceptance Rate (FAR).

Table 3. Comparison of Recognition Accuracy (ARR, EER) Performance with Other Methods.

Method	PolyUII		PolyUM_B		HFUT		HFUT-CS		TJU		MPD	
	ARR	EER	ARR	EER	ARR	EER	ARR	EER	ARR	EER	ARR	EER
CompCode [20]	99.73	0.2300	99.87	0.1560	99.38	0.4300	99.22	0.9000	99.30	0.5500	68.22	14.90
PalmCode [27]	97.22	1.301	99.13	0.9670	89.21	3.892	89.92	4.930	96.24	1.796	56.92	18.36
OrdinalCode [26]	100	0.0259	100	0.0036	99.73	0.3819	99.65	0.6537	99.95	0.2378	68.75	12.36
RLOC [7]	99.87	0.1531	99.97	0.0983	99.75	0.3917	99.36	0.7719	99.37	0.5660	67.15	17.52
POC [28]	99.92	0.5303	99.67	0.7738	97.66	1.827	98.12	2.580	99.21	0.6267	50.61	18.55
SMCC [29]	100	**0.0085**	100	0.0020	99.90	0.2635	99.65	0.3580	99.96	0.1530	70.27	12.40
HOL [30]	99.89	0.2938	99.83	0.4416	99.35	1.0375	99.07	1.677	99.06	0.9013	66.23	11.03
LLDP [31]	100	0.2584	100	0.1996	99.77	0.0225	99.67	0.4975	**99.68**	0.1664	85.66	8.564
CompNet [14]	100	0.0103	100	0.0012	99.80	**0.0111**	99.87	0.0368	99.98	**0.0111**	97.46	0.5960
PTF-ViT	100	0.0221	99.97	**0.0009**	**99.93**	0.0583	100	**0.0082**	99.67	0.0440	**99.35**	**0.2579**

The results in Table 3 indicate that PTF-ViT demonstrates outstanding or near-optimal performance across six datasets, with an ARR exceeding 99%, and achieving 100% for both PolyUII and HFUT-CS. Notably, on the cross-device MPD dataset, the ARR reaches 99.35%, significantly outperforming other methods. Furthermore, PTF-ViT exhibits excellent EER performance on the PolyUMB, HFUT-CS, TJU, and MPD datasets, particularly achieving an EER as low as 0.2579% on the MPD dataset.

The MPD dataset presents challenges due to sample variability resulting from differences in acquisition devices. Traditional encoding methods perform poorly on this dataset, reflecting their sensitivity to noise and variations in texture orientation. In contrast, PTF-ViT demonstrates strong robustness through its effective capture of global positional information and texture features. Additionally, both CompNet and PTF-ViT exhibit excellent performance across all datasets, further highlighting the importance of palmprint texture information in enhancing model recognition performance.

As shown in Fig. 5, the closer the ROC curve is to the upper axis, the better the verification performance. It can be observed that our proposed method performs excellently across all test datasets; although it did not achieve the best results on the TJU dataset, it still obtained a second-best performance. Furthermore, according to the EER values in Table 4, our method consistently ranks among the top for PolyUMB, HFUT-CS, MPD, and cross-device MPD datasets, particularly achieving a very low EER of 0.2579% on the cross-device MPD dataset, significantly outperforming other methods. In contrast, traditional encoding methods performed the worst, confirming their sensitivity to variations in illumination, noise, and rotation.

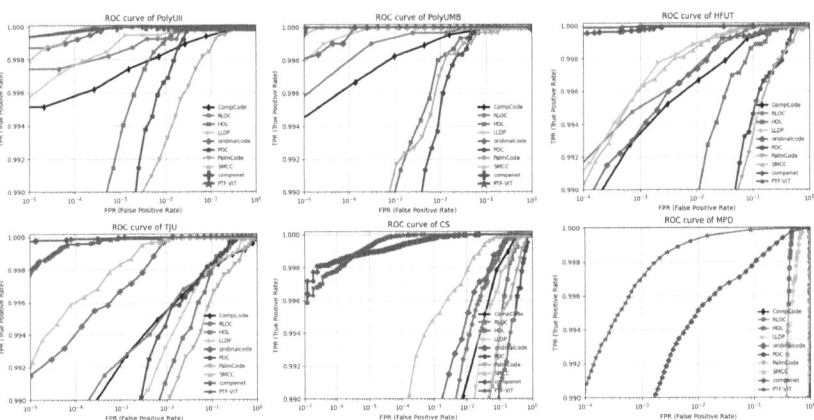

Fig. 5. The ROC curves of the proposed and its competing methods across all datasets.

5 Conclusion

We proposed a texture fusion method based on TinyViT, named PTF-ViT, for palmprint recognition. Leveraging the advantages of Vision Transformer, this approach explores a multi-feature fusion processing technique. By using an attention-based selection fusion method, it integrates both the original palmprint image information and the texture feature information, thereby deeply mining the representation of texture features and effectively enhancing the recognition performance of Vision Transformer on palmprint images, filling a gap in the field of palmprint recognition. Experiments conducted on multiple palmprint datasets demonstrate the effectiveness and advancement of this method compared to traditional and deep learning approaches, thanks to its effective capture of global positional information and texture features in images.

References

1. Hu, B., Wang, J.: Deep learning based hand gesture recognition and UAV flight controls. Int. J. Autom. Comput. **17**(1), 17–29 (2020)
2. Ha, V.K., Ren, J., Xu, X., Zhao, S., Xie, G., Vargas, V.M.: Deep learning based single image super-resolution: a survey. In: Ren, J., et al. (eds.) BICS 2018. LNCS (LNAI), vol. 10989, pp. 106–119. Springer, Cham (2018). https://doi.org/10.1007/978-3-030-00563-4_11
3. Li, C., Wu, X., Zhao, N., Cao, X., Tang, J.: Fusing two-stream convolutional neural networks for RGB-T object tracking. Neurocomputing **281**, 78–85 (2018)
4. Li, C., Liang, X., Lu, Y., Zhao, N., Tang, J.: RGB-T object tracking: benchmark and baseline. Pattern Recogn. **96**, 106977 (2019)
5. Dosovitskiy, A.: An image is worth 16x16 words: transformers for image recognition at scale. arXiv preprint arXiv:2010.11929 (2020)
6. Wu, K., et al.: Tinyvit: fast pretraining distillation for small vision transformers. In: European Conference on Computer Vision, pp. 68–85. Springer (2022)
7. Jia, W., Huang, D.S., Zhang, D.: Palmprint verification based on robust line orientation code. Pattern Recogn. **41**(5), 1504–1513 (2008)
8. Fei, L., Zhang, B., Zhang, L., Jia, W., Wen, J., Wu, J.: Learning compact multifeature codes for palmprint recognition from a single training image per palm. IEEE Trans. Multimedia **23**, 2930–2942 (2020)
9. Fan, D., Liang, X., Jia, W., Zhang, D.: Toward large-scale palmprint image analysis by a rich orientation code. IEEE Trans. Syst. Man Cybern. Syst. (2024)
10. Jia, W., Ren, Q., Zhao, Y., Li, S., Min, H., Chen, Y.: Eepnet: an efficient and effective convolutional neural network for palmprint recognition. Pattern Recogn. Lett. **159**, 140–149 (2022)
11. Fan, D., Liang, X., Zhang, C., Jia, W., Zhang, D.: Amgnet: aligned multilevel gabor convolution network for palmprint recognition. IEEE Trans. Circuits Syst. Video Technol. (2023)
12. Huang, J., Luo, W., Yang, W., Zheng, A., Lian, F., Kang, W.: FVT: finger vein transformer for authentication. IEEE Trans. Instrum. Meas. **71**, 1–13 (2022)
13. Grosz, S.A., Godbole, A., Jain, A.K.: Mobile contactless palmprint recognition: use of multiscale, multimodel embeddings. IEEE Trans. Inf. Forensics Secur. (2024)
14. Liang, X., Yang, J., Lu, G., Zhang, D.: Compnet: competitive neural network for palmprint recognition using learnable gabor kernels. IEEE Signal Process. Lett. **28**, 1739–1743 (2021)
15. Hou, Q., Zhou, D., Feng, J.: Coordinate attention for efficient mobile network design. In: Proceedings of the IEEE/CVF Conference on Computer Vision and Pattern Recognition, pp. 13713–13722 (2021)

16. Zhang, L., Li, H., Niu, J.: Fragile bits in palmprint recognition. IEEE Signal Process. Lett. **19**(10), 663–666 (2012)
17. Ma, F., Sun, B., Li, S.: Facial expression recognition with visual transformers and attentional selective fusion. IEEE Trans. Affect. Comput. **14**(2), 1236–1248 (2021)
18. Xue, F., Wang, Q., Guo, G.: Transfer: learning relation-aware facial expression representations with transformers. In: Proceedings of the IEEE/CVF International Conference on Computer Vision, pp. 3601–3610 (2021)
19. Khosla, P., et al.: Supervised contrastive learning. Adv. Neural. Inf. Process. Syst. **33**, 18661–18673 (2020)
20. Zhang, D., Kong, W.K., You, J., Wong, M.: Online palmprint identification. IEEE Trans. Pattern Anal. Mach. Intell. **25**(9), 1041–1050 (2003)
21. Zhang, D., Guo, Z., Lu, G., Zhang, L., Zuo, W.: An online system of multispectral palmprint verification. IEEE Trans. Instrum. Meas. **59**(2), 480–490 (2009)
22. Jia, W., et al.: Palmprint recognition based on complete direction representation. IEEE Trans. Image Process. **26**(9), 4483–4498 (2017)
23. Jia, W., Hu, R.X., Gui, J., Zhao, Y., Ren, X.M.: Palmprint recognition across different devices. Sensors **12**(6), 7938–7964 (2012)
24. Zhang, L., Li, L., Yang, A., Shen, Y., Yang, M.: Towards contactless palmprint recognition: a novel device, a new benchmark, and a collaborative representation based identification approach. Pattern Recogn. **69**, 199–212 (2017)
25. Zhang, Y., Zhang, L., Zhang, R., Li, S., Li, J., Huang, F.: Towards palmprint verification on smartphones. arXiv preprint arXiv:2003.13266 (2020)
26. Sun, Z., Tan, T., Wang, Y., Li, S.Z.: Ordinal palmprint represention for personal identification [representation read representation]. In: 2005 IEEE Computer Society Conference on Computer Vision and Pattern Recognition (CVPR 2005), vol. 1, pp. 279–284. IEEE (2005)
27. Kumar, A., Shen, H.C.: Palmprint identification using palmcodes. In: Third International Conference on Image and Graphics (ICIG 2004), pp. 258–261. IEEE (2004)
28. Zhu, Y.-H., Jia, W., Liu, L.-F.: Palmprint recognition using band-limited phase-only correlation and different representations. In: Huang, D.-S., Jo, K.-H., Lee, H.-H., Kang, H.-J., Bevilacqua, V. (eds.) ICIC 2009. LNCS, vol. 5754, pp. 270–277. Springer, Heidelberg (2009). https://doi.org/10.1007/978-3-642-04070-2_31
29. Zuo, W., Lin, Z., Guo, Z., Zhang, D.: The multiscale competitive code via sparse representation for palmprint verification. In: 2010 IEEE Computer Society Conference on Computer Vision and Pattern Recognition, pp. 2265–2272. IEEE (2010)
30. Jia, W., Hu, R.X., Lei, Y.K., Zhao, Y., Gui, J.: Histogram of oriented lines for palmprint recognition. IEEE Trans. Syst. Man Cybern. Syst. **44**(3), 385–395 (2013)
31. Luo, Y.T., et al.: Local line directional pattern for palmprint recognition. Pattern Recogn. **50**, 26–44 (2016)

Novel Device Placement Approach with Neighbor Effect Aware Graph Mamba Networks

Hao Shu[1], Wangli Hao[1(✉)], Meng Han[1,2], and Fuzhong Li[1]

[1] Shanxi Agricultural University, Taigu, Jinzhong, Shanxi, China
haowangli@sxau.edu.cn
[2] Hangzhou Dianzi University, Hangzhou, Zhejiang, China

Abstract. For distributed computing, there is an urgent need to solve the problem of device placement, so we propose an innovative graph neural network framework NEAGMamba. It employs a decoupling and importance score evaluation mechanism to prevent excessive smoothing of node encodings during the aggregation process. In addition, a new Graph Mamba Network (Gmamba) is designed to replace the traditional GNN, capable of extracting node characteristics and capturing remote dependencies. Specifically, the features of each node are decoupled into two parts: discriminant features and aggregation features. At the same time, the proposed GMamba is used to obtain the aggregation features. Finally, the importance of adjacent nodes is evaluated by the importance score to realize the integration of adjacent information. The experimental results showed that the execution time was improved by 13.55% using NEAGMamba compared to Placeto. NEAGMamba performed 8.18% better than GraphSAGE and 1.22% better than P-GNN. In terms of computation time, NEAGMamba increased by 96.96% compared with Placeto, 97.34% with GraphSAGE and 96.86% with P-GNN.

Keywords: Distributed Machine Learning · Device Placement · Graph Mamba Networks · Importance Score Evaluation Mechanism

1 Introduction

With the application of Artificial Intelligence (AI) models in Natural Language Processing (NLP) [1], Computer Vision (CV) [2] and Recommendation Systems (RS) [3]. The scale of large language models (LLM) have grown rapidly, with the largest model now being capable of encompassing trillions of parameters [4]. However, training these massive models necessitates harnessing the collective power of thousands of computing devices and ensuring their efficient management. Clearly, traditional single-machine setups cannot meet the demands of large-scale models in terms of training and storage. As a result, both the scientific research and industrial planning groups are becoming more concerned with the efficiently training of large-scale models.

To effectively address this issue, we adopt an automatic parallel training mechanism to identify optimal parallelization strategies. Recently, numerous approaches leveraging graph neural networks (GNN) and reinforcement learning (RL) techniques had been developed, in order to optimize the placement of model tensors across devices for optimal performance. Examples include Placeto [5], GraphSAGE [6] and P-GNN [6], which adopt such strategies. Nevertheless, contemporary graph based methodologies for device placement tend to undervalue the pivotal role of edge information within computational graphs, thereby constraining the overall performance and efficiency achievable through model parallelism.

However, in contrast, our proposed NEAGMamba method is different from traditional node aggregation methods. NEAGMamba not only uses Mamba to integrate into GNN's, but also prevents the over smoothing problem by decoupling and importance score. This new approach not only enhances the node representation, but also significantly reduces the time required for node aggregation (Fig. 1).

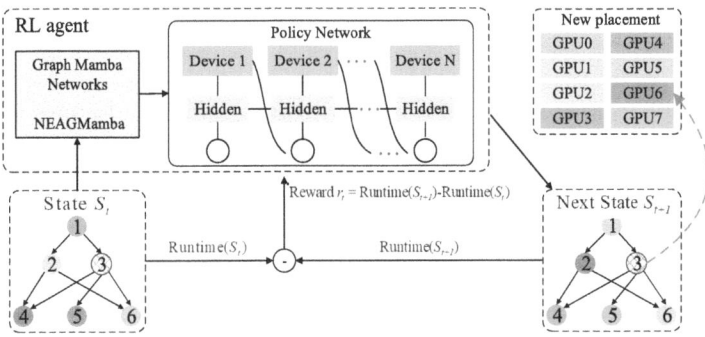

Fig. 1. The overview of the NEAGMamba for device placement.

Following are the key contributions of our research.

(1) A new Neighbor Effect Aware Graph Mamba Networks (NEAGMamba) is designed to replace traditional GNN for node feature extraction and remote dependency capture.
(2) A feature decoupling strategy is proposed, which divides node features into discriminant features and aggregation features, so that the model can process node characteristics and neighbor information more flexibly.
(3) The importance score evaluation mechanism is introduced to quantify the importance of neighbor nodes, which helps in the optimization of the aggregation process for neighbor node information, and also improves the quality of node feature representation.
(4) A large number of experimental results verified the effectiveness of the NEAGMamba in device placement, achieving an improvement of between 10% and 20% in execution time and a 94% improvement in calculation time.

2 Related Works

Recent works portrayed in [5,6] have employed GNNs for node embedding's in the context of device placement. For instance, Placeto [5] iteratively passes messages to capture the graph structure for optimizing the sequential placement of the operators. This approach enables a more effective understanding and utilization of the relationships between nodes in the computation of the graphs, thereby optimizing the allocation of operations across devices (such as GPUs and CPUs) and enhancing computational efficiency.

On the one hand, in the research of GNN [7] model, frequency domain method and spatial domain method are the two main streams. Frequency-domain methods such as GCN [8] proposed by Kipf et al. simplify graph convolution computations by redefining the propagation matrix and stacking layers. Spatial domain methods such as GAT [9] proposed by Veličković et al., use an attention mechanism to learn edge weights and improve the flexibility and accuracy of information interaction between nodes. Moreover, the GIN [10] model proposed by Xu et al. works by introducing graph isomorphic networks.

For the node classification task, Li et al. [11] pointed out that graph convolution of GCN is a special form of Laplacian smoothing, which may lead to fuzzy features of connected nodes and difficulty in distinguishing between different labels. Wang et al. [12] found that multi-layer attention stacking may cause inter-class side information exchange leading to excessive smoothing of node features, affecting classification performance. Therefore, addressing over-smoothing is the main problem faced by current GNN.

On the other hand, Mamba model proposed by Gu & Dao [13] achieves efficient processing of sequence data through selective state transition mechanism and performs well in language modeling tasks, which inspires researchers to explore the application of SSM in GNN [7]. There are transformer-based gt frameworks such as GPS [14] and TokenGT [15], and researchers are trying to replace Transformer with Mamba for more efficient graph data processing and analysis.

3 Methods

In this section, we will introduce the two important modules of NEAGMamba: NEAGNN and GMamba, and will also try to analyze the core elements and ideas the aforementioned process. Which will help in providing the reader with a comprehensive and insightful perspective. For which an illustrative Fig. 2 is made to show the structure of the NEAGMamba model.

3.1 Neighbor Effect Aware GNN: NEAGNN

In this section, the NEAGNN module in NEAGMamba is introduced, which augments GNN by adaptive learning node neighbor effects. So, we decouple node encoding into discriminant features and aggregation features. The discriminant

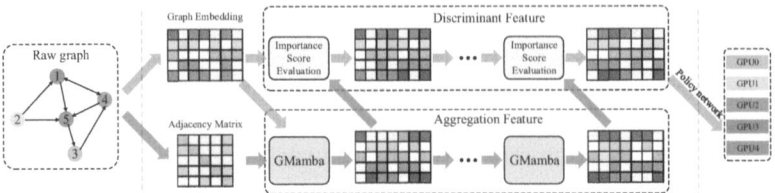

Fig. 2. The pipeline of the NEAGMamba for device placement.

features serves to evaluate the significance of neighbor nodes relative to the current node. The aggregation features are utilized to integrate information from neighbor nodes through graph convolution layers. and use α to evaluate the importance of neighbors to optimize graph convolution.

Let $G(V, E)$ denote a directed graph, where $V \in \mathbb{R}^{1 \times N}$ represents the set of nodes and N denotes the number of nodes. The edge set $E \in \mathbb{R}^{C \times 2}$ contains the set of edges, where C is the number of edges. $X \in \mathbb{R}^{N \times F}$ is encoded as the initial graph, where F is the feature dimension of each node.

$$H^0 = S^0 = Norm\left(Relu\left(Dense\left(X\right)\right)\right) \tag{1}$$

In Formula 1, the initial graph of encoding X is mapped to a higher dimensional space, $X' \in \mathbb{R}^{N \times M}$. And the initial decoupling is obtained as follows: (i) discriminant feature $S^0 \in \mathbb{R}^{N \times M}$ and (ii) aggregation feature $H^0 \in \mathbb{R}^{N \times M}$.

$$H^{t+1} = Norm\left(GMamba\left(H^t\right)\right) \tag{2}$$

In Eq. 2, we proposed a new graph convolution method: GMamba network. And with it denoted the aggregated features as H^{t+1}. The complex theoretical basis and practical implementation details of GMamba are elaborated in Sect. 3.2.

$$\alpha^t = \sigma\left(Dense\left(S^{t-1} \| H^t\right)\right) \tag{3}$$

$$S^t = Norm\left(\left(1 - \alpha^t\right) S^{t-1} + \alpha^t H^t\right) \tag{4}$$

In Eq. 3, 4, the symbol $\|$ denotes the column-wise concatenation operations. Where we used the Sigmoid function σ to normalize the fused features. The discriminant feature S^{t-1} and the aggregation feature H^t are updated by weighted updating to obtain the updated discriminant feature S^t.

$$X' = Fusion\left(Dense\left(S^T\right), X\right) \tag{5}$$

In Eq. 5, at the final time step T, $S^T \in \mathbb{R}^{N \times M}$ is combined with the initial graph coding $X \in \mathbb{R}^{N \times F}$ to ensure that the feature information of each node is not forgotten.

3.2 Graph Mamba Networks: GMamba

The Mamba architecture is known for its superior sequential processing capabilities, but the complexity and nonlinearity of graphs prevent its direct application in graph processing. To overcome this challenge, we creatively integrate Mamba into Graph neural networks (GNN) and proposes a Graph Mamba Networks (GMamba), which open up a new avenue for GNN.

Firstly, in order to further enhance the performance of GMamba, we applied the position encoding and graph encoding techniques in GCN. Inject structure and location information into the input initial graph encoding $X \in \mathbb{R}^{N \times F}$.

$$Y^0 = \sigma \left(\hat{D}^{-\frac{1}{2}} \hat{A} \hat{D}^{-\frac{1}{2}} X \mathbf{W} \right) \tag{6}$$

In Eq. 6, where $\hat{A} = A + I$, A is the adjacency matrix, I is the identity matrix, $\hat{D}$ is the diagonal matrix of node degrees induced by A and I. And σ is the nonlinear transformation function ($ReLU$). Moreover, the purpose of this operation is to symmetrically normalize the matrix A to prevent the changing of the original distribution of features when multiplying with the feature matrix.

$$\mathbf{B} = \mathbf{W_B} Y^t, \mathbf{C} = \mathbf{W_C} Y^t, \Delta = Softplus\left(\mathbf{W}_\Delta, Y^t \right) \tag{7}$$

$$\bar{\mathbf{A}} = exp\left(\Delta \mathbf{A} \right), \bar{\mathbf{B}} = (\Delta \mathbf{A})^{-1} \left(exp\left(\Delta \mathbf{A} \right) - I \right) \Delta \mathbf{B} \tag{8}$$

In Eq. 7 and 8, $\mathbf{W_B}$, $\mathbf{W_C}$ and $\mathbf{W}_\Delta$ are learnable parameterized weight matrices. By encoding Y with their respective weights and input graphs, the projection parameters $\mathbf{B} \in \mathbb{R}^{N \times 1}$, $\mathbf{C} \in \mathbb{R}^{N \times 1}$ and discretization step size $\Delta \in \mathbb{R}^{N \times F}$ are obtained. Through the introduction of discretization operations, the continuous state is transformed into a discrete state, so as to adapt to the data processing in deep learning. The specific implementation is to parameterize Δ as $\mathbf{A}$ function of the input to obtain the evolution parameter $\bar{\mathbf{A}} \in \mathbb{R}^{N \times N}$ and the projection parameter $\bar{\mathbf{B}} \in \mathbb{R}^{N \times 1}$ related to the input Y^t. In addition, the discrete parameters $\bar{\mathbf{A}}$ and $\bar{\mathbf{B}}$ can further control the extent to which Y^t that updates the hidden state, and shows how it affects the input Y^t.

$$Y^{t+1} = SSM_{\bar{\mathbf{A}},\bar{\mathbf{B}},\mathbf{C}} \left(Y^t \right) = \mathbf{C} \left(\bar{\mathbf{A}} Y^{t-1} + \bar{\mathbf{B}} Y^t \right) \tag{9}$$

In Eq. 9, Mamba is used to calculate the selective SSM, it is used to control the flow of information and update the hidden state, and as a result the updated graph embedding $Y^{t+1} \in \mathbb{R}^{N \times F}$ is obtained.

$$X' = Combine\left(Y^T, Y^0 \right) \tag{10}$$

In Eq. 10, the final output graph code $X' \in \mathbb{R}^{N \times F}$ is obtained by fusing Y^T with the initial input graph code Y^0.

4 Experiments and Analysis

In this section, we offer a comprehensive overview of the experimental setup that showcases the results achieved through the utilization of diverse graph embedding models, aiming to enhance the performance of device placement.

4.1 Experimental Setup

To ensure fairness, the Placeto simulator was used in experiments, designed for modern GPUs and accelerators. We assessed NEAGMamba with execution computation time metrics. Experiments used ptb, cifar10 and nmt datasets, with nodes clustered for efficiency. We trained on 3, 5 and 8 devices, randomly selecting 17 graphs/dataset for 20 epochs, aligning with P-GNN benchmarks. The focus is mainly on optimizing the GPU, emphasizing on the ability of our proposed model to be placed on devices.

Table 1. The execution time improvements of various models on three different GPUs across three distinct datasets, compared to the Placeto.

Methods	Datasets	3GPUS		5GPUS		8GPUS	
		Exe_time(s)	Impro(%)	Exe_time(s)	Impro(%)	Exe_time(s)	Impro(%)
Placeto	ptb	5.6592	-	5.5747	-	5.5134	-
GraphSAGE		5.4236	4.16%	5.3365	4.27%	5.3529	2.91%
P-GNN		5.0996	9.89%	4.9096	11.93%	4.7946	13.04%
GNN		5.0664	10.47%	4.9201	11.74%	4.7978	12.98%
GCN		5.0758	10.31%	4.9029	12.05%	4.7973	12.99%
GAT		5.0654	10.49%	4.9156	11.82%	4.8101	12.76%
NEAGMamba		**5.0567**	**10.65%**	**4.9009**	**12.09%**	**4.7862**	**13.19%**
Placeto	cifar10	2.1061	-	2.1649	-	1.9609	-
GraphSAGE		1.9439	7.70%	1.8109	16.35%	1.7763	9.41%
P-GNN		1.7682	16.04%	1.6525	23.67%	1.5991	18.45%
GNN		1.7623	16.32%	1.6669	23.00%	1.5999	18.41%
GCN		1.7682	16.04%	1.6577	23.43%	1.5886	18.99%
GAT		1.7666	16.12%	1.6688	22.92%	1.5926	18.78%
NEAGMamba		**1.7406**	**17.35%**	**1.6502**	**23.77%**	**1.5681**	**20.03%**
Placeto	nmt	2.3136	-	2.0969	-	2.1234	-
GraphSAGE		2.0981	9.31%	2.0392	2.75%	2.0071	5.48%
P-GNN		2.0518	11.32%	1.9869	5.25%	1.9682	7.31%
GNN		2.0570	11.09%	1.9762	5.75%	1.9685	7.29%
GCN		2.0579	11.05%	1.9978	4.73%	1.96	7.70%
GAT		2.0784	10.17%	1.9796	5.59%	1.9698	7.23%
NEAGMamba		**2.0248**	**12.48%**	**1.9226**	**8.31%**	**1.8646**	**12.19%**

4.2 Evaluate the Effectiveness of NEAGMamba in Device Placement

Execution Time. Table 1 compares models like GraphSAGE, P-GNN, GNN, GCN, GAT and NEAGMamba to the Placeto baseline. NEAGMamba tops in reducing execution time, with max gains of 13.19% (ptb), 20.03% (cifar10) and 12.48% (nmt). Its feature tackles the decoupling and over-smoothing, which helps in preserving the key information during propagation. And the importance scoring mechanism precisely assesses the influence of neighboring nodes, allowing the model to focus on relevant information and ignores noise. This is evident from the consistent performance gains across all datasets and GPU configurations.

Table 2. The computation time improvements of various models on three different GPUs across three distinct datasets.

Methods	Datasets	3GPUS		5GPUS		8GPUS	
		Comp_time(s)	Impro(%)	Comp_time(s)	Impro(%)	Comp_time(s)	Impro(%)
Placeto	ptb	430.8948	-	459.0664	-	437.4294	-
GraphSAGE		478.8761	-11.14%	495.6021	-7.96%	553.9057	-26.63%
P-GNN		433.7599	-0.66%	448.3053	2.34%	451.5618	-3.23%
GNN		5.84480	98.64%	5.8515	98.73%	5.8360	98.67%
GCN		**5.3585**	**98.76%**	**5.3867**	**98.83%**	**5.4459**	**98.76%**
GAT		10.6206	97.54%	8.9339	98.05%	10.8413	97.52%
NEAGMamba		**10.3281**	**97.60%**	**12.1111**	**97.36%**	**10.8461**	**97.52%**
Placeto	cifar10	122.7012	-	117.9207	-	119.0673	-
GraphSAGE		137.0734	-11.71%	128.6956	-9.14%	136.168	-14.3%
P-GNN		93.9383	23.44%	85.8227	27.22%	90.3976	24.08%
GNN		2.41261	98.03%	2.4046	97.96%	2.4141	97.97%
GCN		**2.3516**	**98.08%**	**2.3447**	**98.01%**	**2.3686**	**98.01%**
GAT		5.0706	95.87%	3.806	96.77%	5.039	95.77%
NEAGMamba		**4.3618**	**96.45%**	**4.5284**	**96.16%**	**4.621**	**96.12%**
Placeto	nmt	64.6854	-	63.4991	-	63.6237	-
GraphSAGE		72.162	-11.56%	73.073	-15.08%	73.8744	-16.11%
P-GNN		68.1539	-5.36%	67.0998	-5.67%	79.8831	-25.56%
GNN		2.0174	96.88%	2.0120	96.83%	2.0211	96.82%
GCN		**1.7244**	**97.33%**	**1.7512**	**97.24%**	**1.7727**	**97.21%**
GAT		3.7789	94.16%	3.2788	94.84%	3.765	94.08%
NEAGMamba		**3.7618**	**94.18%**	**3.7987**	**94.02%**	**2.7932**	**95.61%**

Computation Time. Table 2 demonstrates that for the ptb dataset, Placeto takes 437.4294 s using 8 GPUs. In comparison, GCN, GAT and NEAGMamba reduce the computation time by 98.76%, 97.52% and 97.52%, respectively. The NEAGMamba achieves 97.36% improvement with 5 GPUs. On 8 Gpus of cifar10, the time is reduced by 98.01%, 95.77% and 96.12%, respectively, while P-GNN

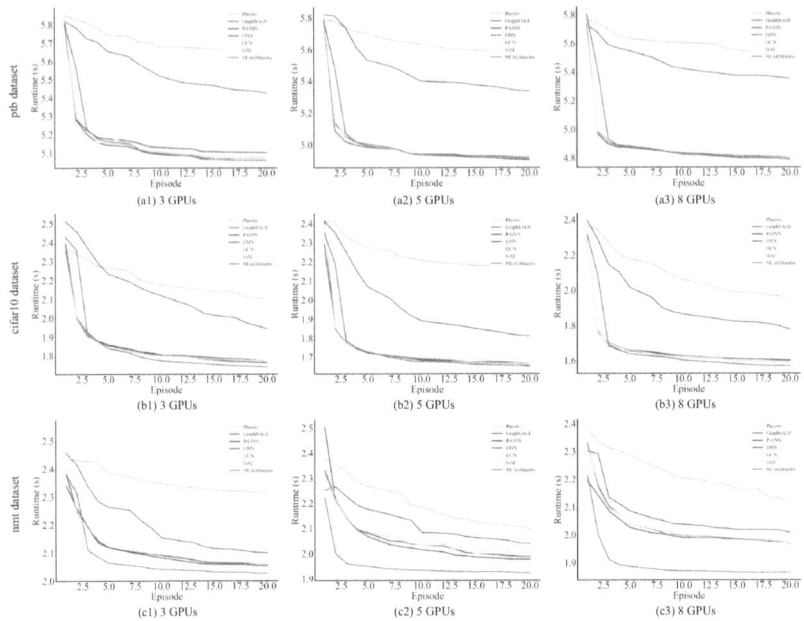

Fig. 3. The execution time of various comparison models under distinct episodes.

is only reduced by 27.22% on 5 GPUs. Finally, for the computation time of nmt on 8GPUs, NEAGMamba is also between GCN and GAT. And NEAGMamba maintains over 94% speedup across all datasets and configurations. This also shows that NEAGMamba as an advanced graph neural network optimization framework, its computing time is not only comparable to GCN, GAT, etc., but also much better than other methods such as Placeto, P-GNN and GraphSAGE.

The Fig. 3 compares the execution times of various models using 3, 5 and 8 GPUs on the ptb, cifar10 and nmt datasets over 20 episodes. NEAGMamba consistently achieves the lowest running time across all GPU configurations, while Placebo consistently achieves the longest running time. This pattern is evident on the ptb datasets (a1, a2, a3), cifar10 datasets (b1, b2, b3), and nmt datasets (c1, c2, c3). For other models, such as GNN, GCN and GAT models show performance close to NEAGMamba, but not beyond, which also further highlights the advantage of NEAGMamba in terms of running time reduction.

4.3 Comparing Models with and Without the NEAGNN Module

Execution Time. Table 3 show the execution time of GNN, GMamba and NEAGMamba on ptb, cifar10 and nmt datasets with 3, 5 and 8 GPUs configurations.GMamba consistently shows improvements over traditional GNNs, with notable reductions in execution time: (0.61–0.78% for ptb, 1.18–0.83% for cifar10, 0.46–5.30% for nmt). However, NEAGMamba stands out as the ultimate optimizer, outperforming both GNN and GMamba across the board. Its

Table 3. The execution time improvements of GMamba and NEAGMamba compared to Placeto on three different GPUs across three distinct datasets.

Methods	Datasets	3GPUS		5GPUS		8GPUS	
		Exe_time(s)	Impro(%)	Exe_time(s)	Impro(%)	Exe_time(s)	Impro(%)
Placeto	ptb	5.6592	-	5.5747	-	5.5134	-
GNN		5.0664	10.47%	4.9201	11.74%	4.7978	12.98%
GMamba		5.0939	9.99%	4.9438	11.32%	4.8684	11.70%
NEAGMamba		**5.0567**	**10.65%**	**4.9009**	**12.09%**	**4.7862**	**13.19%**
Placeto	cifar10	2.1061	-	2.1649	-	1.9609	-
GNN		1.7623	16.32%	1.6669	23.00%	1.5999	18.41%
GMamba		1.7416	17.31%	1.6536	23.62%	1.5821	19.32%
NEAGMamba		**1.7406**	**17.35%**	**1.6502**	**23.77%**	**1.5681**	**20.03%**
Placeto	nmt	2.3136	-	2.0969	-	2.1234	-
GNN		2.0570	11.09%	1.9762	5.75%	1.9685	7.29%
GMamba		2.0641	10.78%	1.9397	7.50%	1.8648	12.18%
NEAGMamba		**2.0248**	**12.48%**	**1.9226**	**8.31%**	**1.8646**	**12.19%**

Table 4. The average computation time of various models on three different GPUs across three distinct datasets.

Methods	Datasets	3GPUS		5GPUS		8GPUS	
		Comp_time(s)	Impro(%)	Comp_time(s)	Impro(%)	Comp_time(s)	Impro(%)
Placeto	ptb	430.8948	-	459.0664	-	437.4294	-
GNN		**5.84480**	**98.64%**	**5.8515**	**98.73%**	**5.8360**	**98.67%**
GMamba		10.7251	97.51%	10.7918	97.65%	10.7827	97.53%
NEAGMamba		10.3281	97.60%	12.1111	97.36%	10.8461	97.52%
Placeto	cifar10	122.7012	-	117.9207	-	119.0673	-
GNN		**2.41261**	**98.03%**	**2.4046**	**97.96%**	**2.4141**	**97.97%**
GMamba		3.4257	97.21%	3.3942	97.12%	3.4794	97.08%
NEAGMamba		4.3618	96.45%	4.5284	96.16%	4.621	96.12%
Placeto	nmt	64.6854	-	63.4991	-	63.6237	-
GNN		**2.0174**	**96.88%**	**2.0120**	**96.83%**	**2.0211**	**96.82%**
GMamba		3.1948	95.06%	3.1422	95.05%	3.1717	95.01%
NEAGMamba		3.7618	94.18%	3.7987	94.02%	**2.7932**	**95.61%**

Sequential State Management (SSM) mechanism efficiently manages information flow and minimizes redundant computations, driving remarkable execution time improvements.

Computation Time. Table 4 shows the comparison of computation time and improvement rate of three models (GNN, GMamba and NEAGMamba). In general, compared with the traditional GNN, GMamba has a slight increase in computation time, which also shows that the improvement of GMamba's execution time in the device placement task does not cause a lot of computation overhead. In addition, NEAGMamba adds the NEAGNN module on the basis of GMamba, but with a small increase in computing time, as shown in the cifar10 datasets on 8GPUs, the time is 4.621 s and 3.7494 s. Despite this slight increase, the overall execution time efficiency has been improved.

5 Conclusions

The NEAGMamba model integrates feature decoupling, importance scoring and a selective state transition mechanism (SSM) to enhance GNN efficiency. It tackles over-smoothing by decoupling node features, precisely assesses neighbor influence via scores and optimizes sequential data processing with SSM. Experiments prove NEAGMamba boosts execution by over 10% across datasets, with a 94% acceleration ratio. This model pioneers new GNN research avenues, fostering potential, applications and future technological advancements.

Funding Information. This work was supported by the GHfund D [ghfund202407042032]; the Shanxi Province Basic Research Program [202203021212444]; Shanxi Agricultural University Science and Technology Innovation Enhancement Project [CXGC2023045]; Shanxi Province Higher Education Teaching Reform and Innovation Project [J20220274]; Shanxi Postgraduate Education and Teaching Reform Project Fund [2022YJJG094]; Shanxi Agricultural University doctoral research start-up project [2021BQ88]; Shanxi Agricultural University Academic Restoration Research Project [2020xshf38]; Young and Middle-aged Top-notch Innovative Talent Cultivation Program of the Software College, Shanxi Agricultural University [SXAUKY2024005].

References

1. Brown, T.B., et al.: Language models are few-shot learners, CoRR, vol. abs/2005.14165 (2020). https://arxiv.org/abs/2005.14165
2. Dosovitskiy, A., et al.: An image is worth 16x16 words: transformers for image recognition at scale, CoRR, vol. abs/2010.11929 (2020). https://arxiv.org/abs/2010.11929
3. He, X., Liao, L., Zhang, H., Nie, L., Hu, X., Chua, T.: Neural collaborative filtering, CoRR, vol. abs/1708.05031 (2017). http://arxiv.org/abs/1708.05031
4. Fedus, W., Zoph, B., Shazeer, N.: Switch transformers: scaling to trillion parameter models with simple and efficient sparsity, CoRR, vol. abs/2101.03961 (2021). https://arxiv.org/abs/2101.03961
5. Addanki, R., Venkatakrishnan, S.B., Gupta, S., Mao, H., Alizadeh, M.: Placeto: learning generalizable device placement algorithms for distributed machine learning, CoRR, vol. abs/1906.08879 (2019). http://arxiv.org/abs/1906.08879
6. Mitropolitsky, M., Abbas, Z., Payberah, A.H.: Graph representation matters in device placement. In: Proceedings of the Workshop on Distributed Infrastructures for Deep Learning, ser. DIDL 2020, pp. 1–6. Association for Computing Machinery, New York (2021). https://doi.org/10.1145/3429882.3430104

7. Hamilton, W.L., Ying, R., Leskovec, J.: Inductive representation learning on large graphs. In: Proceedings of the 31st International Conference on Neural Information Processing Systems, ser. NIPS 2017, pp. 1025–1035. Curran Associates Inc., Red Hook (2017)
8. Kipf, T.N., Welling, M.: Semi-supervised classification with graph convolutional networks, CoRR, vol. abs/1609.02907 (2016). http://arxiv.org/abs/1609.02907
9. Veličković, P., Cucurull, G., Casanova, A., Romero, A., Liò, P., Bengio, Y.: Graph attention networks (2018). https://arxiv.org/abs/1710.10903
10. Xu, K., Hu, W., Leskovec, J., Jegelka, S.: How powerful are graph neural networks? CoRR, vol. abs/1810.00826 (2018). http://arxiv.org/abs/1810.00826
11. Li, Q., Han, Z., Wu, X.: Deeper insights into graph convolutional networks for semi-supervised learning, CoRR, vol. abs/1801.07606 (2018). http://arxiv.org/abs/1801.07606
12. Wang, G., Ying, R., Huang, J., Leskovec, J.: Improving graph attention networks with large margin-based constraints, CoRR, vol. abs/1910.11945 (2019). http://arxiv.org/abs/1910.11945
13. Gu, A., Dao, T.: Mamba: linear-time sequence modeling with selective state spaces (2024). https://arxiv.org/abs/2312.00752
14. Yun, S., Jeong, M., Kim, R., Kang, J., Kim, H.J.: Graph transformer networks, CoRR, vol. abs/1911.06455 (2019). http://arxiv.org/abs/1911.06455
15. Kim, J., et al.: Pure transformers are powerful graph learners. In: Proceedings of the 36th International Conference on Neural Information Processing Systems, ser. NIPS 2022. Curran Associates Inc., Red Hook (2024)

Research on Improved PointPillars Algorithm Based on Attention Mechanism and Feature Fusion

RunMei Zhang[1], AnLong Zhang[2], and Lei Yin[2(✉)]

[1] School of Mechanical and Electrical Engineering, Anhui Jianzhu University, Hefei 230601, China
[2] School of Electronics and Information Engineering, Anhui Jianzhu University, Hefei 230601, China
leiyin@mail.hfut.edu.cn

Abstract. 3D point cloud-based object detection is a crucial topic in the field of computer vision. This paper addresses the issue of low detection accuracy in 3D object detection using LiDAR during autonomous driving by proposing an improved 3D object detection method based on PointPillars. Firstly, in the feature extraction stage of the PointPillars model, the CA attention mechanism is embedded in the backbone network to learn similarity information and focus on important features. By calculating attention in two spatial dimensions (height and width) separately, the spatial distribution characteristics in the image are captured more accurately, thereby more comprehensively capturing the dependency relationships between features. Secondly, when performing feature fusion on the extracted information, instead of simply adding or concatenating the feature maps directly, suitable weights are learned for each position to select features of different scales, enabling more effective fusion. This allows the model to more efficiently detect feature information at different scales. To verify the performance of the improved algorithm, experiments were conducted on the publicly available KITTI dataset. The experimental results show that the proposed optimization algorithm achieves certain improvements over other publicly available algorithms. Compared to the baseline algorithm PointPillars, the average precision was improved by 4.48%.

Keywords: PointPillars · 3D Object Detection · Feature Fusion · Attention mechanism

1 Introduction

3D object detection algorithms are a key research focus for achieving autonomous driving path planning and safe obstacle avoidance. With the advancement of LiDAR manufacturing technology and the development of deep learning algorithms, the accuracy of 3D laser measurements is becoming increasingly high, and deep learning detectors that take point clouds as input are also gradually maturing. In this context, methods based on 3D point cloud object detection are being increasingly applied in fields such as intelligent driving [1].

Compared to image-based 2D detection methods, point cloud data can provide the 3D coordinates of objects, whereas 2D object detection can only provide bounding box information on the image plane. Additionally, 3D detection methods are better at understanding the interactions between objects in complex environments, such as the relative positions and movement directions of pedestrians and vehicles on the road. Lastly, 3D object detection can better handle various perspectives and target size variations, and it is relatively less affected by lighting and occlusion.

Currently, there are still several challenges that need to be overcome in the field of 3D object detection. 3D point cloud data is usually sparse and irregular, especially at long distances or at the edges of the sensor's field of view, making it more difficult to construct efficient and accurate networks. The features of small and medium-sized objects are usually more sparse to being overlooked or misdetected. Additionally, high-quality 3D detection networks often require complex structures and substantial computational resources, which conflicts with real-time requirements.

This paper proposes an improved algorithm to address the issues of insufficient feature extraction and inadequate feature fusion in the PointPillars algorithm. In the original algorithm, the generated pseudo-image is processed with 2D convolutions for feature extraction without considering the abundant redundant information present in the space, which not only wastes considerable computational resources but also affects the accuracy of subsequent detection. By introducing the CA attention mechanism, the model can more effectively focus on local effective feature information through the interaction relationships between features in different regions [2]. Additionally, the feature fusion utilizes an FPN with a top-down approach, transmitting high-level semantic features to lower-level features layer by layer. This method does not account for the adaptive needs of feature maps under different tasks or scenarios, leading to insufficient feature fusion. Moreover, due to the unidirectional propagation from top to bottom in FPN, the detailed information of the lower-level feature maps is not fully utilized when transmitted upward, resulting in inadequate information transfer between feature hierarchies. By incorporating ASFF (Adaptively Spatial Feature Fusion), the model adaptively merges feature maps from different network layers, helping it achieve a balance between different scales, semantic complexities, and spatial precision [3].

2 Related Work

Scholars both domestically and abroad have proposed numerous efficient algorithms in the field of 3D object detection. In 2017, Charles et al. introduced the PointNet algorithm [4], which pioneered a deep learning method that directly processes point cloud data, significantly improving the ability to extract features from point clouds. That same year, Zhou et al. proposed VoxelNet, which discretizes point cloud data into a 3D voxel grid and then uses a 3D convolutional neural network to extract features from the voxel grid [5]. In 2018, Yan et al. introduced the SECOND algorithm, which uses 3D sparse convolution, significantly improving the algorithm's running speed [6]. In 2020, the PV-RCNN algorithm proposed by Shaoshuai Shi et al. combines point features and voxel features. Point features capture fine local geometric information, while voxel features help capture global structure [7]. The PointPillars algorithm introduced by Lang et al. in their work proposes a novel feature representation method called Pillar, which greatly speeds up the algorithm's runtime and addresses the real-time issue of 3D detection networks [8].

In recent years, 3D point cloud-based object detection technology has made significant progress in the field of autonomous driving. For example, PointPainting introduces RGB image information into point cloud detection for multi-sensor fusion. It first performs semantic segmentation on each pixel of the image, then projects the semantic information onto the point cloud, and the fused point cloud features are used for object detection [9]. 3DETR, a transformer-based 3D object detection model, inputs the 3D point cloud into a Transformer architecture, leveraging the self-attention mechanism to capture global contextual information, enabling efficient 3D object detection [10].

Here have been many improvements made to PointPillars. By improving the feature extraction and encoding methods, the dynamic voxelization adaptively adjusts the size and shape of the voxel grid, enhancing the accuracy and computational efficiency of feature extraction [11]. The introduction of attention mechanisms enhances the effectiveness of feature extraction, allowing the model to focus better on important areas of the point cloud [12]. Multimodal fusion is also an approach, such as the fusion of LiDAR point clouds and camera images. By leveraging texture information from images and depth information from point clouds, the accuracy of object detection can be improved [13]. Such as the Frustum-PointPillars [14], the ExistenceMap-PointPillars [15].

3 PointPillars Detection Algorithm Based on Attention Mechanism and Feature Fusion

3.1 Detection Network Model Framework

The PointPillars algorithm first transforms the point cloud into a sparse pseudo-image through a point cloud encoding network, then performs subsequent 2D feature extraction and predicts the location and category of targets based on the extracted features. In the original process, only 2D convolution was used for feature extraction, with simple feature concatenation and without fully utilizing multi-scale features. The original 2D convolutional network is improved based on the concepts of attention mechanisms and feature fusion. The structure of the improved model is shown in the Fig. 1.

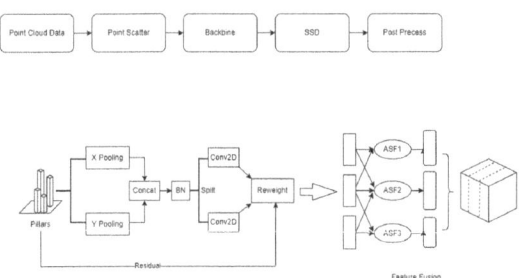

Fig. 1. The main model framework and the redesigned backbone network.

3.2 Attention Mechanism

The Attention Mechanism is a technique widely used in deep learning, initially introduced in natural language processing (NLP) tasks and later expanded to other fields such as computer vision [16].

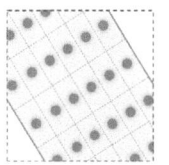 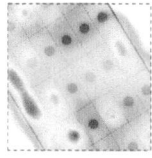

Fig. 2. The visual representation of features after being processed by the attention mechanism.

In visual models, attention mechanisms help deep learning networks focus on the most informative parts, thereby improving overall prediction accuracy and computational efficiency [17] in the Fig. 2.

The commonly used SE attention module explicitly models the importance of each feature channel and adjusts the weights of feature channels [18]. However, the SE attention mechanism only focuses on relationships between channels while ignoring spatial information, resulting in a loss of spatial details. Researchers have proposed several more advanced attention mechanism modules, such as CBAM (Convolutional Block Attention Module) [19], and Cross-Attention [20].

The CA (Coordinate Attention) mechanism is an innovative attention mechanism that, compared to traditional global attention mechanisms, considers not only the content of feature maps but also their spatial coordinate information. It captures information along horizontal and vertical paths separately, then combines them to produce an attention map that incorporates spatial position and channel information. In PointPillars, the pseudo-image essentially projects 3D point cloud data onto a 2D bird's-eye view (BEV) plane. By capturing the spatial position information of the feature map, the CA mechanism enables more precise feature localization, enhancing the accuracy of feature representation. The application of the CA mechanism in pseudo-images effectively integrates the spatial coordinate information of 3D point clouds, providing higher-quality feature maps—an innovative approach for handling 3D information on a 2D plane.

3.3 Feature Fusion

Feature fusion is a technique that combines features from different sources or stages to enhance model performance. Classical feature fusion methods include concatenation, addition, weighted sum, and multiplication.

ASFF (Adaptive Spatial Feature Fusion) is a method designed to optimize multi-scale feature fusion by adaptively selecting and combining features to improve object detection performance. PointPillars converts point cloud data, which often contains more noise and irregular distributions. ASFF needs to discern and handle these features on this basis, conducting more meticulous multi-level feature fusion. This includes not only the fusion of 2D feature maps but also the effective integration of spatial, density, and other 3D information, resulting in more precise and accurate object detection.

By generating weights for feature maps at each scale, the model can effectively allocate these weights between different feature maps. Each scale's feature map is multiplied by its corresponding weight and then fused into a weighted average feature map. This enables the model to adaptively select and utilize the most significant features, thereby enhancing the accuracy of object detection.

3.4 Detection Algorithm

Pillar Feature Network. First, the original point cloud data needs to be processed. For the original point cloud $L(x, y, z)$, a grid of fixed size is defined in the XY plane. The collection of pillars is formed by extending these grids along the Z-axis in space. Each point in the point cloud contains information in four dimensions (x, y, z, r), where (x, y, z) represents the spatial coordinates of the point and r denotes the reflectance.

Let there be a total of P point cloud pillars at this time, then the dimensionality of the point cloud data $D = 9$. This data is raised to 64 dimensions through a fully connected layer, followed by max pooling, which compresses the number of points in each pillar to 1, representing the most representative point within that pillar. The point cloud data is represented as a tensor of shape (C, P), where C is the number of channels. Finally, these features are stacked according to the positions of the original point cloud data to obtain a pseudo-image of size (C, H, W), where (H, W) s the height and width of the pseudo-image. At this point, the point cloud data has been transformed into a tensor of shape (C, H, W), which is then input to the backbone network.

Backbone Network. Referencing the principles of CA Attention, a feature weighting equation is designed to extract features from pseudo-images. For the input feature map of shape (C, H, W), a global pooling operation is first performed on the height and width dimensions, resulting in two feature maps of shapes $(C, H, 1)$ and $(C, 1, W)$, respectively.

$$z_c^h(h) = \frac{1}{W}\sum_{0<i<W} x_c(h, i) \qquad z_c^w(w) = \frac{1}{H}\sum_{0<i<H} x_c(j, w) \qquad (1)$$

Then, the two feature maps (width and height) are transposed to the same orientation for concatenation. After concatenating the feature map, dimensionality reduction and activation operations are performed. Here, a scaling factor k can be introduced to reduce the number of network parameters.

$$f = \delta(F_1([z^h, z^w])) \qquad (2)$$

Next, f is split to obtain $[\frac{C}{k}, H, 1]$ and $[\frac{C}{k}, 1, W]$, which are then upsampled and activated to produce the attention vectors.

$$g^h = \sigma(F_h(f^h)) \qquad g^w = \sigma(F_w(f^w)) \qquad (3)$$

Finally, the attention output is obtained by multiplying with the original features.

$$y_c(i, j) = x_c(i, j) * g_c^h(i) * g_c^w(j) \qquad (4)$$

For the feature $X \in R^{(H*W*C)}$ from the previous step, channel and spatial information are aggregated by applying global average pooling to the feature map of each channel.

$$z_c = \frac{1}{H*W}\sum_{H}^{i=1}\sum_{W}^{j=1} x_{ij} \qquad (5)$$

The global average values z_c of the c channels obtained through global pooling are passed through several fully connected layers to generate the importance weights $W \in R^{(H*W*C)}$. The specific steps involve first using a $1*1$ convolutional layer, followed by a $softmax$ function for normalization.

$$w = \sigma(Wz + b) \qquad (6)$$

Here, w is a weight vector of length c (the number of channels), W and b represent the weights and biases, respectively, and σ is the sigmoid activation function used for normalizing the weights.

$$w_c = \frac{exp(w_c)}{\sum_{k=1}^{C} exp(w_k)} \qquad (7)$$

The generated weights are used for feature selection and suppression. For the important features,

$$X'_c = X_c * w_c \qquad (8)$$

For the features that need to be suppressed,

$$X''_c = X_c * (1 - w_c) \qquad (9)$$

Finally, the two are fused to obtain the final features.

Detection Head. SSD (Single Shot MultiBox Detector) is a deep learning model used for object detection. Compared to other object detection methods, SSD offers high detection speed and good accuracy, making it outstanding for real-time detection tasks. The core idea is to perform object detection using multi-scale feature maps and default boxes. At each position on the feature map, multiple default boxes with different aspect ratios and scales are used to cover a variety of shapes and sizes of objects.

Loss Function. Referring to the loss function in the original text, the ground truth bounding box is represented as $(x, y, z, w, l, h, \theta)$, where (x, y, z) are the coordinates of the center of the bounding box, (w, l, h) are its length, width, and height, respectively, and θ is the yaw rotation angle around the z-axis. The localization regression residual between the ground truth and the predicted box (anchor) is calculated as follows:

$$\begin{cases} \triangle x = \frac{x_{gt}-x_a}{d_a}, \triangle y = \frac{y_{gt}-y_a}{d_a}, \triangle z = \frac{z_{gt}-z_a}{d_a} \\ \triangle w = \log \frac{w_{gt}}{w_a}, \triangle h = \log \frac{h_{gt}}{h_a}, \triangle l = \log \frac{l_{gt}}{l_a} \\ \triangle \theta = \sin(\theta_{gt} - \theta_a) \end{cases} \qquad (10)$$

Here, $\triangle x, \triangle y, \triangle z$ represent the differences between the ground truth and predicted boxes in (x, y, z), $d_a = \sqrt{w_a^2 + l_a^2}$ is the distance, and $\triangle w, \triangle h, \triangle l, \triangle \theta$ are the differences between the ground truth and predicted boxes in (w, l, h, θ). Localization loss function of the model.

$$L_{loc} = \sum_{b \in x,y,z,w,h,l,\theta}[SmoothL1(\triangle b)] \qquad (11)$$

Here, $SmoothL1$ is a smooth version of the $L1$ function, b represents the difference between the ground truth and predicted values, $\triangle b$ is the localization regression residual, and β is a parameter, where $\beta = 1$. The classification loss uses the Focal Loss function.

$$L_{cls} = -\alpha_a(1 - P_a)^\gamma \log P_a \tag{12}$$

Here, P_a is the class probability of the predicted box, α_a is the modulation factor, set to $\alpha_a = 0.25$, and γ is a constant, set to $\gamma = 2$. Since the localization loss cannot distinguish whether the bounding box is flipped, it is necessary to use the cross-entropy loss function L_{dir} to detect the orientation of the predicted box. Therefore, the overall prediction function can be expressed as:

$$L = \frac{1}{N_{pos}}(\beta_{Loc}L_{Loc} + \beta_{Lcs}L_{Lcs} + \beta_{dir}L_{dir}) \tag{13}$$

Here, N_{pos} is the number of correctly predicted anchors, and β represents the coefficients for each loss function, with $\beta_{Loc} = 2$, $\beta_{Lcs} = 1$, $\beta_{dir} = 0.2$.

4 Experiments

4.1 Experimental Equipment

The hardware configuration of the computer includes: CPU Intel i5 13400F, GPU NVIDIA GeForce RTX 4060 Ti, and 16GB of RAM. The software configuration consists of the Linux operating system Ubuntu 22.04 and Python 3.9. The dataset used is the open-source KITTI dataset. The KITTI dataset is a widely used benchmark dataset for autonomous driving research, jointly released by the Karlsruhe Institute of Technology (KIT) and Daimler AG. The dataset includes real driving images from urban, rural, and highway environments, with each image containing up to 15 cars and 30 pedestrians, as well as varying degrees of occlusion. It consists of 7,481 frames of training data and 7,518 frames of testing data, with each frame containing RGB images, LiDAR point clouds, and camera annotations.

4.2 Experimental Results

In the KITTI dataset validation, the model's performance is quantitatively analyzed by comparing it with other algorithms, using the average precision (AP) as the evaluation metric. The results are shown in Table 1 (Fig. 3).

Additionally, to verify the detection efficiency of the improved model, the inference speed of the model was compared with mainstream algorithms. The results are shown in the Fig. 4.

From the Table 2 above, it can be seen that the improved model shows a significant enhancement in detection accuracy compared to the baseline PointPillars model. In the easy detection category, improvements were 2.59 To validate the impact of each module on the overall model, ablation experiments were conducted. Experiment one was the basic PointPillars, with the experimental results evaluated using mean average precision (mAP) as the metric. The results are shown in Table 2.

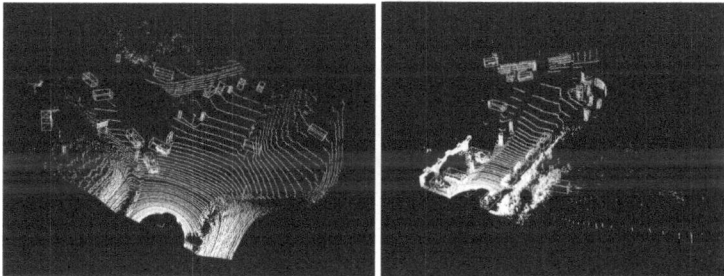

Fig. 3. Visualized Detection Results, the green boxes represent detected vehicle positions, the yellow boxes indicate bicycle positions, and the blue boxes show pedestrian positions. It is clear from the image that the model can effectively detect the objects in the scene. (Color figure online)

Table 1. Comparison of Average Precision in 3D Point Cloud Object Detection.

	Car			Cyclist			Pedestrian		
	easy	moderate	hard	easy	moderate	hard	easy	moderate	hard
SECOND	87.74	78.62	77.93	80.55	61.83	61.76	60.87	55.72	55.25
VoxelNet	85.72	75.41	73.69	77.49	60.62	58.61	55.71	51.02	50.21
PointRCNN	88.21	82.16	80.66	81.62	67.30	65.03	59.31	56.94	52.59
PointPillars	87.59	81.51	77.42	80.35	65.28	63.56	59.18	57.33	52.96
Ours	**90.18**	**83.77**	**82.97**	**85.34**	**71.75**	**68.34**	**63.88**	**60.83**	**55.44**

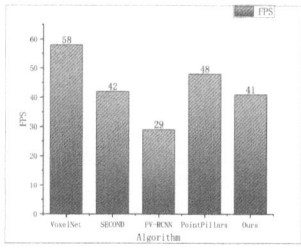

Fig. 4. Comparison of network model reasoning speed.

Table 2. The impact of different modules on detection accuracy.

	CA Attention	ASFF	Car/mAP	Cyclist/mAP	Pedestrian/mAP
Exp1			81.17	69.73	56.49
Exp2	√		83.58	72.08	58.87
Exp3		√	80.31	67.42	51.74
Exp4	√	√	85.64	75.14	60.05

From the table above, it can be seen that both the attention mechanism module and the adaptive feature fusion module have a positive effect on improving the overall model detection accuracy. By incorporating the attention mechanism, CA attention helps the model to more accurately identify target boundaries and details by effectively distributing attention to spatial and channel information within the feature maps, thus reducing false positives and false negatives. This incorporation leads to an overall detection accuracy improvement of approximately 2.38.

5 Conclusion

This paper improves the PointPillars model based on the CA attention mechanism and feature fusion methods, aiming to address the issue of insufficient detection accuracy due to inadequate feature extraction in the PointPillars model. The CA attention mechanism enhances the model's ability to capture spatial and channel information more effectively by separately processing these two types of information, thereby strengthening feature representation. Experiments show that after incorporating the CA attention mechanism, the overall accuracy of the model improves by approximately 2.38 At the same time, feature fusion combines features from different levels and scales, allowing for better capture of the diversity and complexity of the targets. By fusing features of different scales, the model can focus on both the global information of large targets and the local details of small targets. Experimental results indicate that after introducing the new feature fusion method, the overall accuracy improves by about 2.1.

Acknowledgments. Provincial and Ministerial Laboratory Development Fund: FZ2021KF10, SGCZXZ-D2101, University-Level Reserve Fund Project: 2022XMK03. National Natural Science Foundation General Program: 52378001, Construction of a Drone Safety Knowledge Base Based on Knowledge Graphs: FZ2021KF10. Intelligent Building System Integration and Intelligent Operation and Maintenance Research Oriented Towards Big Data:2019QDZ38.

References

1. Liang, Z., Huang, Y., Song, Z., Ding, J.: A review of 3D object detection methods based on deep learning in autonomous driving. J. Univ. Shanghai Sci. Technol. **46**(2), 103–119 (2024)
2. Hou, Q., Zhou, D., Feng, J.: Coordinate attention for efficient mobile network design. In: Proceedings of the IEEE/CVF Conference on Computer Vision and Pattern Recognition, pp. 13713–13722 (2021)
3. Liu, S., Huang, D., Wang, Y.: Learning spatial fusion for single-shot object detection. arxiv 2019, arXiv preprint arXiv:1911.09516 (1911)
4. Qi, C.R., Su, H., Mo, K., Guibas, L.J.: Pointnet: deep learning on point sets for 3D classification and segmentation. In: Proceedings of the IEEE Conference on Computer Vision and Pattern Recognition, pp. 652–660 (2017)
5. Zhou, Y., Tuzel, O.: Voxelnet: end-to-end learning for point cloud based 3D object detection. In: Proceedings of the IEEE Conference on Computer Vision and Pattern Recognition, pp. 4490–4499 (2018)
6. Yan, Y., Mao, Y., Li, B.: Second: sparsely embedded convolutional detection. Sensors **18**(10), 3337 (2018)

7. Shi, S., et al.: PV-RCNN: point-voxel feature set abstraction for 3D object detection. In: Proceedings of the IEEE/CVF Conference on Computer Vision and Pattern Recognition, pp. 10529–10538 (2020)
8. Lang, A.H., Vora, S., Caesar, H., Zhou, L., Yang, J., Beijbom, O.: Pointpillars: fast encoders for object detection from point clouds. In: Proceedings of the IEEE/CVF Conference on Computer Vision and Pattern Recognition, pp. 12697–12705 (2019)
9. Vora, S., Lang, A.H., Helou, B., Beijbom, O.: Pointpainting: sequential fusion for 3D object detection. In: Proceedings of the IEEE/CVF Conference on Computer Vision and Pattern Recognition, pp. 4604–4612 (2020)
10. Misra, I., Girdhar, R., Joulin, A.: An end-to-end transformer model for 3D object detection. In: Proceedings of the IEEE/CVF International Conference on Computer Vision, pp. 2906–2917 (2021)
11. Li, W., Chen, Z., Qu, J., Cui, L., Chu, W., Gao, H.: 3D object detection based on fusion sampling and graph network. J. Harbin Univ. Sci. Technol. 1–13
12. Liu, M., Yang, Q., Hu, G., Guo, Y., Zhang, J.: 3D point cloud object detection algorithm based on transformer. J. Northwestern Polytech. Univ. **41**(06), 1190–1197 (2023)
13. Tao, L., Wang, H., Cai, Y., Chen, L.: Multi-target point cloud detection algorithm for autonomous driving scenarios. Autom. Eng. **46**(07), 1208–1218+1238 (2024)
14. Paigwar, A., Sierra-Gonzalez, D., Erkent, Ö., Laugier, C.: Frustum-pointpillars: a multi-stage approach for 3D object detection using RGB camera and lidar. In: Proceedings of the IEEE/CVF International Conference on Computer Vision, pp. 2926–2933 (2021)
15. Hariya, K., Inoshita, H., Yanase, R., Yoneda, K., Suganuma, N.: Existencemap-pointpillars: a multifusion network for robust 3D object detection with object existence probability map. Sensors **23**(20), 8367 (2023)
16. Vaswani, A., et al.: Attention is all you need (NIPS) (2017). arXiv preprint arXiv:1706.03762, vol. 10, p. S0140525X16001837 (2017)
17. Alexey, D.: An image is worth 16x16 words: transformers for image recognition at scale, arXiv preprint arXiv: 2010.11929 (2020)
18. Hu, J., Shen, L., Sun, G.: Squeeze-and-excitation networks. In: Proceedings of the IEEE Conference on Computer Vision and Pattern Recognition, pp. 7132–7141 (2018)
19. Woo, S., Park, J., Lee, J.-Y., Kweon, I.S.: CBAM: convolutional block attention module. In: Proceedings of the European Conference on Computer Vision (ECCV), pp. 3–19 (2018)
20. Vaswani, A.: Attention is all you need. In: Advances in Neural Information Processing Systems (2017)

Correction to: Intensity Controllable Emotional Speech Synthesis Based on Valence-Arousal-Dominance

Guoping Li and Yanxiang Chen

Correction to:
Chapter 4 in: A. Hussain et al. (Eds.): *Advances in Brain Inspired Cognitive Systems*, **LNAI 15497, https://doi.org/10.1007/978-981-96-2882-7_4**

In the original version of this chapter, the name of the first author and affiliation of all the authors were wrong. This has been corrected. Correctly it should read as name: "Guoping Li", affiliation: "Hefei University of Technology".

The updated version of this chapter can be found at
https://doi.org/10.1007/978-981-96-2882-7_4

Author Index

B
Bruevich, Maria II-178

C
Chen, Jiacong I-177
Chen, Jie II-1
Chen, Jinlan II-104
Chen, Rongjun I-20, II-273, II-283
Chen, Weiqi I-136
Chen, Yanxiang I-30
Chen, Yuan II-1, II-11

D
Dai, Xiaodong I-104
Dashtipour, Kia II-159, II-178
Deng, Shunan II-273
Diao, Liangjin I-177
Ding, Jonathan II-117
Ding, Zhuanlian II-41
Dong, Xingbo I-166
Dong, Yiting I-249

E
Efunwoye, Ibukunoluwa Oluwabusayo II-159

F
Fan, Chen I-1
Feng, Guanyuan I-20
Feng, Mingchen II-201
Fu, Yanping I-126

G
Gao, Fei I-1
Gao, Yuefang II-62
Gogate, Mandar II-159, II-178
Gunathilake, Nilupulee A. II-178

H
Han, Meng I-269, II-82, II-188, II-233
Hao, Li II-62
Hao, Wei I-82
Hao, Wangli I-269, II-82, II-233
He, Jian I-259
He, Yuan II-93
Hou, Wanzhen I-93
Hou, Yanqi I-177
Hu, Guyue II-93
Hu, Jingfei I-197
Hu, Wei I-10, I-136
Hu, Xiaojing I-62
Hu, Xinyuan II-233
Huang, Chengchuang II-188
Huang, Fuchuan I-259
Huang, Lili I-208
Huang, Yuesheng I-20
Hussain, Adeel II-159, II-178
Hussain, Amir I-1, II-159, II-178, II-273, II-283

J
Jia, Congcong I-166
Jia, Wei I-259
Jiang, Bo II-1, II-11
Jiang, Fengling II-159, II-178
Jiang, Hongfan II-138
Jiang, Longteng I-218
Jiao, Hui II-117
Jin, Xin I-218
Jin, Yuguang I-52
Jin, Zhe I-166
Jing, Yang I-41

L
Li, Anzhen II-201
Li, Ao II-168
Li, Chenglong I-155
Li, Chuanfu I-208

Li, Dazhi II-62
Li, Fuzhong I-269, II-233
Li, Guoping I-30
Li, Huihui I-136
Li, Jiawen I-20
Li, Ping II-222
Li, Qilang II-104
Li, Shiqi I-208
Li, Weiyi I-249
Li, Xin II-188
Li, Xinhui II-168
Li, Xinwei II-117
Li, Xuetao II-52
Li, Yi I-155
Li, Yiming I-197
Li, Yufei I-187
Li, Ziyu II-21
Ling, Chen I-20
Lingkang, Gu I-41
Liu, Fengxiang II-252
Liu, Lei I-155
Liu, Shixi I-62
Liu, Tianjin I-1
Liu, Wenhao II-41
Liu, Xiaoyong I-136
Liu, Xinyu II-72
Liu, Yijian I-218
Liu, Yudian I-62
Liu, Yuqing II-1, II-11
Liu, Zhenfei II-211
Lu, Enmeng I-229
Lu, Feixiang I-218
Lu, Jianwei II-72
Lu, Xiaofei II-82
Lu, Yongyi II-243
Luo, Bin II-159, II-178
Luo, Ye II-72
Lv, Jujian I-20
Lv, Zhao II-168

M
Ma, Leilei I-52
Ma, Qingchuan I-104, I-115
Ma, Yong II-31
Ma, Yuanmin II-1
Mai, Kaizhan II-62
Mengqi, Wu I-41
Mi, Pingping II-62
Mi, Xin II-222

Mu, Chaofan II-52

P
Pang, Mengyin II-211

Q
Qin, Jinghui II-243
Qin, Liwen II-201
Qin, Sizhe I-239
Qin, Weijie II-201
Qing, Shufan II-201

R
Rao, Ji II-72
Ren, Jie II-222
Ren, Jinchang II-159, II-178, II-201, II-273, II-283
Ren, Ximing I-20
Ren, Yongjian II-82

S
Shang, Dongxu I-72
Shang, Shoulai I-82
Shen, Guobin I-249
Shen, Longfeng I-177
Shen, Zhiya II-104
Sheng, Cunyu I-259
Shi, Chao II-222
Shi, Meilin II-201
Shu, Hao I-269, II-82, II-233
Su, Deyu I-239
Sun, Dengdi II-41, II-52
Sun, Lingma I-208
Sun, Meijun II-211
Sun, Yinqian I-229

T
Tan, Wenhao II-148
Tang, Jin I-115, II-262
Tao, Liang I-72
Tao, Wan II-127
Tu, Zijian I-239

W
Wang, Futian II-252
Wang, Hongbo II-222
Wang, Hua I-197
Wang, Huabin I-72
Wang, Jianfei II-72

Author Index

Wang, Jing I-10
Wang, Jun I-1
Wang, Leijun I-20
Wang, Lina II-211
Wang, Linbo II-148
Wang, Peixian I-20
Wang, Qianhao I-229
Wang, Shiao I-104, I-115
Wang, Wenzhong II-262
Wang, Xiao I-104, I-115, II-252
Wang, Xiaoran II-127
Wang, Xiuli II-72
Wang, Xuzhen II-262
Wang, Yifeng I-104, I-115
Wang, Zheng II-211
Wang, Zi I-239
Wei, Yubin I-187
Wu, Chenming I-218
Wu, Xia II-21
Wu, Xiaojun II-138
Wu, Xinyu I-20
Wu, Yanqiang I-10, I-136
Wu, Yuntao II-283
Wu, ZeHong II-243

X

Xi, Shanlin II-21
Xiang, Dawei II-31
Xiao, Zhanhao I-136
Xu, Guosheng I-115
Xu, Minghui I-126
Xu, Song II-211
Xu, Tianyang II-138
Xu, Wantong II-72
Xu, Youle I-177

Y

Yan, Ning I-115
Yan, Zheng II-148
Yang, Lianqiang I-166
Yang, Qingquan I-104, I-115
Yang, Tianxia I-126
Yang, Yiming II-31
Yang, Zhijing II-243
Yin, Lei I-280
Yu, Shan II-93
Yuan, Hongrui I-72

Z

Zeng, Yan II-82
Zeng, Yi I-229, I-249
Zhang, AnLong I-280
Zhang, Jiajing II-104
Zhang, Jiawei II-201
Zhang, Jicong I-197
Zhang, Jilin II-82, II-188
Zhang, Jingjing I-72
Zhang, Linglin II-72
Zhang, Mengya I-155
Zhang, Qian I-229
Zhang, Qiang II-127
Zhang, Qintao II-188
Zhang, RunMei I-280
Zhang, Wannan I-147
Zhang, Xiaojun II-117
Zhang, Yepeng I-82
Zhang, Yiwen I-187
Zhao, Dongcheng I-249
Zhao, Haifeng I-52, I-93, I-126
Zhao, Huimin II-273, II-283
Zhao, Wei I-177
Zheng, Tong I-104
Zhou, Die II-11
Zhou, Shiwei I-93
Zhouxiang, Xia I-41
Zhu, Xuefeng II-138
Zhu, Ying II-188

The manufacturer's authorised representative in the EU is Springer Nature Customer Service Centre GmbH, Europaplatz 3, 69115 Heidelberg, Germany. If you have any concerns regarding our products, please contact ProductSafety@springernature.com

Printed and bound by CPI Group (UK) Ltd, Croydon, CR0 4YY

26/03/2026

02078935-0012